THE CASTLE OF KNOWLEDGE

ROBERT RECORDE

THE CASTLE OF KNOWLEDGE

A FACSIMILE OF THE
FIRST EDITION
IMPRINTED AT LONDON BY
REGINALDE WOLFE
1556

RENASCENT BOOKS

THE CASTLE OF KNOWLEDGE

First imprinted by Reginalde Wolf in 1556
Second Edition
Imprinted by Valentine Sims
Assigned by Bonham Norton
1596

Hardback facsimile edition published by
TGR Renascent Books
27 Springdale Court
Mickleover, Derby DE3 9SW
United Kingdom
2009

Paperback edition first published 2012

ISBN 978-1-4783559-8-4
www.renascentbooks.co.uk

Printed and bound in the United States of America by
CreateSpace, Charleston, South Carolina, U.S.A.

for
STUART

INTRODUCTION

This book is a facsimile of the first edition of Robert Re-corde's *The Castle of Knowledge*, originally printed in London by Reginalde (or Reynold) Wolfe in 1556. Recorde was a typical Renaissance man, a physician at the courts of Henry VIII, Edward VI and Mary I, a mathematician and a very learned scholar. This book was the fourth in a series of justly famed mathematical works with the following titles: *The Grounde of Artes, The Pathway to Knowledge, The Gate of Knowledge, The Castle of Knowledge, The Treasure of Knowledge* and *The Whetstone of Witte*. Of these six, *The Gate* and *The Treasure* are no longer extant. The fourth book in the series (the one you now hold) deals with cosmography, which may be defined as a description or mapping of the general features of the universe. This 'universal world' as discussed by sixteenth-century authors, including Recorde, consisted of the spheres of the earth and the heavens. Progress in the study of cosmography, devoted to the charting of that world on globes and plane maps, was attended during the century by increasing exploration of the celestial as well as of the terrestrial sphere. The great gains made by English astronomers and geographers during the Tudor period are well known. Recorde was in the forefront of those seeking to instruct ordinary Englishmen, often mariners, not in scholarly Latin but in their mother tongue. A mark of the respect with which Recorde's writings were held is evidenced by the selection of *The Castle of Knowledge* for inclusion in the small library which Martin Frobisher purchased for the first voyage in search of the Northwest Passage. Included with William Cunningham's rather less famous sequel, the *Cosmographical Glass*, the ship's inventory shows that the two books together cost ten shillings. It is probable that both texts served as training manuals for the ship's officers. The *Castle* is written in dialogue form, carried on between a very knowledgeable master and a quite brilliant pupil. The book deals chiefly with elements of Ptolemaic

Introduction

astronomy but includes some geographical information. It is an original and exhaustive study intended to modernise Proclus and Sacrobosco.

BRIEF BIOGRAPHY

Robert Recorde was born between 1510 and 1512 in Tenby, Pembrokeshire. He was the second son of Thomas Recorde and Rose Jones. Practically nothing is known of his childhood and the first thing in his life about which we can be certain is his entrance into Oxford University in about 1525. It is not known what he studied but we may assume it was the usual (for the time) *trivium* of grammar, logic and rhetoric, followed by the *quadrivium* of arithmetic, geometry, music and astronomy. He graduated with a B.A. in 1531 and was elected a Fellow of All Souls College in the same year. He may have taught at Oxford for a few years but the evidence for this is scanty. At some time he moved from Oxford to Cambridge, where he studied for an M.D. and graduated in 1545 at the age of 35. He then moved to London, where for a few years he practised medicine. In later years he was always to describe himself as 'physician'. A defining moment in his life occurred in 1549 when he was appointed Controller of the Bristol Mint. It was during his time there that he made a very powerful and ruthless enemy. Sir William Herbert was sent by Edward VI to help suppress a revolt by John Dudley, Earl of Warwick, in the west country. Herbert demanded that Recorde divert funds from the mint to pay and support his army, but Recorde refused on the grounds that the order did not come from the king. Herbert countered and accused Recorde of treason. He was lucky to incur the mild penalty of confinement to court for 60 days. However, apparently all was later forgiven because in 1551 he was appointed general surveyor of Mines and Monies in Ireland. He was placed in charge of the Wexford silver mines and also became the

Introduction

technical supervisor of the Dublin mint. In the meantime, Sir William Herbert was created Earl of Pembroke for his services to the crown during the rebellion, and there was continued animosity between him and Recorde. Although the silver mines at Wexford had great potential, the enterprise was largely unsuccessful, mainly due to a lack of royal investment and the imperfect state of mining technology. The mines closed in 1553 and Recorde was recalled to England. Upon the accession to the throne of Mary, the daughter of Henry VIII, Recorde's old enemy the Earl of Pembroke was made a privy councillor for his support of Mary's claim to the throne. For some strange reason, Recorde chose the moment when Pembroke was strongest to try and get his revenge, charging him with misconduct in gaining his court positions. The allegation was probably true, but Pembroke was in favour with the monarchy and so had almost perfect immunity. He responded by suing Recorde for libel. There was a hearing in January 1557 and Recorde was ordered to pay the huge sum of £1000 compensation. He either could not or would not pay and so was sentenced to imprisonment in the King's Bench Prison in Southwark, for debt. Whilst in prison he made his will, leaving small sums of money to various people, including £20 to his mother. The date of his death is not known with any certainty, but is generally supposed to have been in the later part of 1558, only a short time after making his will.

The following notes are provided as a guide to reading, understanding and enjoying this facsimile edition of Recorde's treatise.

PAGINATION

Page numbering begins at the start of the first treatise and continues orderly into the second treatise, until page 53 is reached. Thereupon the first error in pagination occurs – 53

Introduction

is followed by 34 which is obviously a simple mistake in type-setting. From this point on similar errors occur with increasing frequency. Many pages which should bear numbers in the sequence 1xx are numbered 2xx, likewise some pages in the proper sequence 2xx are numbered 5xx. Caution is therefore required when page referencing, since the errors of the sixteenth century compositors are faithfully reproduced in this facsimile reprint. However, for the benefit of modern day readers who expect consistent pagination, modern page numbering is applied over a line drawn above Recorde's text.

SPELLINGS AND PUNCTUATION

Dictionaries and standard spellings did not really exist when *The Castle of Knowledge* was first printed and many words are not even spelt phonetically. Therefore many spellings in the book appear peculiar to the modern reader, but a little practice at reading Early Modern English soon renders the text intelligible. Many familiar words look strange simply because, unlike modern spellings, they end with the silent letter e and the last consonant might or might not be doubled, hence mane or manne (man), and rune or runne (run). The letter y is often used in place of i, for example fynde (find) or fyrste (first). Note that early printing conventions were to use the terminal letter s at the end of words, as today, but the long form everywhere else, for example poſſeſs (possess). The letters u and v were not considered to be two distinct letters, but different forms of the same letter. Typographically, v was often used at the start of words and u elsewhere, hence vnmoued (unmoved) or vnloued (unloved). But conversely, the letter v was often used where today we would expect the letter u, as in, for example, thervnto (thereunto). Neither were the letters i and j considered distinct, so that the planet Jupiter would appear spelt as Iupiter. In short, expect to read modern spellings such as sum, divisor and just for example, as ſumme, diuiſor and iust.

Introduction

CONTRACTIONS

Often the words 'the' and 'that' are contracted to yᵉ and yᵗ respectively, with the small letters e and t placed directly above the y. Of course, these contractions should be read in the text with their full pronunciation – 'the' and 'that'.

DIACRITICAL MARKS

Diacritical marks have been used to abbreviate printed words ever since Gutenberg and early English printers adopted the same conventions that Gutenberg did for Latin texts (which he copied, in turn, from the handwritten texts of medieval scribes). Diacritical marks are used on many pages of *The Castle of Knowledge* to indicate the omission of the consonant m or n where this follows a vowel. The missing letter is indicated by placing the mark (a bar) over the vowel. Instances are exāple (example), quotiēt (quotient), ī (in), nōbre for nombre (number) and chaūce for chaunce (chance). All such abbreviated words with diacritical marks should be read of course with their full pronunciation.

TYPOGRAPHICAL FEATURES

All the contractions and abbreviations found in the pages of *The Castle of Knowledge* are compositors tricks to help in the justification of entire paragraphs – something that was considerably easier in the days before standard spellings and orthography. Justification of paragraphs was not then merely a cosmetic feature (as it is today). Early printers would be laying out movable metal type into a square wooden frame and if the frame was not completely filled, the types would move under the action of the press and smudge the ink. In other words, each line of each paragraph had to extend fully from left to right, or the page would be unprintable. One way for Renaissance printers to do this might have been by inserting blank spaces of suitable lengths between the words of each line, but this is not a satisfactory solution. The result is usu-

Introduction

ally 'rivers' of blank space flowing down the page, which seriously interrupt reading and which was recognised as a problem from the very earliest days of printing. Hence the use of aggressive hyphenation, contracted words, diacritical marks and variant spellings like hed (which has three letters), head (which has four), or hedde (which has five), all very useful when striving to obtain justification and spelling is not a problem. The compositor would use any or all of these tricks at will in order to obtain a solid block of text on each page.

FAULTS

There are errata listed at the end of *The Castle of Knowledge,* supplemented by a smaller list before the first treatise, which describe those faults recognised by the author and his printer 450 years ago. However, the attentive reader will find a few more faults and errors besides those listed and, of course, no attempt has been made to indicate or remedy these in this facsimile reprint. The erratic page numbering has already been mentioned, and sometimes spelling mistakes occur despite the relaxed phonetic spelling. An example here is 'eihhte', which should be 'eighte' (eight) and is obviously wrong. In a number of cases the running head at the top of each page is not consistent in reflecting the content of the page beneath it. When readers encounter these typographical errors in the book, remember that they occur in the original printing and are faithfully reproduced in this facsimile reprint.

TITLE PAGE ICONOGRAPHY

On the right side of the title page is shown 'The wheele of Fortune whose ruler is Ignorance'. The wheel is supported by a female figure, blindfolded, holding a bridle in her left hand and standing on a globe. These are attributes associated with the goddess Fortuna: the bridle signifies restraint, the globe signifies instability, or possibly (in the Renaissance period) the world over which Fortune holds sway. The legend on the

Introduction

rim of the wheel reads, '*Qui modo scandit corruet statim*' (Who climbs at all will promptly fall). This was a familiar sentiment in the literature of the time and is perhaps expressed most memorably by Sir Walter Raleigh in an exchange with Elizabeth I: 'Fain would I climb, yet fear I to fall'. 'If thy heart fails thee, climb not at all'. On the left side is shown 'The Sphere of Destinye whose governour is Knowledge'. Here, too, the supporter is a female figure who may reasonably be identified as Virtue, often contrasted with Fortune and standing on a solid block, a cube, signifying stability, unlike Fortune's unsteady globe. In Virtue's right hand are a pair of dividers, a mathematicians instrument, and she holds aloft an armillary sphere. Above the Sphere of Destiny stands the sun; above the Wheel of Fortune is the moon. The allusions are clear. They combine the Aristotelian/Ptolemaic distinction between the sublunary sphere where all is perishable and the supralunary where all is stable. This is contrasted with the Neoplatonic distinction between the world of ideas or forms which is illuminated by the sun and is ultimately intelligible to the true philosopher, and the world of shadows, hazards and uncertainty illuminated only indirectly by the moon. In the centre is the Castle of Knowledge, from whose towers two astronomers survey the heavens. The one on the left measures angular distances in the starry sky with a quadrant, the one on the right holds aloft an astrolabe to measure the sun and the moon. Atop all sits Ptolemy, recognisable by his robes and the crown on his head. It was a common misunderstanding in medieval times to confuse this Ptolemy – Claudius Ptolemy of Alexandria, the renowned astronomer and geographer – with the one-time ruling Ptolemy dynasty in Egypt. Hence the crown!

THE DEDICATORY EPISTLE AND PREFACE

In the dedicatory epistle addressed to Queen Mary, Recorde announces his purpose and explains his title: 'to buylde a

Introduction

Castle for Knowledge to reste in, after hir longe banishment & tedious exile'. There is also a Latin dedication to Cardinal Pole, Mary's newly appointed Catholic archbishop of Canterbury. In the preface to the reader Recorde extols the heavens as the theatre of God's mighty power, and pronounces the celestial realm the chief spectacle of all his divine works.

NOTABLE PAGES

The Castle of Knowledge contains the celebrated reference to the heliocentric theories of Nicolas Copernicus (on page 165, or page 181 according to the modern numbering), written just thirteen years after publication of '*De revolutionibus Orbium Coelestium*' (On the revolutions of the Celestial Spheres). This is the first mention ever in an English book of Copernicanism. The startling doctrines of Copernican theory, that the sun and not the earth is at the centre of the world and that the earth moves, were apparently so contrary to English common sense and so upsetting to the theologians, Protestant as well as Catholic, that they cannot be said to have been welcomed in England. Recorde's discussion is brief and cautious, but seems to imply approval. His promise to set forth the Copernican theories at a later time, in such a way that the scholar will be as eager to believe them as he is now to condemn them, makes it seem probable that Recorde himself not only rejected the Aristotelian arguments against the possibility of the earth's rotation, but was also prepared to accept all the implicit details of a heliocentric cosmos. But he may well have had good reason to proceed with caution in introducing or developing discussion of Copernican astronomy. The case of Richard Eden, a contemporary of Recorde, illustrates why this may have been so. In 1555 a large collection of the best and latest lore about the new geographical discoveries available on the Continent was published. Richard Eden, the editor and translator, prefaced the volume with a long and fulsome praise of Mary and Philip, obviously with

Introduction

the intent of ingratiating himself with their majesties. Unfortunately, an allegedly objectionable passage contained in the preface caught the eye of the Church. Eden was promptly dismissed from a position in King Philip's English treasury, and was hauled just as promptly before Stephen Gardiner, Bishop of Winchester, on heresy charges. Further difficulty was avoided by Gardiner's timely death, and nothing more was made of the charges. Because of this disturbing incident it is reasonable to assume that the lesson was not lost on Robert Recorde. *The Castle of Knowledge* closes poignantly with an autobiographical allusion on the very last page (page 284, or page 300 according to the modern numbering): 'Master. ...but nowe farewell for a time: I am dryven to omytte teachinge of Astronomye, and muste of force go learne some lawe'. Recorde was committed to the King's Bench Prison, for reasons that are not entirely clear, at some time during 1556 and he may have completed *The Castle* while imprisoned there. Recorde's mention of Pluto's forge in the introductory verses of *The Castle* no doubt refers to actual imprisonment and he was to die, whilst still a prisoner, two years later.

SOURCES
Readers wanting to know more about Robert Recorde and his famous series of mathematical books should consult the following: For an easily accessible biography visit the MacTutor History of Mathematics, 'Robert Recorde', [online] http://www-history.mcs.st-andrews.ac.uk/Biographies/Recorde.html. Written sources are:

Stephen Johnston, 'Recorde, Robert (c1512–1558)' *Oxford Dictionary of National Biography*, Oxford University Press, 2004.

Howell Lloyd, 'Famous in the Field of Number and Measure: Robert Record, Renaissance Mathematician', *Welsh History Review*, Vol. 2 (2000), pp. 254-282.

William Barr, 'A World View of Robert Recorde: A Brief Study of Tudor Cosmology, *Albion: A Quarterly Journal Concerned with British Studies*, Vol. 1, No. 1 (1969), pp. 1-9.

Introduction

Joy B. Easton, 'The Early Editions of Robert Recorde's Ground of Artes', *Isis*, Vol. 58, No. 1 (Winter 1967), pp. 515-532.

Joy B. Easton, 'On the date of Robert Recorde's birth', *Isis*, Vol. 57, No. 1 (Spring 1966), p. 121.

Margaret E. Baron, 'A Note on Robert Recorde and the Dienes Blocks', *The Mathematical Gazette*, Vol. 50, No 374 (Dec 1966), pp. 363-369.

Louise Diehl Patterson, 'Recorde's Cosmography, 1556', *Isis*, Vol. 42, No. 3 (Oct 1951), pp. 208-218.

E.R. Sleight, 'Early English Arithmetics', *National Mathematics Magazine*, Vol. 16, No. 4 (Jan 1942), pp. 198-215 and Vol. 16 No. 5 (Feb 1942), pp. 243-251.

Francis R. Johnson & Stanford V. Larkey, 'Robert Recorde's Mathematical Teaching and the Anti-Aristotelian Movement', *The Huntingdon Library Bulletin,* No. 7 (Apr 1935), pp. 59-87.

David Eugene Smith & Frances Marguerite Clarke, 'New Light on Robert Recorde', *Isis*, Vol. 8, No. 1 (Feb 1926), pp. 50-70.

David Eugene Smith, 'New Information Respecting Robert Recorde', *The American Mathematical Monthly*, Vol. 28, No. 8/9 (Aug–Sep 1921), pp. 296-300.

Frank V. Morley, 'Finis Coronat Opus', *The Scientific Monthly*, Vol. 10, No. 3 (Mar 1920), pp. 306-308.

HERE BEGINS
THE CASTLE OF KNOWLEDGE

The Castle of Knowledge.

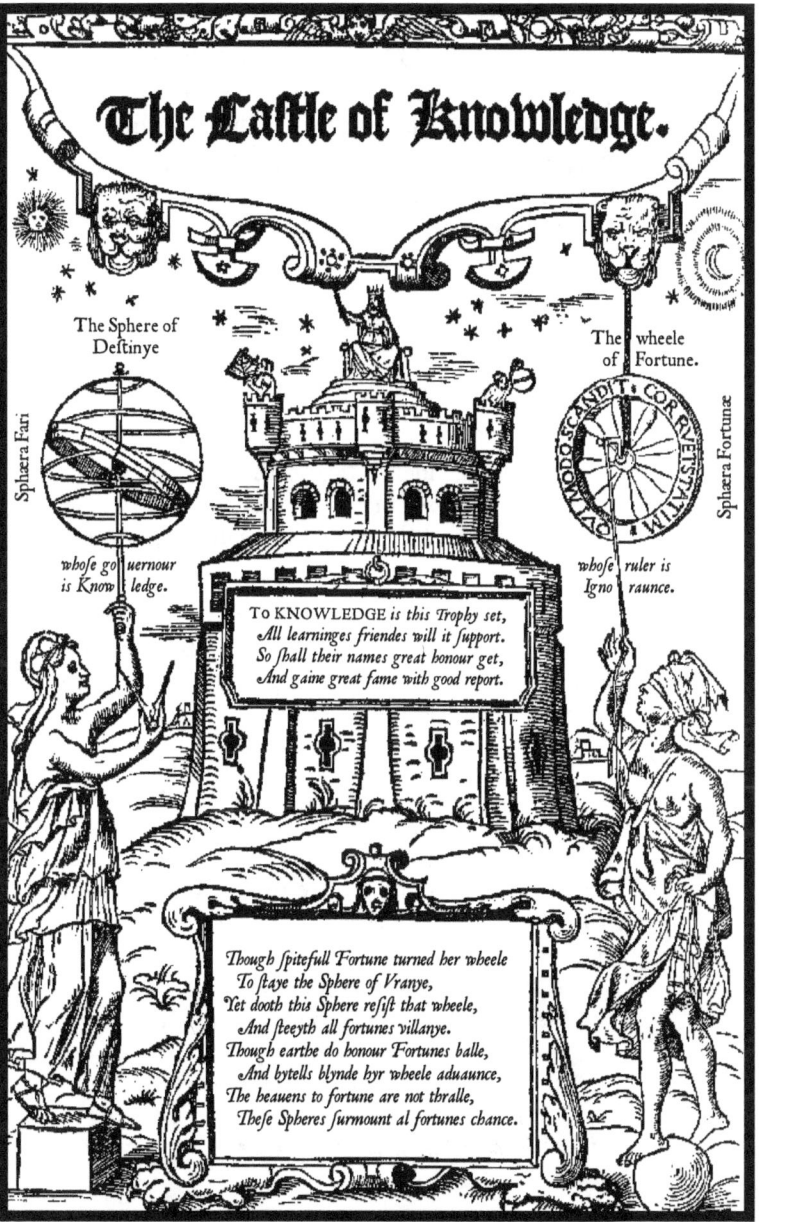

The Sphere of Destinye

Sphæra Fari

The wheele of Fortune.

Sphæra Fortunæ

whose gouernour is Knowledge.

whose ruler is Ignoraunce.

To KNOWLEDGE is this Trophy set,
All learninges friendes will it support.
So shall their names great honour get,
And gaine great fame with good report.

Though spitefull Fortune turned her wheele
To staye the Sphere of Vranye,
Yet dooth this Sphere resist that wheele,
And steeyth all fortunes villanye.
Though earthe do honour Fortunes balle,
And bytells blynde hyr wheele aduaunce,
The heauens to fortune are not thralle,
These Spheres surmount al fortunes chance.

The contentes in briefe of the 4 Treatiſes of

THE CASTLE OF KNOWLEDGE

CONTAINING THE EXPLICATION OF THE SPHERE

bothe celeſtiall and materiall, and diuers other
thinges incident therto. With ſundꝛy plea=
ſaunt pꝛoofes and certaine newe demon=
ſtrations not wꝛitten befoꝛe in any
vulgare wooꝛkes.

The firſt treatiſe is an introduction into the Sphere, declaringe the neceſſarye partes of it, as well for the materiall
Sphere, as for the celeſtiall : And that no partes of it are
admitted without profitable vſe.

The ſecond treatiſe doothe teache the makinge of the
ſphere, as well in ſound and maſſye forme, as alſo in Ringe
forme, with hoopes : And the proportions of eche of them
iuſtly deſcribed.

The thyrde treatiſe dooth briefly declare certain thinges
appertaininge to the vſe of the Sphere, and other matters
thervnto incidente : without proofe or demonſtration : and
that briefly, for eaſineſſe in learninge and remembringe.

The fourthe treatiſe doth approue manye thinges, that
were noted in other partes before : and beſide then addeth
diuers other maters, concerninge the neceſſarye vſe of the
ſphere, whiche were not touched before, and doth bring demonſtration or other certaine proofe for the perſwadinge
of them : wherein are many Tables ſet forth very pleaſaunte
and profitable.

If ought here want, that you deſire,
Remembre where this woorke was wrought :
In Plutos forge with ſcarſe good fier,
This ruſtye Sphere to eande was brought.
But if I may it fyle agene,
The ruſte I truſte to ſcour of clene

TO THE MOSTE MIGHTIE AND
MOST PVISSANT PRINCESSE MARYE, BY

the grace of God Queene of England, Spain,
bothe Siciles, Fraunce, Jerusalem, and Irelande :
Defendour of the faithe : Archeduchesse of Austria : Duchesse
of Millayne, Burgundye, and Brabaunt. Countesse of
Habspurge, Flaundres, and Tyroll. ꝛc.

AS LOVE OF LEARNYNGE AND
zeale vnto knowledge (most dradde so=
ueraine Ladye) dyd prouoke me to at=
tempte an enterprise farre aboue myne
habilitie, that is, to buylde a Castle for
knowledge to reste in, after hir longe
banishment & tediouse exyle. Althoughe
I could not be permitted by disturbaunce of cruell For=
tune, to accomplish now my buyldyng as I had drawen
the platte : yet in despite of Fortune, thus muche haue I
doone. which is more then euer was done in this tonge
before, as farre as I can heare. But considering by mis=
fortune this Forte lacketh sence, and needeth som good
gouernoure to supplye that that wanteth, that Know=
ledge maye reste vnder safe protection, I thought it my
duetye to make moste humble sute vnto your excellente
Maiestie, that it might please your highnes to accepte
this poore Castle into your gracious tuition : that not
only in time of your Maiesties raign, but by your high=
nes speciall defence, knowledge myght bee maintained
and reuoked fró exyle. Vnto whiche sute I am the more
boldened, throughe remembraunce howe Godde in de=
spite of cancred malyce and of frowninge Fortune, dyd
exaulte your maiestie to that throne royall, whiche of
iustice dyd belonge vnto your highnes, althoughe the
musers of mischief wrought muche to the contrary. In
whiche matter as knowledge did detect the malyce of
other, and taught your true subiectes their duty to their
Soueraine, so knowledge yet diuers waies shall fur=
ther your Maiestie. And therefore am I encouraged to

a.ij.   sue

sue to your royall excellencye, not onlye for to take into
your highnes protection this Castle of Knowledge, but
all Knowledges friends, which in hir maintenaunce do
keepe continuall warre againſt peſtilente Ignoraunce,
the ſubuerter of Realmes : which knoweth no vertu, ho=
neſty, nor duety, and therefore meaneth no truthe, how
ſo euer ſhe flatter. yet doth ſhe often tymes ſhewe great
countenaunce of friendſhip, when ſhe meaneth nothing
leſſe. Here coulde J paint forth Ignoraunce in hir right
colours, but vnto your Maieſtie it is needleſſe, whome
God not only hath endewed with excellent knowledge,
but alſo hath ayded with ſuch prudent Coūcellars, that
it maye ſeeme arrogancy in any ſuche as J am, to make
explication, or in manner more then onlye inſinuation of
anye doubtefull matters. Jt maye therefore pleaſe your
Maieſty, for loue vnto Knowledge, and fauour to your
highnes ſubiectes, to accept this ſimple Caſtle into your
graces defence, and ſo ſhall J bee animated to fy=
niſhe the reſt, and to publiſh it vnder your Ma=
ieſties name. whome God of his mercy in=
creaſe in all honour royall, and true fe=
licity, and continue proſperouſlye
and longe amongeſt vs.
Amen.

Your Maieſties moſte humble ſubiecte,

Roberte Recorde Phyſicion.

INCLITISSIMO CARDINALI

POLO, CANTVARIENSI ARCHIEPISCOPO &C.

Reuerendiſſimo Archiepiſcopo Eboracenſi, Nico-
lao, ſummo Angliae Cancellario. ac vniuerſo ſacrae
Regiae Maieſtatis Conſiliariorum Praeclariſſimorum
Senatui, dominis maxime ſuſpiciendis.

POLLOPHANES clarusille ſophiſta, qui in He
liopoli Aegypti ciuitate vna cū Dionyſio Areo-
pagita eo ipſo tempore fortè degebat, quo Serua
tor hominum Chriſtus crucis mortem ſuſtinuit,
quum admirandam illam eclipſim conſpexiſſet,
reſpōdiſſe dicitur : Θἐωρ ἀμοιβαὶ πραγμάτων. Dio-
nyſius verò altius quodāmodo adſpirans, ἤ τό θἐι-
ορ (inquit) πάσχϳ,ἤ τῷ πάσχοντι συμπάσχϳ. Adeo cer
ta quidem ratio eſt coeleſtium motuum, vt li quid præter conſuetum
in coelo eluceat, noui cuiuſdam ac inſoliti euentus indicium certiſ-
fimum eſſe conuincatur. Adde quod qua eſt benignitate Deus opti-
mus maximuſcꝫ, non vult homines inaduertentes opprimi, niſi eo-
rum ſupina admodum inertia, aut cōtumax planè malitia diuinas eas
admonitiones vecordius aſpernetur. Erunt (inquit Chriſtus) ſigna in
Sole & Luna. diuinae quidem in nos philanthropiae certiſſima teſti-
monia, ac noſtrae, ſineglexerimus, veſanie argumenta irrefragabilia. Si
ingratiigitur in deum dici horreamus, praeſertim in noſtra ipſorum
cauſa : imò ſi in ipſos nos iniurijeſſe, quod vitium naturæ aduerſiſsi-
mum cenſetur, nolimus, coelum aſſidue contemplemur, diuinam in
eo potentiam ſuſpiciamus, prouidentiam admirantes amplectamur,
ſapientiarn adoremus & exoſculemur. ſiquidem dicente Propheta,
ὁι ὀυρανὸι δικγουῦται Δόξαν θἐꝹ. atqui ne quis ad formam coeli, & motus
tantum referat, ημἐρα (inquit) τῆ ἡμἐρᾳ ἐριύγκται ῥῆμα,καὶ νὑξ τῆ νυκτῒ
ἀναγγἐλἊ γνῶσιψ. Serenitatem itaꝗ veſgtram rogo, ac per pietatem ob-
teſtor, per celſitudinis apices, honorumꝗtitulos, quos diuina fauen-
te clementia adepti eſtis, obſecro : vt quod alij multiex ſumma pru-
dentia in vobis probant, id vos viciſſim in alijs exoptetis. adꝗ ea
ſtudia alios, ingenua precipuè indole præditos, à vanis ludicriſꝗ ex-
ercitijs, ne dicam improbis planecꝗimpijs, reuocetis. Penes celſitudi-
nem excellentiasꝗ veſtras eſt, ſubditorum ſtudia moderari, exercitia
preſcribere, impetus eſſrenatos coercere. Vos oculi, aures, adeocꝗ mès
ipſa Regiae Maieſtatis eſtis. Vos regniſydere poſt ſolem ac lunam ip-
ſam ſplendidiſſima collucetis. Vos omnes probitanquam patrie pa-

         aiij.   rentes

EPISTOLA.

rentes, imo terreſtres deos cernui adorant : veſtris veſtigijs aduoluun-
tur : opem veſtram niſi aſſidue ſenſerint, actū plane de ſeiure optimo
putant. At haec ſtudia fortaſſis quibuſdam male feriatis ingenijs pa-
rum reipublice commoda, eoꝗ veſtro fauore aut ſubſidio indigna
videri poſſunt. Aliter longe exiſtimauit Atlas rex, qui inde ſibl æter-
nitatis nomen meruit, coelumꝗhumeris ſuſtinere praedicatur, quod
Aſtronomiae ſtudioſiſſimus, ſydera obſeruarit ſedulò. Hūc Euſebius
Enoch eſſe arbitratur. Hic inter Titanos praecipuus erat. quos ſi
recte intucamur, veneratione, nedum admiratione dignos cenſebi-
mus : quod induſtria maxima altiſſimos montes ſcandentes, ibiꝗ in
deſeſsi pernoctantes, ſydera obſeruando, munia cuiusꝗ vero animad
uerterint, primiꝗ oſtenderint ea vnius ſummi Dei imperio parere,
nec deos eſſe : vanamꝗ grioilium deorum opinionem arguerint. eoꝗ
Iouem coelo deturbare conatos eols poëtæ aſſerunt. quo nomine quā
tum illis debeat ſyncerior religio, pij omnes agnoſcūt. Liceret hic, ni
longioris commemorationis tedium vitarem, referre Orionem, Hy-
perionē, Endymionem lunae amaſium, Ganymedem, Adonim, Aeo-
lum, Phaëtontem, & Ptolemæos, omnes principes viros, & aſtronomiae ſtudioſos, vt qui obſeruationibus inuigilarint, motuscꝗ ſyderum
notarint. Alfonſiueròregis præclariſsimi non vnquam intermoritu-
ram famam, ex hac arte multo celebriorem redditam, omnes norunt.
Quin ceſſo artem omni laude maiorem amatoribus eius ſummis eni
    xius obtrudere? Hæc eſtilla maxima ſecundum Theologiam ſci-
       entia, ſolo ſilentio predicanda. Veſtrae itaꝗcelſitudini tam
          eam quam alumnos eius omnes, precipuè verò Recor-
             dum, ſupplex commendo. Deus vobis omnia ſe-
                cunda donet, ex animi ſententia.

                    Celſitudini excellentieꝗveſtrae deditiſſimus

                    Robertus Recordus Medicus.

THE PREFACE TO THE
READER.

If reaſons reache tranſcende the Skye,
Why ſhoulde it then to earthe be bounde?
The witte is wronged and leadde awrye,
If mynde be maried to the grounde.

THEREFORE,

HEN SCIPIO BEHELDE OFTE *of the high heauens the ſmallenes of the earth with the kingdomes in it, he coulde no leſſe but eſteeme the trauaile of men moſte vaine, which ſuſtaine ſo muche grief with infinite daungers to get ſo ſmall a corner of that lyttle balle. ſo that it yrked him (as he then declared) to conſidre the ſmalnes of that their kingdom, whiche men ſo muche did magnifie. Who ſoeuer therefore (by Scipions good admoniſhment) doth minde to a-uoide the name of vanitie, and wiſhe to attayne the name of a man, lette him contemne thoſe trifelinge triumphes, and little e-ſteeme that little lumpe of claye : but rather looke vpwarde to the heauens, as nature hath taught him, and not like a beaſte go po-ringe on the grounde, and lyke a ſcathen ſwine runne rootinge in the earthe. Yea let him think (as Plato with diuers other philoſo-phers dyd trulye affirme) that for this intent were eies geuen vnto men, that they might with them beholde the heauens : which is the theatre of Goddes mightye power, and the chiefe ſpectakle of al his diuine workes. There are thoſe viſible creatures of God, by which many wiſe philoſophers attainted to the knowledge of his in-uinſible power. There are thoſe ſtraunge conſtellations, by whiche Job doth prooue the mightye Maieſtie and omnipotency of God. There are thoſe pure creatures, whiche waxe not werye with la-boure, nother growe olde by continuance, but are as freſhe nowe in beutye and ſhape, as the firſte daye of their creation . and as apte nowe to perfourme their courſe, as they were the firſte hower that*

a.iiij.　　　　　　　　they

THE PREFACE

they began. *And thoughe time wholly depend of it, yet time cannot vtter anye force in it, yea thoughe all other thinges in the worlde by tyme be confumed, and euen the mofte harde metals freted into droffe, yet the liquide heauens not only gouerne time it felfe, but vtterly ftande cleere from all corruption of time. Oh woorthy temple of Goddes magnificence : Oh throne of glorye and feate of the lorde : thy fubftaunce moft pure what tonge can defcribe? thy beuty with ftarres fo garnifhed and glytteringe : thy motions fo meruailous, thine influence ftrange, thy tokens fo terrible, to ftonifhe mennes hartes. thy fignes are fo wonderous, furmountinge mannes witte, the effects of thy motions fo diuers in kinde : fo harde for to fearche, and worfe for to fynde. Thy greatnes fo huge, thy compaffe fo large, thy rollyng fo fwifte, and yet feemeth flowe : thy ftaye fo vnknowen, thy place without name : thy fpheres are mere wondres, and fo is thy frame. Thy lyghtes are fo lykinge to comforte mennes myndes, no beafte is fo brutifhe, but that hee ftyll fyndes, thy warmenes to woorke him greate folace and eafe : thy coloure to comforte his fight and his braine. Thy ftarres in fuche ordre, thy circles fo fine : thy platte forme is painted with manye a figne. Oh meruailous maker, oh God of good gouernaunce : thy woorkes are all wonderous, thy cunning vnknowen : yet feedes of all knowledge in that booke are fowen. The fignes of the tymes who can them comprife? the tokens of troubles what man could de uife? And yet in that boke who rightly can reade, to all fecrete knowledge it will him ftraighte leade. The ftarre in the eafte dyd gouerne the Wifemen, and taughte them the very region where Chrifte fhould be borne. And farther by it they vnderftode, that he was the true kynge of Jewes, and fauiour of Ifraell. And thoughe manye fawe the ftarre as well as they, yet fewe or none knewe the fignification but they. yet dyd God at the beginning ordaine the ftarres to be as fignes and tokens of times alteration : and namely of fuche ftraunge effectes as feldome come in vre, and therefore are knowen but to fewe men. Thefe woorkes the more*

ftrange

THE PREFACE

strange they be, the more oughte men to esteeme the frute of them :
to magnifie the knowledge of them, and to studye to vnderstande
the mean to attaine them, but most of all to honour, praise and glo-
rifie the author of them. who willeth nothinge to happen so so-
denly on the moste wicked, but by som signes and tokens hee giueth
warnyng of them. of which thing who so euer standeth in doubt,
let him peruse the state of tymes, and hee shall see wonderouse
thinges. Before the floude of Noe althoughe God did by speciall
reuelation vtter his mynde to his seruaunte Noe, yet dyd hee also
by wondrefull signes and straunge coniunctions, expresse the same
to the whole world. for all the Planetes were in coniunction in wa-
terye Signes. so that no nation might excuse them selues, for that
they were so farre distaunte from Noe, that they could not heare
his preachinge sith all nations myght see the heauens and the to-
kens in it, althoughe but fewe in euery nation coulde skyll of them.
And thoughe Noe coulde not in person go into all partes of the
worlde, yet was that office supplied by the heauens, of whose re-
uolutions it is written by Dauid the prophet : They haue no speach
nor language, so that their voice can not bee hearde. yet did their
course extende into all the earthe, and their woordes into the ex-
treame boundes of the worlde. So was there neuer anye greate
chaunge in the worlde, nother translations of Imperies, nother
scarse anye falle of famous princes, no dearthe and penurye, no
death and mortalitie, but GOD by the signes of heauen did pre-
monishe men therof, to repent and beware betyme, if they had any
grace. The examples ar infinite, and all histories so full of them,
that I thinke it needeles to make any rehersall of them now : espe-
cially seeyng thei appertain to the Iudicial part of Astronomy,
rather then to this parte of the motions, yet shall it not bee preiu-
diciall anye waies, to repeate an example or twoe. As namelye
before the buildinge of Rome, there was a verye notable eclipse
of the Sonne, declaringe that the libertye of the worlde beganne
then to decay, whē Rome began to rise: which shuld subdue all the

a.v.   worlde

TO THE READER

world neare hand : as in effect afterwarde it dyd succeede, increa-
singe styll by lytle and little, and continuynge for a longe tyme, tyll
the Gothes in the time of Arcadius and Honorius, did spoile that
citye, and subdue their power. At which time also straunge signes
dyd appeare in the ayer, and in the skye : whiche seemed not onlye
to signifie the deuastation of the Imperye of Rome, but also the
subduyng of all the weste prouinces, by straunge inuasion of bar-
barous nations. Many other straunge eclipses both of Sonne and
Moone, beside the appearing of sondrye Sonnes, and straunge
shapes of the Moone, and the starres diuerselye disordered, with
Rainbowes of meruailous formes, Cometes of diuers kindes, and
other wonderfull signes, whiche euer were messangers of as won-
derfull effectes, of newe innouations, straunge transmutations,
and sometime vtter subuersions, not onlye of small prouinces, but
also of great kingdomes, yea and of many regions at ones. And
therefore sayth M. Manilius.

Nunquam futilibus excanduit ignibus aether.
The earthe doth euer feele griefe and teene,
When those straunge syghtes in heauen be seene.

But who that can skyll of their natures, and coniecture rightlye
the effect of them and their menacynges, shall be able not only to
auoide many inconueniences, but also to atchiue many vnlikelye at-
temptes : and in conclusion be a gouernoure and rulare of the stars
accordynge to that vulgare sentence gathered of Ptolemye :

Sapiens dominabitur aftris.
The wife by prudence, and good skyll,
Maye rule the starres to serue his will.

I mynde not to discourse in declaringe the profite and commodity
of Astronomye, but only to admonishe briefly the reader, that hee
maye thinke the study woorthye his trauaile, and to knowe it to be
the most necessary studye that can be, for anye man that desireth
perfection of wisedome. What benefite doth come by it to the true
knowledge of husbandrye and nauigation, I am assured the verye
simplest in those artes do partlye perceaue : and the cunningest

THE PREFACE

in the fame do fo fullye vnderftande, that they iudge them felues
naked and bare without it, and vtterlye deftitute of all excellency
in their arte. In phyficke the vfe of it is fo large in iudginge duely of complexions, in prefcribinge righte ordre of diete and con-
uerfation, in gouernaunce of healthe, for iufte miniftration of me-
dicines in time of fickenes, and in righte iudgement of the Criticall daies, that without it phyficke is to be accompted vtterlye imperfecte. For proofe wherof althoughe there be infinite places in
Hippocrates and Galene, and diuers other good writers, yet hee
that hathe readde in Hippocrates but that one booke of Ayer,
water, and Regions, and Galen his third boke of Criticall daies,
can not be ignoraunte howe neceffarye an inftrument Aftronomy
is vnto Phyficke, as bothe thofe bookes do teftifie at large. But
omittinge the teftimonies of famous wryters (whiche would make
a wonderfull volume of them felues, if they were written only together) I wyll vfe a fimple plaine proofe manifeft to all men, and
therefore mofte apte for to perfwade all men. Firfte to begin
with fowinge of graine, with graffynge and plantinge, who is fo
rude, but knoweth that without thefe be dulye doone, and in their
feafonable time, men can not conueniently lyue on the earthe? And
howe are their times knowen, but by the rifinge and fetting of cer
taine notable ftarres? Peraduenture fome man will anfwere, that
by the monethes of the yeare all men do know their times without
farther Aftronomy. whiche anfwere is fuche, as if a carpentar
or mafon fhoulde faye, that he can woorke with his compaffe, ru-
lar, fquire, plumbe rule, and fuche like inftrumentes, without any
knowledg in Geometrye. but how ridiculous an anfwer this were,
all men can iudge. Likewaies, if a mafter of a fhippe would fay,
that he can faile and gouerne his courfe by his compaffe and his
carde, with his quadrante and his other inftrumentes, without any
knowledge in Cofmographye or Aftronomye, would not all men
that heare him, deryde him, or thinke him madde, for fpeaking fo
vndifcreatly, efpecially fuch as know (as few ar ignorant therein)
                                                                    that

TO THE READER

that all thofe inftrumentes are made by thofe artes, and appertain
to them? So if the diftinction of times do depende of Aftronomy all toguther, and the monethes woulde foone runne out of their
courfes, if the ayde that it hathe by that arte were neglected, fo
that Michelmas day wold happen in the Spring time, and the An
nunciation of our Ladye would fall after harueft (as the truthe is,
it could do, if Aftronomic all accompte were not) who can fhew
him felfe fo madde as to denye the neceffarye vfe of Aftronomye,
in due keping the times of the yeares? The ecclefiafticall hiftorye
dothe declare at large, and other writers in greate numbre do teftifie, that greate controuerfye hath beene in the church, for the
righte obferuation of Eafter, which controuerfye could neuer be
decided but by the knowledge of Aftronomye. And of late
yeares in diuers councelles redreffe hath beene fought for the iufte
obferuation of it : confideringe that if errour be in it, all other moueable feaftes, are wrongly kepte by that occafion, and Lente dif-
placed fo, that fome tyme it hath beene kepte fooner then it ought,
and at other times later then it oughte. whiche faulte can neuer bee
redreffed but by aftronomy. Whereby it appeareth alfo manifeftly,
that in ecclefiafticall maters Aftronomy hath a great vfe. but that
is fo well knowen, that euerye man almofte doth confeffe it. And
generally who fo euer dothe take benefite by the dewe diftinction
of the yeare, he can not chofe but acknowledge that the fame com
moditie doth come by Aftronomy. If I fhould fpecially and per
ticularlye difcourfe in euerye kinde of fcience and artes, and fhewe
how they are ayded by aftronomye, I fhould make my preface ouer
longe, and repeate thinges that all men doth knowe. In lawe for
contractes and bargaines the time is mofte neceffarye to be obferued : but efpeciallye if they depende of moueable feaftes, wherein
aftronomy muft difcuffe the doubte. In Grammar, Logike and
Rhetorike howe needefull it is, and in hiftories alfo, I neede fay
nothinge, but remitte all men to the readinge of thofe bokes, which
are vfed in thofe artes, whereby it fhall appeare, that without the

                                                                          prin-

THE PREFACE

principles of Aſtronomye thoſe bookes can not bee vnderſtande.
Then for vulgare artes how the knowledge of ebbes and fluddes
doth profite, manye men, but ſpeciallye mariners can teſtifie : and
namely ſuche as vnderſtande, what errour commeth by the diffe-
rence of the true accompte therein and the vulgare accompte. A-
gaine for loppinge of trees and wodde fall, and diuers other ob-
ſeruations in huſbandry, the conſideration of the ſonne and com-
monlye of the moone doth greatly healpe. Wherfore I maye con-
clude, that in all artes and ſciences, in lawe, phyſicke and diuini-
tie, in mariners arte and huſbandrye, the profite of Aſtronomye is
exceding neceſſarye. But aboue all other thinges the teſtimonye of
Chriſte in the ſcripture doth moſt approue it, when he doothe de-
clare that ſignes of his coming, and of other ſtraunge effectes
ſhall be ſeene in the Sonne, Moone and Starres. Alſo for alte-
ration of wether he teſtified that many did marke the face of hea-
uen, and pronounced truly of the wether, and therefore blameth
them that thei coulde not marke and iudge the ſignes of the com-
ming of the Sonne of man. But here poſſiblye ſome men will ob-
iecte the ſaynge of the prophete : Feare not the ſignes of heauen
whereunto I maye duelye anſwere : that thoſe woordes of Hie-
remye do forbidde honouringe of them as goddes, as the texte
is plaine. for oftentimes in the ſcriptures fear of God is taken for
honoure of God, and ſo is it here. els other wayes might I anſwer
that the true ſeruauntes of God whiche haue repoſed the loue and
feare of God in their heartes, are neuer aferde of any tokens that
God ſendeth, but reioyce to ſee them, and glorifie God for them.
But bicauſe in this caſe there be manye diuines that can better de-
clare thoſe thinges then I whiche am a man of an other profeſſi-
on, I will remitte that matter to them. only admoniſhing all men,
that the Sonne, the Moone and the Starres, were ordained of
God to ſerue all nations that be vnder the heauens, as Moſes
dooth teſtifie. Then ſeynge God hath made them for mannes com-
moditie, and to be diſtincters of times, and for ſignes and tokens,
for

TO THE READER

for aide of mennes knowledge, let not men be vnkinde to God a=
gain, but lyfte vp their eies to heauen and beholde the good guiftes
of God : Note diligently their meruailous motions, and studiouslye
confidre their wondrefull alterations, with perpetualle conftancye
and inuiolable ordre : fo fhall men neuer bee doubtfull of Goddes
prouidence towarde them, of his daylye prouifion for them, when
they fee that he hath made fuche an vnexplicable frame to ferue
onlye for mannes vfe, for whofe fake all other creatures alfo were
made. In token therfore of thankfulnes, let vs finge an Hymne vn
to that God, praifinge his name, and magnifynge him foreuer and
euer.

> The worlde is wroughte righte wonderouflye,
> Whofe partes exceede mennes phantafies:
> His maker yet mofte meruailouflye
> Surmounteth more all mennes deuife.

> No eye hath feene, no eare hath hearde
> The leafte fparkes of his Maieftie:
> All thoughtes of heartes are fullye barde
> To comprehende his Deitye.

> Oh Lorde who maye thy power knowe?
> What mynde can reache the to beholde?
> In heauen aboue, in earthe belowe
> His prefence is, for fo hee woulde.

> His goodnes greate, fo is his power,
> His wyfedome equalle with them bothe:
> No wante of will fith euerye hower
> His grace to fhewe he is not lothe.

> Beholde his power in the fkye,
> His wifedome eche where dooth appeare:
> His goodnes dooth grace multiplye,
> In heauen, in earthe, bothe farre and neare.

FINIS.

AN ADMONITION FOR THE

orderly trade of studye in the Authors woorkes, appertainyng
to the mathematicalles.

> *The grounde is thought that steddye staye,*
> *Where no foote faileth that well was pyghte:*
> *Whereon who walketh by certaine waye,*
> *His pase is lyke to prosper ryghte.*

1.   *The* Grounde of Artes *who hathe well tredd,*
    And noted well the slyppery slabbes,
    That may him force to slyde or falle,
    He hathe a staffe to staye withall.

2.   *Then if he trade that* Pathwaye *pure*
    That vnto Knowledge leadeth sure:
    He maye be bolde t' approche The Gate
3.   Of Knowledge *and passe in thereat.*

    Where if with Measure he doo well treate:
4.   *To* Knowledges Castle *he maye soone get.*
    There if he trauaile and quainte him well.
5.   The Treasure of Knowledge *is his eche deale.*

5.   *This* Treasure *though that some wold haue,*
3.   *Whiche* Measures *friendshippe do not craue,*
2.   *Nor walke the* Patthe *that leadeth the waye,*
1.   *Nor in* Artes *grounde haue made their staye,*
    Thoughe bragge they maye, and get false fame,
4.   *In* Knowledges *courte thei neuer came.*

Certaine faultes omitted out of the corrections.

10.29,proofe of my woordes. And in the meane ceason to procede as I began: you must. 212.1, differeth not. In thi table the fyrste. 279.17 defer-entes. 280.28, within the waddowe. 281.15, in euery common almanach. 283.21,alwaye runneth. 284.10. And the rather.

THE FYRST TREATISE OF
THE CASTLE OF KNOWLEDGE.

Whiche is an induction to the neceſſary partes of the Sphere,

as well celeſtiall as materiall.

SCHOLAR.

HE TIME SEMETH *The deſire*
longe (bee it neuer ſo *of know-*
ſhorte in deed) to hym *ledge.*
that deſirouſly looketh
for any thing : for as the
obtainĩg of it bringeth
great pleaſure, namelye
the thinge it ſelfe being
profitable, ſo the wante
therof cauſeth diſplea-
ſure and cõtinuall grief
tyll the deſire be eyther
fully ſatiſfied , other
partly (at the leaſt) accompliſhed.

Maiſter. And ſometimes we ſee, that when the deſire is
partly perfourmed, and the pleaſantnes of the ſame ones ta-
ſted of, the deſire therby nothinge aſſwageth, but contrarye
ways greatly increaſeth : and the more it getteth, the more
is deſireth. ſo that in this point may knowledge well be cõ-
pared to couetouſnes : for as the couetour mynd with get-
tyng is neuer ſatiſfied, ſo knowledge by knowing doth couet ſtyll more : And as it increaſeth, ſo doth it ſtill learne the
vilenes of Ignorance, and profite of Sciences, and therfore
can not reſt from ſearching more knowledge, as long as it
ſpyeth any ſpot of ignorance.

Scholler. This oftentymes as I haue conſidered, maketh
me to muſe what mynd is in them, which care for no knowledge, nor eſteeme any ſcience. *The groſe-*
nes of i-
Maiſter. This is the greateſt pointe of all ignorance, not *gnorance.*

A.i.   to

to know the groffenes of ignorance, and not to vnderftand
the benefite of knowledge, and with this faulte are a greate
numbre fpotted. The nexte is their faulte, whiche perceaue
fufficientlye what vilenes is in ignorance, and what profite
in knowledge, and yet of a certaine negligence partelye,
and partlye for other pleafures, they omytte to trauaylea-
nye whitte for knowledge, and contente them felues wyth
wilfull ignoraunce : but as thefe men do trouble the good
ftate of the worlde, fo the talke of them wyll hynder the
talke of the worldes knowledge, whiche is the thinge that
you fo muche longe after : and therefore befte it is, that wee
let them lye ftill tomblinge in the dyche of ignoraunce, and
that wee trauaile forward towarde the Caftle of knowledge.
But firft let me heare what is your chief defire.

The occa-
fion of this
booke.

　　Schollar. Syth my lafte talke with you aboute the
knowledge of the worlde and the partes of it, I haue readd
dyuers bookes that intreate of that matter, as namelye
Proclus fphere, Ioannes de Sacro bofco, Orontius cofmographye, and diuers other, whofe woordes in manye
thinges I remember, but of the matter I haue fondry doubtes, and therefore, defire muche your healpe therein. For
althoughe I haue confulted with diuers men therein, yet
me thynketh they tell me but the fame woordes in lyke forte
as I readde theym before, or lyttle other wayes altered,
but lyghte of vnderftandynge, I haue gotten lyttle
yet.

　　Mafter. Then proue againe, peraduenture your chaunce
may be better : that whiche at the fyrfte femeth harde, maye
at lengthe become eafy : for Vfe maketh mafterye, all men
confeffe. And, The beft thynges are not mofte ea-
fieft to attayne. begynne in that ordre as youre Authors doo.

The diuer-
fitye of
writers.

　　Scholar. Theyr ordres bee as dyuers as theyr names
be, fo that I knowe not whofe ordre is beft. For Proclus in
treatinge of the Sphere, defineth firfte the Axe tree of
                                                          the

the worlde, before hee had ſhewed other what the worlde is, or what hee calleth a Sphere, or what neede the worlde hathe of anie Axe tree. Therfore I tourned to Ioannes de Sacro boſco our contry man, whiche beginneth firſte with the definition of a ſphere, but nothinge lyke to that ſphere, whiche I before had bought, as an apt inſtrument to learne by. Then ſee I Orontius diſagree from them bothe : and generallie, euerye one from other, ſo that I know not wher to beginne.

Maſter. As touchynge thoſe writers, I will ſaye no more nowe, but although euerye one of them haue ſome thinges that exactlie ſcanned may be misliked, yet he that hath doone worſte, is woorthie of thankes, for his ſtudious paines in furtheringe of knowledge. And ſeyng you doubte of their ordre, lette the thinge it ſelfe miniſter ordre. What is it that you deſire to knowe?

Scholar. I ſee in the heauen meruailous motions, and in the reſte of the worlde ſtraunge tranſmutations, and therfore deſire muche to know what the worlde is, and what are the principall partes of it, and alſo how all theſe ſtraung ſightes doo come. *The argument of this booke.*

Maiſter. Then is the worlde the thinge that you woulde knowe firſt, ſyth all theſe other things are incident to it. What doo your authors call the Worlde?

Scholar. Orontius defineth the worlde to be the perfect and entiere compoſition of all thinges : a diuine worke, infinite and wonderfull, adorned with all kindes and formes of bodies, that nature coulde make. *What the world is.*

Maſter. This definition doth muche agree with thoſe that bee writen by aunciente authors, and namely Ariſtotle whiche defineth it thus.

κόσμΘ· ἐsὶ σύsεμα ἐξ οὐρανοῦ Ⰽϳ γῆς, Ⰽⰹ τῶῤὲῤ τούτεισ περιχομἐῳῤ φύσεῳῤ.

Mundus eſt compages ex coelo & terra, & reliquis in ijſdem contentis naturis.

                    A.ij.   The

4   THE FIRST TREATISE OF

The worlde is an apte frame of heauen and earthe, and all other naturall thinges contained in them. The like wordes hath Cleomedes and others. So that the worlde is that en-tiere body, whiche containeth all thinges that euer God made, and man can fee, nothinge excepted but God himſelf only, whiche is not comprehenſible by any worldly meanes. This worke is ſo pure and wonderfull in beauty, that it bea-

Wherof the worlde is named. reth the name of cleannes, bothe in Greke and Latine, that is κόσμ☉ in Greeke, and Mundus in Latine. and thereto allu-deth Sibyll in her verſes ſpeakinge of the diſſolution of the worlde, ſaying :

τοτται κόσμ☉ ἄκοσμ☉ ἀπολλυμβίων αὐθρώπων.

Erit mundus immundus, pereuntibus hominibus. The worlde (ſaith ſhe) ſhalbe vnclean, or leeſe his beuty, whē all mē ſhal periſh.

Schollar. And ſo dooth that ſentence leeſe his beautye by the tranſlation, for there canne bee no ſuche alluſion of woordes in the englyſhe of that ſentence, as there is in the other tongues.

Maſter. You ſay truthe, except a man wold rather allude at the woordes, then expreſſe the ſentence, for ſo might it be tranſlated thus : It ſhall bee an vnwordlye worlde, when all men ſhall periſhe : But here the ſenſe is loſte : for this name

Divers ſi-gnificatiōs of that worde worlde. Worlde, hath not the like deriuation of cleannes in englyſh, as the Latine and Greeke names haue in their tongues : no-ther can I well tell wherof this englyſhe name is deriued, al-though I remember ſom other ſignifications of this worde, as firſte it is vſed in Scripture for a name of long continu-ance of tyme, when we ſay : Worlde without ende. and, tho-rough worlde of worldes : whiche ſignifieth for euer. Alſo this name dooth ſignifye ſometymes a greate wonder, as when wee ſaye : It is a worlde to ſee the crafte that ſome menne vſe vnder colour of ſimplicitye. Nowe if anye man wyll contende, that this worde Worlde dooth principal-lye betoken a wonder, and that the worlde for the won-derfull ſhape of it, tooke that name, as the chieffe won-

der

THE CASTLE OF KNOWLEDGE 5

der of all wonders, I will not greatelye repine, but then
mufte I needes wonder, to fee the chieffe worldely men
to wonder fo lyttle at this wonderfull wonder, and to bend
all theyr ftudye to the centre of the worlde, I meane the
Earthe, whiche in comparifon to the whole worlde is not
onlye a parte without all notable quantitye, but alfo leafte
adourned with meruailous woorkes, and mofte fubiecte
to all frayle tranfmutation and chaunge, ftyll repleni-
fhed with continuall corruption. And yet on it only doth
the greateft numbre fet all their ftudye. For it they fu-
ftaine great trauaile and toyle : for yt they chide, quarrell
and fyghte : to gette it they venter lyfe and lymme, and
when they thynke mofte affuredlye that they haue gotten the Earthe, then in deede the earthe hathe gotten them,
and mofte commonlye then doothe the earthe confume
them, when they thinke theym felues fulle maifters
of yt.

Schollar. By thefe mennes trauaile (I thynke) it came
to paffe, that the earthe doothe vfurpe the name of the
Worlde, as thoughe it were all, and that befides it were
nothinge.

Mafter. Thereof commeth that common Prouerbe
of a couetous manne : All the worlde is to lyttle for him.
where he in deede feeketh nothynge but the earthe, whiche *The fmale-
earthe in comparifon to the whole worlde beareth no grea- nes of the
ter vewe, then a muftarde corne on Malborne hylles, the whole
or a droppe of water in the Occean fea. for of all the par- worlde.
tes of the worlde, the earthe is the leafte, and that withoute comparifon, as hereafter I fhall not onlye tell you,
but alfo prooue it by inuincible reafon. And therefore to proceede in oure matter, I thynke it befte not
onlye to make difcourfe lyghtlye of the principall partes of the worlde, but to dooe it in fuche a brief forte, *The beft
as the mynde maye conceaue it fooneft, and the memo- ordre in
rye alfo retaine it longeft : and therefore will I omytte teachinge.

A.iij. all

6   THE FIRST TREATISE OF

all proofes, tyll we haue ones generally drawen the ymage
of the whole worlde, so shall not your memory be troubled
with sundrye thinges at ones, as in learnyng a science whiche seemeth sumthing straunge, and in conceauyng the reasons of it, whiche in declaring, seeme much more straunge.

Scholar. In deed I haue felt the discommoditie of suche
hasty desires : for where I haue sought reason, before I vn-
derstoode, whereto that reason tended, I haue troubled
my mynde, and hyndred my knowledge. wherefore it may
please you in your ordre to procede.

The ordre of the ele- ments.

Master. I haue all ready sayd, that of all the partes of the
worlde the Earthe is the leaste : wherby you may conceaue,
that within it is nothyng : for so should that (what so euer
it were) be lesser then the earthe. but without the earthe,
dooth the Water lye, whiche couereth a great parte of the
same : about them bothe, dooth the Ayer run, and occupieth (as we maye easilye consider) muche more roome, then
bothe the sea and the londe : aboue the ayer, and rounde a-
bout it, (after the agreement of the moste wise men) dooth the
Fyer occupye his place. And these foure that is, earth, water
ayer and fyer, are named the foure elementes, that is to say

All thinges compounde ar made of the foure elementes.

the fyrste, symple and originall matters, whereof all myxt
and compounde bodies be made, and into whiche all shall
tourne again.

Scholar. Oftentimes haue I heard it, that bothe man and
beastes are made of earthe, and into earthe shall retourne
againe : but I thought not that they had been made of water, and muche lesse of ayer or fyer.

Master. Of earthe only, nothinge is made but earthe :
for an herbe or tree can not growe (as all men confesse) excepte it be helped and nourished with ayer conuenient, and
due wateringe, and also haue the heat of the Son. and generally, syth all thynge is maintained by his lyke, and is de-
stroyed by his contrarye, than if man can not be maintained without fyer, ayer and water, it must needes appeare,
                                                                    that

THE CASTLE OF KNOWLEDGE   7

that he is made of them, as well as of earthe, and fo likewaies all other thinges that be compounde.

Scholar. This talke delyteth me meruailoufly, fo that I
can not bee wearye of it, as longe as it fhall pleafe you to
continue it.

Maifter. This talke is not for this place, partly for that
it is more phyficall then aftronomicall : and partly bicaufe
I determined in this firfte parte, to omitt the caufes and reafons of all thinges, and brieflie to declare the partes of the
worlde, whereof thefe foure elementes, beinge vncom-
pounde of them felfe, that is fimple and vnmixt, are accōp-
ted as one parte of the worlde, whiche therfore is called the
Elementarie parte, and bicaufe thofe elementes do dailye increafe and decreafe in fome partes of them (though not in
all partes at ones) and are fubiecte to continuall corruptiō,
thei are diftinct from the reft of the worlde, which hath no
fuche alteration nor corruption, whiche parte is aboue all
the foure elementes, and compaffeth them about, and is called the Skie, or Welkin, & alfo the Heauens : this part hath
in it diuers leffer or fpecial parts, named cōmonly Spheres :
as the fphere of the Moone which is loweft, and nexte vnto
the elementes : then aboue it, the fphere of Mercury : and
nexte to it the fphere of Venus : then foloweth the Sonne,
with his fphere : and then Mars in his ordre : aboue him, is
Iupiter : and aboue him, is Saturne. Thefe feuen, are named
the feuen Planetes, euery one hauinge his fphere by himfelfe
feuerallie, and his motion alfo feuerall, and vnlike in time to
anie other. But aboue thefe feuen planetes, is there an other
heauen or fkie, whiche commonly is named the Firmament, and hath in it an infinite numbre of ftarres, wherof it
is called the Starrye fkie. and bicaufe it is the eighte in ordre of ẙ heauēs or fphers, it is named alfo the Eight fphere.
This heauen is manifeft inough to all mennes eies, fo that
no man needeth to doubte of it, for it is that fkie, wherein
are all thofe ftarres that we fee, except the fiue leffer planets,
                              A.iiij.   whiche

*The elemē-
tes are fim
ple.*
*The elemē-
tes do alter
dailye in
their parts.*
The fkye.
*The ordre
of the
fpheres.*
*The feuen
Planetes.*

whiche I dyd name before, that is Saturnus, Iupiter, Mars,
Venus and Mercurye.

Schollar. The Sonne and Moone alſo muſt bee excepte
oute of that numbre, for they haue their ſpheres by them
ſelues, as well as the other Planetes.

Maſter. Truthe it is. but bicauſe no man dooth accompte them as ſtarres, therefore they neede none exception, where mention is made of ſtarres onlye, where
as the other fiue ſmaller Planetes (which I named before) ar
ſo like to other ſtarres, that no manne, but ſuche as are
of good experience in Aſtronomy, can diſcerne them from
the other ſtarres, although manye men doo make a difference of them by twinkelinge, affirming that the Fixed ſtarres doo twinkle, and not the Planetes, with other differē̄ces
difficult to obſerue, and ſcarſe certeine in diſtinction. But
this is their moſte certaine difference, that all thoſe ſtarres,
whiche be in the firmament, do ſtande and continue in one
forme of diſtaunce eche from other, and chaunge not their
places in their ſphere, and therefore be they called Fixed ſtarres : for althoughe thei go rounde aboute the worlde in 24.
houres, that is euerye day ones, yet they keepe their places
in their ſphere, and tourne onlye with their ſphere : or (as
Aratus ſayth) thei be drawen with their heauen, wher as the
ſeuen Planetes are not only carried round about the earthe
with the like motiō of heauen euery day, but they do moue
of them ſelues, and doo chaunge their places in their owne
ſpheres, and for that cauſe are they called Planetes, that is
to ſay, Wanderynge ſtarres.

Scholar. Oftentimes haue I hearde this, but yet can I not
tell howe to perceaue it.

Maiſter. That ſhall be referred to the fourth treatiſe, wher
I wyll ſhewe you the proofe of all that you ſhall thinke
doubtfull.

Scholar. Yet I beſeche you lette me knowe this, Whye
are thoſe heauens called Spheres? for (in my phantaſye)
they

Howe the Planets are knowen from other ſtarres.

THE CASTLE OF KNOWLEDGE 9

they are nothinge like that inftrument of fundrye cirkles, whiche is commonly called the Sphere, fyth neither can I fe in them fuche cyrkles as are in that materiall fphere : nother is there in the materiall fphere anye fuche reprefentation of fuche dyuers heauens, nother of fuche varietie of ftarres.

Maifter. This doubte was moued before nowe, by Ioa-chim Ringelbergh, in a treatife that he wrote of the Sphere, but it fhall be anfwered eafily by your felfe, after a lyttle de-claration of the celeftiall fpheres. And for that caufe, I wyll omitte it tyll anone, and will firfte declare certaine other ac-cidentes of the heauens, and of the other partes of the worlde.

Hitherto you haue hearde onlye the names of the partes of the worlde, and of their fituation, howe they be placed in ordre. Nowe for the forme and fhape of them, you fhall vnderftande, that the whole worlde is rounde exact-lye as anye ball or globe, and fo are all the principall par-tes of it, euerye fphere feuerallye and ioyntlye, as well of the Planetes, as of the Fixed ftarres, and fo are all the foure Elementes. And they are aptly placed togither, not as a numbre of rounde balles in a nette, but euery fphere indu-

The forme of the world and his partes.

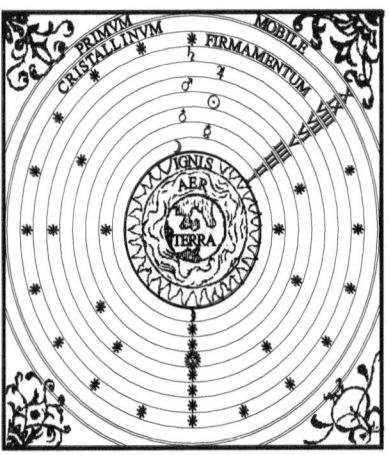

deth other, as they be in ordre of great-nes, beginning at y̆ eighte fphere or fir-mamente, and fo de-fcending to the lafte and loweft fphere, is the Sphere of the Mone : vnder which the foure elementes fuccede : firft the fier, then the ayer : nexte foloweth the water : which with the earth
ioyntlye

ioyntlie annexed, maketh as it were, one ſphere only.

Scholar. This I do well vnderſtande in wordes, and the eaſier by this picture, whiche I finde in euerie booke of the Sphere, but that I ſee there more ſpheres, then you ſpeake of : for in ſome bookes mention is made of nyne ſpheres : and in other are ten ſpheres named, where you ſette foorthe but eighte.

Maſter. The cauſe of this diuerſitie will I in the fourthe treatiſe declare : in the meane ſeaſon, I thinke it beſt to tell you of no mo ſpheres, then are perceptible by ſighte, for ſo manye are we certaine of. And therefore vnderſtande you thus, that as the eihhte ſphere is the greateſt, and hath none other without him that may be ſeene, ſo the earthe is the leaſte, and hathe none other within hym, but it ſtandeth in the middle and in the centre of the whole worlde, and of euery one of theſe ſpheres, and therfore it is called the Centre of the worlde : ſo that although the earthe in itſelfe haue a greate and notable quantity, yet in compariſon to the firmament, it is to bee eſteemed but as a centre or little pricke, yea in deed muche leſſe than any notable ſtarre that you ſee, & if I ſhall ſpeak boldly that which I intend herafter to proue certainly, the earthe is leſſer then the leaſte ſtarre in the firmament whiche is commonly ſeen, but yet is it greater thē Venus or Mercury, yea greater then the Moone.

Schollar. This affirmation ſeemeth to me impoſſible, or at the leaſt contrary to ſence : for the Mone ſeemeth bygger muche then any ſtarre, yea ſomwhat bigger then the Sonne.

Maſter. Content your ſelfe to credite me, tyll tyme ſerue for the proof of my woordes, and in the meane ſeaſon, to procede as I began. You muſt thinke, that the earth and the water annexed togither in one globe, are of no notable quantitye, in compariſon to the firmament, and that it ſtan deth as the centre of the worlde, and hath no motion out of his place, nother yet circular mouyng about his owne centre, but reſteth (as we may ſay) quiete without all ſuch mouyng,

The earthe is the cētre of the worlde.

The earthe hath no quantity in reſpecte to the world.

The earthe hath no motion.

uynge. Lyke wayes muſt you thinke of the other elementes, whiche of their owne nature haue none other motion then a ſtone or a lyghte fether, ſo that they may be accompted all four to be without naturall motion.

Scholar. Yet in the water and in the ayer we ſee euerye day notable mouynge. and ſometime I haue hearde of mouynge of the earthe, by earthquakes : and as for the fyer that we ſee, it alwaies moueth and flyckereth in burninge.

Maſter. And ſo you haue ſeene a ſtone moue ſwiftelye, when it fell from anye hyghe place. but theſe motions haue an ende quicklye, excepte they be continued with violence, as hereafter I will ſufficietlye declare. But as the ſtone although it wyll moue in fallinge, yet in his place lyeth quiete without motion : ſo the earthe of it ſelfe, and the other elementes muſte be accompted quyete by nature, and without motion.

¶The heauens contrarye wayes haue ſuche a naturall motion that neuer reſteth nythte nor daye, nother can be ſtaied by any violence. This motion wee ſe in the heauens daylye by their mouinge from the eaſte to the weſte, and from the weſte to the eaſte againe, aboute the earthe, ones euerye 24. howers, and therfore is thys motion named the Daily motion, for it is the meaſure of a Naturall day, commonly accompted. and this motion is lykewayes called of aunciente writers the motion of the Firſt firmament, accordynge to whiche motion you ſee the Sonne in the daye tyme, and the ſtarres in the nyghte tyme, and the Moone both in the day and the nyghte, to paſſe from the eaſte into the ſouthe, and ſo into the weſte, and at the ende of 24. houres to come againe into the eaſte : wherby you may eaſily vnderſtand, that this motion is common to all the ſpheres of heauen. *The motions of the heauens.* *A Daye.*

Scholar. This maye all men ſee, that can ſee any thing. yet haue I heard of ſome ſo groſſely witted, that they doubted which way the Son and the Moone dyd come into the eaſt agayne, as though they did not thinke that the ſkye dydde moue

moue about the earthe.

Mafter. Suche groffe ignorance happened fomtymes to famous men, for lacke of due confideration of that, whiche all men maye fee, as I will in place conueniente more largelye note.

Schollar. Yet one doubte I haue, of whiche I wolde gladly be rydde, and that is of the Moone : for as you faye, and by fyghte wee perceaue, all the ftarres with the Sonne and Moone go round about the earth in 24. houres, faue that the Moone is flacker then all the reft, for fhe is euerye daye later in ryfynge by an hower, then fhe was the daye before : but howe that cometh to paffe, I doo not vnderftande.

A diuers motion in the Mone.

Ma. This doubt is well moued, and in good tyme, for by it will I take occafion to inftruct you not only in the true knowledge of it, but alfo of other fondrye motions in all the heauens : for in euery one of them dooth there appeare a lyke motion, contrarye to the dailye mouinge of the Firmament, whiche in the Moone is mofte fwifteft, and therefore may be perceaued daylye of all men : but in the Sonne it is not fo fwifte, and therfore not fo eafilye perceaued : yet all men fee a greate alteration in the mouynge of the Sonne in one yeare : for fomtimes he is hygher and nearer ouer our headdes, and fometime farther from our headdes, and lower in the fouthe : yea fometime he fhineth with vs almofte 18. howers, (as in the middle of the Sommer) and in the middle of Winter hee fhineth but 6. houres or lyttle more : this euerye childe dooth fee, althoughe they knowe not the reafon thereof.

A feuerall mouing in the Sonne.

Scholar. Yet the reafon of that is eafy inough to be conceaued, for when the daye is at the longeft, the Sonne mufte needes fhine the more tyme, and fo muft it needes fhine the leffer tyme, when the day is at the fhorteft : this reafon I haue hearde many men declare.

Mafter. That may well be called a crabbed reafon, for it goeth backward lyke a crabbe. The day maketh not the fon

to

to fhyne, but the Sonne fhynynge maketh the daye. And
fo the lengthe of the daye maketh not the Sonne to fhine
longe, nother the fhortenes of the day caufeth not the Son
to fhyne the leffer tyme, but contrarye waies the longe fhyninge of the Sonne maketh the longe daye, and the fhorte
fhyning of the fonne maketh the leffer daye : els anfwere me,
what maketh the dayes longe or fhorte?

Schollar. I haue heard wife men fay, that Sommer maketh
the longe dayes, and Wynter maketh the longe nyghtes.

Mafter. They myghte haue fayde more wifelye, that long
dayes make fommer, and fhorte dayes make winter.

Schollar. Why, all that feemeth one thing to me.

Maifter. Is it all one to fay God made the earth. and the
earthe made God. Couetouifnes ouercometh all men. and
all men ouercome couetoufnes.

Schollar. No not fo, for heere the effecte is tourned
to bee the caufe, and the agente is made the paciente.

Mafter. So is it to faye, Sommer maketh longe dayes,
where you fhoulde faye : Longe dayes make fommer.

Schollar. I perceaue it nowe, but I was fo blynded
with the volgare erroure, that if you hadde demaunded
of me farther what dydde make the Sommer, I hadde
beene lyke to haue aunfwered, that greene leaues doo
make Sommer : and the fooner by remembraunce of an
olde fayinge : that a yeare fhoulde come, in whiche the
Sommer fhoulde not bee knowen, but by the greene
leaues.

Mafter. Yet this fayinge dooth not importe that greene
leaues do make fommer, but they betoken fommer : fo are
they the figne and not the caufe of fommer.

Schollar. So I perceaue nowe that the longe fhinynge of
the Sonne doth make the dayes longe. But nowe can I not
tell what caufeth the Sonne to fhine longer one tyme of the
yeare, then an other.

Mafter. That is it that draue wife menne to fearche, and

                          B.i.   marke

marke the motions of the Sonne, whereby at lengthe they founde, that the Sonne hathe an other courſe, contrarye to the daylye motion of the ſkye. And as the Moone doth accomplyſhe her propre courſe (whiche is from the weſt into the eaſte, contrarye to the daylye motion) euerye mo-

A yeare. neth in the yeare, ſo the Sonne dothe ende his courſe, in his propre motion, but ones in the yeare. And to expreſſe it aptlye, I muſte ſaye, that the true terme of a yeare is no-thynge els, but the verye tyme of the courſe of the Sonne from a certaine pointe in heauen, tyll his retourne to the

A moneth. ſame pointe againe. And a Moneth is the iuſte time of the propre courſe of the Moone, from chaunge to chaunge :

A weeke. and euerye quarter of the Moone maketh a Weeke. of whiche I will ſpeake more in the nexte treatiſe, with the declaration of the diuerſitye for the begynninge of Mo-nethes and Yeares. But nowe to contynewe oure princi-pall matter the more ordrelye, I woulde haue you repeate the chieffe articles of our talke hitherto.

The fyrſte Schollar. This is the ſumme of all your doctrine hy-
repetition. therto.

1. That the worlde is that entiere body, which containeth in it all the heauens and the elements, with all that in them is.

2. The partes of the world ar two eſpecial, the heauens whi-che are eighte in numbre, and the elemenents whiche are .iiij. in kinde.

3. The ordre and ſituation of all theſe partes, as well ele-mentes as heauenly ſpheres, beginning at the higheſt, and proceding to the loweſt, is this. the Firmanent, Saturne, Iupiter, Mars, the Sonne, Venus, Mercury, and the Moone.

THE FOVRE ELEMENTS.

Fyer,   Ayer,   Water,   and Earthe.

and euer the hygher incloſeth all that is vnder it.

4. The worlde and all his principall partes are rounde in fourme and ſhape, as a globe or ball.

5. The

5. The earthe is in the middle of the worlde, as the centre of it : & beareth no vewe of quãtitye in cõparifon to the worlde.

6. The earthe hathe no motion of it felfe, no more then a ftone, but refteth quietly : and fo the other elementes do, except they be forceably moued.

7. The heauens do moue continually from the eafte to the weft, and that motiõ is called, The dayly motion : and is the meafure of the Common day.

8. The Mone hath a feuerall motion from the weft toward the eafte, contrarye to that mouyng of the dailye courfe, and that motion is ẙ iuft meafure of a moneth, and euery quarter dooth make a weeke.

9. The Son alfo hath a peculier motion from the weft toward the eafte, whiche he accomplifheth in a yeare, and of that courfe the yeare taketh his meafure and quantitye.

Now then it may pleafe you to procede to farther explication of the apparaunces which are noted in the heauens, and to fhew the manner of their motions.

Mafter. To the intent that you may vnderftand all thinges the more eafilye, I thinke it good to defcribe vnto you *A material* a Materiall fphere, whiche fhall containe in it fuche nota- *all fphere.* ble cyrcles only, as haue fpeciall vfe in the declaration of the heauenly motions, and fuche as reafon fhall driue a man to appointe, as certaine boundes of the motions in the heauens : yea fuche I faye, as your felfe fhall by interrogatories be conftrayned to confeffe needfull to that knowledg which you defire.

Schollar. If nothinge bee placed in that fphere but that which muft needes be had, then can I not accompt any part of it fuperfluous. And againe, if it ferue fufficiently to inftructe me in that I defyre to knowe, I canne not iuftlye blame it in anye pointe as infufficiente, fo mufte it needes be a perfect inftrument, voyde of defaulte, and without fuperfluitye.

Mafter. So fhall it be, for fo much as this parte of knowledge

B.ij.

ledge requireth. Now then to begin. ye doo beleue that the worlde is rounde.   Schollar. Yea for foothe.

The makig of a Globe.   Mafter. Then muft that inftrument alfo be round, which fhall aptelye expreffe the forme of the worlde.

Schol. Truth it is. Maft. Can there be any thinge more round then a circle? Schollar. No trulye.

Maifter. And dooth not twoo halfe cyrdes make a whole cirde?   Schollar. It can not be denayed.

Mafter. Then take halfe a cirde, and faften it on an axtre or on any diameter, and then tourne it rounde about, fyrfte lettyng the halfe cyrde hang downward vnder the diameter

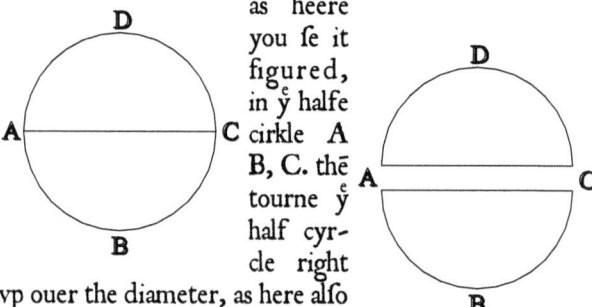

as heere you fe it figured, in ỹ halfe cirkle A B, C. thē tourne ỹ half cyrcle right vp ouer the diameter, as here alfo is reprefented in the halfe cyrcle A, D, C. do not thefe two pofitions make a whole cyrcle?   Scholar. Yes furely.

Mafter. Then fet the halfe cirde fo, that the diameter may ftande ftyll firmelye fixed, and the halfe cyrcle maye tourne rounde about. Do not you imagin nowe that euery dyuers pofition of this halfe cyrcle with the contrary place againft it, dooth make a whole cyrcle? Schollar. Yes verelye.

Mafter. And bycaufe there is no place round aboute that diamenter, within the reache of that halfe cirde, but that half circle hathe paffed it, there can no voyde place be affigned but it is occupied and fylled with halfe a cyrcle, and euerye halfe cyrcle with his contrarye dooth make a whole cyrde, fo doth this whole reuolution of the halfe circle make a iuft cyrcular bodye.   Scho-

THE CASTLE OF KNOWLEDGE   17

Here is the lyke fourme
of that worke.

Schollar. So it appeareth trulye.

Maifter. This circular body is na- *A Sphere is*
med a fphere, as it may appeare by the *defined.*
defcription that Euclide maketh of a
fphere : whiche is this in greeke, as him
felfe wrote it, in his eleuenth booke of
Geometrye.

Σφαιρα ɛsιψ οταψ ἡμικυκλίου μβεθίσης τῆσ διημε-
τρου , πιρινιγθιψ τα ἡμικύκλιου είς το αὐτο παλιψ
αϗ ϗϛισαθη, οθιψ ἤρϗαϗ φιρ.αϗῆ, το πιρμκρθιψ χἡμα

Whiche into Latine may well be translated thus.

Sphaera eft figura comprehenfa ex circumductu femicirculi, donec eo redeat,
vnde moueri incoepit, manente interim immota femicirculi eius diametro.

And thus it foundeth in englifhe.

A Sphere is a found figure, made by the tournynge of half
a circle, tyll it end where it began to be moued, the diameter
of that halfe circle continuyng fteddye all the meane whyle.
This defcription dooeh Ioannes de Sacro bofco expounde
thus : that a fphere is a rounde and found body made by the
tournynge of halfe a circle.

    Schollar. So that a fphere is nothinge els but a rounde
and maffye bodye dofed with one platforme, whiche you in
your Pathwaye doo call a Globe.

    Mafter. You take it ryghte. But nowe muft you marke, *The centre*
that as a circle is made about his centre, fo a globe alfo hath *of a Globe*
his centre, as you may eafilye vnderftande, from which cen- *or Sphere.*
ter all the lynes that may be drawen to the platforme, or vtter parte of the globe, are all equall togither, accordyng to
Theodofius definition, whiche faythe thus : A fphere is a
maffye bodye, indofed with one platforme, and in the middle of it there is a pricke, from which all lynes drawen to the
fayde platforme, are equall eche to other, and that pricke is
the centre of the globe and fo fayth Euclide alfo.

Κἰντρον δὲ τῆς ντπαίρασ ɛsῖ το αυτο, ὁ χαὶ του ἡμικυκλιου.

Idem centrum fphaeræ eft, quod & femicirculi.

                    B.iii.   The

The centre of a globe is the fame centre that a femicirde
hath, by whiche the globe was made.

Schollar. It mufte needes bee fo : and lykewaies the diameter of them bothe mufte needes be all one, as I thynke.

Maifter. You faye not muche amyffe. Yet muft you put
A Diame- a difference in a globe, betwene a Diameter and an Axe tre.
ter and an For euery right lyne that paffeth frō fide to fyde in a globe,
Axe tree and touches the centre, is aptely called a diameter. fo that
differ. as ther may be many diameters in a cyrkle, fo may ther be as
many alfo in a Globe : But of all that multitude, one only is
called the Axe tree, and that is it on whiche the globe tourneth. This difference did Ioannes de Sacro bofco ouerpaffe
not ignorantly, but negligently, or els wittingly : but fo dyd
not Euclide, whiche defineth them bothe thus.

An axr tre   ἄξων δὲ τῆσ σφαίρασ ἐςὶν, ἡ μένουσα ἐυθᾶα, περὶ ἥν τὸ ἡμικύκλιον ςρέφεδὴ.

Axis Sphaerae eft, recta illa ftabilis linea, circa quam femicirculus rotatur.

The Axe tree (faith he) is that righte
lyne whiche moueth not, but the halfe
cirkle moueth aboute it. Thefe wordes
haue refpect not only to the makynge
of a Globe or Sphere, but alfo to the
vfe of it. But now the diameter is defined by him thus :

                    διάμετρῶ- δὲ τῆσ σφαίρασ ἐςὶν ἐυθᾶα τίς διὰ τοῦ
A diameter   κἰντρȣ ἀγμένη,καὶ περαὃυμένη ἐφ' ἑκάτερα τὰ μέρη, ὑπὸ τῆσ ἐπιφανέασ Τῆσ
σφαίρασ.

Dimetiens vero Sphaerae eft recta quaeq; linea per centrum acta, & vtrinque
definens in fphaerae fuperficie.

The diameter of a Sphere, is anye ryghte
lyne that is drawen by the centre, and ended
in the plat forme of the fphere.

Schollar. This difference mufte needes
feeme reafonable, fyth there maye be fo many diameters drawen as a man lyfteth, but
Axe trees there can be but one in one globe.

                                                          Ma-

THE CASTLE OF KNOWLEDGE   19

Maifter. When a globe tourneth rounde, are there anye mo poyntes then twoo in that globe, on whiche it doothe tourne?

Schollar. By proof it appeareth, that all partes of the globe moue, excepte the two endes of that Axe tree, wheron it mooueth, and they mooue not out of their place.

Mafter. Thofe twoo pointes are named the poles in a fphere, wherby alfo you may vnderftande, that there can be but two poles in one fphere : marke this well, for it will ferue your turne in place conueniente. Nowe applye all thefe to the worlde, whiche in his whole fubftaunce is rounde, and therefore aptelye maye bee called a fphere : yow fee it tourne aboute rounde, and therefore muft it haue twoo poles, on whiche it tourneth fo. Alfo bicaufe it is rounde, it mufte haue a centre (whiche I dyd affirme before to bee the earthe) and by this centre, we may imagine a right line to run from the one pole to the other, whiche righte lyne mufte be called the Axe tre of the worlde. *Poles of a Sphere.*

Schollar. For the centre of the worlde, it mufte needes be fomthinge : for I perceaue a globe can not be, but it muft neceffarily haue a middle pricke or centre, no more then a lyne maye be made whiche hath no myddell, or a circle that hathe no centre : whiche bothe appeare vnpoffible. Alfo for the pooles, they appeare needefull, or rather of neceffity to folowe the mouinges of heauen. For in all rounde thinges that mooue roundly, there be fuche two pointes that feeme not to moue : but why there fhoulde by any axe tree requyred in the worlde, I fee no reafon : for if the myghtye power of God dyd not ftaye the worlde, there coulde bee no Axe tree able to beare it.

Mafter. Your imagination in this pointe is to groffe. I fayde not that the Axe tre was made to ftay the worlde, but that it paffeth as a lyne only from the one pole to the other : and is not without greate and profitable vfe, bothe in doctrine, and alfo in practife, for placynge of inftruments, as

B.iiij.   you

you fhall know better hereafter. But nowe heare howe Proclus dooth applye thefe to the worlde.

ἄξων καλᾶται ϙῦ κόσμου ἡ διάμετρ῀ αὐϙῦ, πεϱὶ ἥν ϛϱ ἐϙϛνϙμι, ϙὰ δὲ πέϱαϛ
ϙκ ϙῦ ἄξων῀ πόλοι λέϡονϙμ ϙῦ κόσμου·Ϊῶμ δὲ πόλωμ,ὁ μϣϛ λὲϡϛϙμ-ϐόϱϛϛϛ
ὁ δὲ νόϟι῀﹢   Whiche wordes our worthye contrye man
D. Linaker, tranflateth thus.

Axis mundi vocatur dimetiens ipfius, circa quam voluitur. Axis extrema,
poli mundi (feu vertices) funt nominati horum alter Septentrionalis, alter
Auftrinus dicitur.

The Axe tree of the worlde, is named the Diameter of it,
aboute whiche it tourneth and the endes of that Axe tree,
are called the Poles of the world. of whiche poles one is named the Northe pole, and the other the South pole. The
North pole is alwaies feene of vs where as we dwell, and
the Southe pole is neuer feene in this oure contrye, but is euer
more vnder our Horizonte, and that as lowe, as the Northe
pole is highe aboue our Horizonte.

The north and fouthe Poles.

Schollar. I haue beene taughte to knowe the Northe pole,
and I haue marked it oftentimes, wherby I perceaued a great
numbre of ftarres to moue aboute it, and were fometymes
higher then it, and fometymes lower then it : nowe on the
eafte fyde of it, and nowe on the weft fyde : but that pole
ftarre feemed not to fturre oute of his place at anye tyme :
whereby I gather, that he is neuer oute of fighte to vs, when
the ftarres appeare, and that is all the nyghte. but what be-
commeth of him in the daye tyme, I cannot tell.

Mafter. I wyll cleere you of all fuche doubtes before I
leaue you : but in the meane tyme I meruaile you founde no
doubte at the name of the Horizonte.

The Hori-zonte.

Schollar. That name I learned to fignifye that cyrcle,
whiche goeth along by the edge of the ground, and parteth
that parte of the worlde whiche we fee, from that part which
we fe not : & when the Son rifeth, then is he in our horizonte,
& fo is he, when he is goyng downe as lowe as we can fee him.

Mafter. This is not greatlye amiffe. the lyke expreffynge
of it

Here the Horizonte is reprefented by
the lyne A. C.

of it dooth Hyginius vfe
in his fyrfte booke, and in
the .iiij. alfo of his aftronomye : but Proclus in
his Sphere, dooth define
it thus.

ὁρίζων δὲ ἐςὶ κύκλος ὁ διορίζων
ἡμῶν τὸ, τε φανερὸν καὶ τὸ ἀφανὲσ
μέρῷ· Τῦ κόσμϛ , καὶ διχοτομῶν
τὴν ὅλην σφαῖραν Τῦ κόσμϛ, ὥςε
ἡμισφαίριον μὲν ὑπὲρ γὴν καταλαμ
βάνεδδι, ἡμισφαίριον δὲ ὑπο γὴν,

Horizon vero circulus eft, qui confpectam mundi partem ab incon-
fpecta dirimit: itaq in duas partes vniuer fam Sphaeram fecat, vt alterū
hemifphaerium fupra terram, alterum fub terra relinquat.

The Horizonte is a cyrcle whiche parteth that parte of the
worlde that wee fee, from that whiche wee fee not : and it de-

And here the Horizonte is the edge betwene the
lyght parte (whiche ftandeth for that whiche wee
fee) and the darke part whiche dooth fignifie that
whiche wee can not fee of the fkye.

uideth the whole
fphere of ẙ world
into twoo equall
partes, in fuche
forte, that half of
that fphere is e-
uer abooue the
grounde, & halfe
alwaies vnder the
earthe. This cyr-
cle you perceaue
to be neceffary in

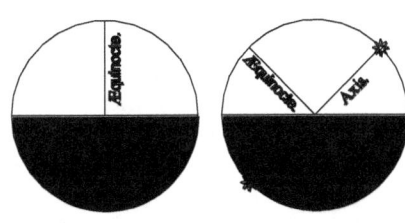

the materiall fphere, feynge it hath fo great vfe in the heauenly motions, that by it we iudge the rifynges and fettings
of the Sonne and the Moone and all other ftarres. what fay
you then for the noone fteede of the day, from whiche you
recken all your houres, as it appeareth both by the clockes
and dyals? for as the clocke ftriketh one nexte after noone,

The meri-
dian circle.

and

and fo increafeth forward in the numbre of houres, fo likewaies are your howers marked in the dialles.

Schollar. I thinke it very meete to haue the fouth pointe
well knowen, as well for this, as for ftandynge dialles, and
for knowledge of the tyme of the nyght by the moone, and
by other ftarres.

Maifter. Then mufte there be a cirde appointed for that
vfe, whiche is called therfore the Meridiane circle, and may
The None- be named well the Noone fteede cyrcle. This circle is thus
fteed circle. defined by Proclus.

κιωσιμζρισα· ἀ ὁϊι κυπλΘ· βαⱮ τὼγ κυῦ κόσμχ πόλαψ καὶ τξ̄ κατπκτωρυφὴψ
σημαίω χραπὶμιΘ· κὑμλΘ· λϕ' ου γρόμιΘ· ὁ ἡλιΘ· τὶ μίστι τὼγ ἡμιρῶγ, κὴ
τὶ μίστι τξυ νυκτῶγ σκιϐθκαι.

Meridianus circulus eft, qui per mundi polos & punctum, quod nobis
fupra verticem eminet, ducitur in quem cum folincidit, medios dies,
medias qi noctes efficit.

The Meridian is a cyrde drawē by the poles of the world, &
the point right ouer our heads. in which cirde whē the Son
is, he maketh the myddle of ẙ day, & the middle of ẙ nyghte.

Nowe farther to procede to other partes needfull in the

The Meridiane cyrde here is refem-
bled to the circle A, B, C, D.

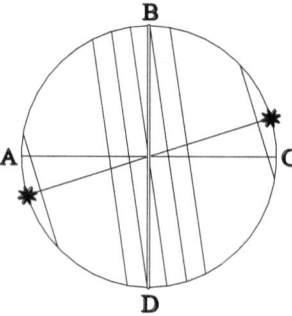

fphere you do fe, that twife
in the year the daies & nights
ar equall, & the Son rifeth in
the iuft eaft, & goeth doune
in ẙ full weft, wher as in ẙ fõ
mer ẙ Son rifeth northeaft,
and fetteth northwefte : & at
nonetide is very high ouer
our heds : but in ẙ winter, cõ
trary ways ẙ fon rifeth fouth
eaft, & fetteth fouthweft : & at
nonetide is very low. thynk
you not that thefe thre boū
des of the courfe of the Son would be well noted, and haue
their peculiar circles, for diftinction of thofe tymes?

                                                        Schol-

Schollar. I thynke nothinge more needefull then that.

Mafter. Thefe thre cirdes (with two other that I will next fpeake of) are named the fiue Paralelles, and the middle cirde of thofe, is named the Equinoctiall, bicaufe that when the Sonne is vnder it, the dayes and nyghtes are equall in all the worlde, except only twoo places. This cirde is thus defined by Proclus.

The Equi-noctial circle

ἰσημερινὸσ δὲ κύκλ⊙- ἐστὶν ὁ μέγιστος τῶν πέντε παραλλήλων κύκλων, ὃ διχοτί-μέμλν⊙- ὑπὸ τοῦ ὁρίζοντος, ὥστε ἡμικύκλιον ὑπὲρ γῆν ἀπολαμβάνεϑϑ ἡμικύκλι-ον δὲ ὑπὸ τὸν ὁρίζοντα, ἐφ᾽ οὗ χρόμλν⊙- ὁ ἥλι⊙- τὰσ ἰσημερίασ ποιεῖται : τήν τ᾽ ἐαρινὴν καὶ τὴν φθινοπωρινήν.

Aequator, circulus is eft, qui maximus aequidiftatiem circulorum ftatuitur, ita nimirum ab Horizonte diffectus, vt alter eius femicirculus fupra terram, alter fub terra condatur : in hoc fol duplex aequinocrium, vernum autumnaleq facit.

The equinoctiall cirde is the greateft of the fiue Parallele circles, and is deuided fo equallye into two partes, by the Horizonte, that the one halfe of it is aboue grounde, and the other is vnder the horizonte : and when the Sonne is in this cirde, he maketh the daies equall with the nightes, ones in the Springe tyme, and againe in the Harueft. This equinoctiall cirde and the other feuen that folowe, to be declared, doo moue all as the fkye moueth but the Horizonte and the Meridian doo not moue with the heauen, but ftand ftedye, and keepe their places.

Schollar. That feemeth reafonable, els coulde not men knowe the rifyng, fetting, and noonefteed of the Sonne. but howe fhall I knowe this equinoctiall cirde in heauen, feynge I can not fee any fuche cirde there?

Mafter. Marke the courfe of the Sonne aboute the eleuenth daye of Marche, or els about the fourtenth daye of Septembre, and fo may you beft vnderftande the place of this cirde, for at thofe two tymes the Sonne runneth directly vnder the equinoctiall cirde, and dothe (as it were) defcribe it by his motion in four and twenty howers. And if

Howe to knowe the place of the circle equinoctial

you

you fyrſte do marke the ryſinge of the ſonne that daye, you maye know the preciſe pointe of the eaſte, and at nyghte he ſetteth in the iuſte poynt of the weſte.

Schollar. I woulde I knewe as good markes of the other cyrdes.

Maſter. So wyll I geue you in their conuenient places and times good orders to know them al : and firſt I muſt tel you, that theſe other two cyrdes, which I named before (with the *The know ledge of the ij tropikes* equinoctiall) are called the twoo Tropike cyrdes after the greeke deriuation, and maye be called in englyſhe the Sonne boundes, bycauſe the Sonne doth neuer paſſe them, nother towardes the northe, nor yet toward the ſouthe : but when he toucheth any one of them, he doth tourn his courſe toward the other, as for example : All the tyme from the myddle of December vntill the eleuenth daye of Iune, you maye per-ceaue the Sonne to ryſe hi-gher and hygher, and that daye hee is at the hygheſt that hee canne go towardes our heads, and then dooth hee by his courſe deſcribe that Sommer tropike, after whiche daye hee draweth agayne lower and lower e-uerye daye, tyll the twelfte daye of December, for then he is at the loweſt, and that daye he doth deſcribe the Winter tropike. Nowe marke howe Proclus defi-neth them.

Examples of thoſe circles and other that foloweth
A, C, the Horizonte.
✳ ✳ The poles of the worlde.
G, H, The Equinoctiall cirde.
B, F, one tropike, and
E, D, the other tropike.
The Sōmer tropike. A, I, the artic cirde,
C, K, the antartike cirde.

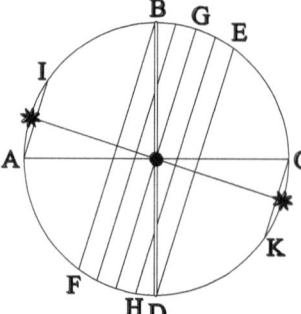

The win-ter tropik

θερινὸσ δ̀ τροπικὸς κύκλος ἐςὶν ὁ Βορειότατος, ἴϖ ὑπὸ τὸ ἥλιου γρα-φομένων κύκλων, ἐφ' οὗ γινόμενO ὁ ἥλιO- τὴν θερινὴν τροπὴν ποιεῖται, ἐν ᾧ ἂ μιγίςη μὲν ποιεῖ τὴν ἑμέ-ραν τῇ ἐνι-αυτᾶ

THE CASTLE OF KNOWLEDGE　　25

αυτῇ ἡμέρα, ἰλαχίσυ δὲ ἡ νὺξ γίνεται. μετὰ μὲν ͦ τὴν θερινὴν τροπὴν ὀκ ἔτι
πρὸσ τὰσ ἄρκτους παροδεύον ὁ ἥλι☉ θεωρᾶται, ἀλλ᾿ ἐπὶ θάτερα μέρη τρέπτ᾿)
τῦ κόσμυ, διὸ κέκληται βορπικόσ.

*Solſtitialis autem circulus is eſt, qui omnium, qui à ſole deſcribuntur
maximdè ſeptemtronionalis habetur in quem quum ſe ſol receperit, aeſtiua
ceciprocationem peragit, longiſſimuſq totius anni dies, breuiſſimaq
nox erit poſt hanc autem reciprocationem, nequaquam vltra verſus ſep
temtriones ſolem progredi, quin potius ad diuer ſa mundi regredi cer-
nas vnde & Tropico graece nomen.

Linacer
nimiū co
actè com-
mune no
mē utriq
tropico æ
ſtiuo vni
tribuit,
Pliniū im
portunè
ſecutus.

The Sommer tropike is the moſte northerlye circle of all
thē that the Sonne deſcribeth : in the which when the Sonne
is, he maketh his Sommer turne, at which time is the lōgeſt
day of al the year, and the ſhorteſt night : for after this Sommer turne, you ſe the Sonne go no more toward the north,
but turneth to the contrary coaſte of the worlde, and therof
is that circle named (in greeke) a Tropike : that is to ſaye, a
Returninge circle, or a circle of Returne.

The Sonne aftter he beginneth to turne, maye be perceaued
euery day, or at the leaſt euery weeke, and chiefly at nonetide
to waxe lower & lower, vntill he come to the Winter tropike,
and there he turneth againe, as by the definition of that tro
pike you may vnderſtande.

χαιμερινὸς δὲ βορπικὸς κύκλος ἐςὶν ὁ νοτιώτατος ͦ ͦ ἀπὸ τῦ ἁλίε γραφομένων κύκλων κατὰ τὴν ἕωυ τῦ κόσμυ γινομένην περιςροφὴν. ἐφ᾿ ὃν γενόμεν☉ ὁ ἥλι☉
τὴν χαιμερινὴν τροπὴν ποιεῖται. ἐφ᾿ ᾗ ἡ μεγίςη μὲν πασῶν ͦ͜ ἐν τῷ ἐνιαυτῷ νύξ
ἐπιτελεῖται, ἰλαχίςη δὲ ἡμέρα. μετὰ μὲν ͦ ͦ τὴν χαιμερινὴν τροπὴν ἐκ ἔτι πρὸσ
μεσημβρίαν παροδεύων ὁ ἥλι☉ θεωρᾶται, ἀλλ᾿ ἐπὶ θάτερα μέρη τρέπεται τοῦ
κόσμυ, διὸ κέκληται καὶ οὗτ☉ τροπικός.

Brumalis circulus is eſt, qui omnium circulorum qui à Sole circumactu munde deſcribuntur, maxime ad auſtrum pertinet : in quo ſol bru-
malem reciprocationem facit, maximaq totius anni nox, minimuſq
dies efficitur. poſt hanc metam nequaquam vltra progreditur * ſol, ſed
ad alteras mundi partes reuertitur : vnde tropicus hic quoque, quaſi ver-
ſilis, appellatur.

Intellige
verſus au
ſtrū quod
& græcè
additur.

The winter tropike, ſayth Proclus, is the moſte ſoutherlye

The win-
ter tropike.

C.i.　　　　circle

circle of all them that the Sonne doth defcribe, by the reuolution of the worlde, in whiche when the Sonne is, hee maketh his Winterly tourne, and then is the longeft nyghte in
all the year, and the fhorteft day, for after this Winter turn,
the Sonne is not feene to go any farther towarde the fouth,
but tournith to the contrarye coaftes of the worlde, and
thereof is this cyrcle alfo named a Tropike or cyrde of
Retourne. And thus haue we the three circles that are principallye noted for the courfe of the Sonne. Nowe are there

*The fouthe
and northe
circles.*

other twoo whiche be Paralleles with thefe thre, whereof
the one is more foutherlye (to vs) then in the Winter tropike,
and the other is more northerly, thē is the Sommer tropik,
whiche whether they be needfull or not, their vfe maye declare. I remember, that you fayd, you had oftentymes be-
holden the Northe pole, where you myghte fee manye ftarres about it, that neuer go vnder our Horizont. do you not
thinke it good that all thofe ftarres were inclofed in a circle
to be difcerued from al other, which rife fomtime aboue the
Horizont, and fomtime againe do fet vnder the fame?

Schollar. Yes verilye, it were pleafaunt to know.

Maft. And profitable alfo, as you fhal hereafter perceaue.

*The ufe of
the Arctik
and Antar-
ctik circles.*

Now contrary waies, there are other ftarres, that are neuer
feene of vs in this cuntrye, and yet muche mention is made
of them in writers, were it not good that their bounde were
marked, that all other maye be knowen from them?

Schollar. Els myghte men often looke for fuche ftarres
as they reade of, and fhulde loofe their labour, for they fhall
not fee them.

Mafter. And yet are there goodlye bryghte and notable
ftarres, whiche are not feene here, but in fouthe Spaine, in
Barbary, in Guinea and Calecut, and many other cuntries,
they appeare fayre and pleafaunt to beholde.

Scholar. I pray you, what call you thofe cyrcles that
inclofeth thofe ftarres?

Mafter. They are named after the coafte of the worlde
                                                        where

where they bee. So that the circle whiche inclofeth all thofe
ftarres that be about the Northe pole, is named the Arctyke
circle or Northe circle : and the contrary circle in the fouth,
is called the Antartike circle by the greeke compofition, as
you woulde fay, Contrary or againft the Arctike circle : and
it may well be called the South circle. But nowe heare how
Proclus defineth them.

Ἀρκτικὸν μὲν δὴ κύκλθ· ὁ μέγιϲτος τῶν ἀεὶ θεωϲϲμϲίων κύκλον, ὁ ἐφαπτόμϲθ·   *The Arctik*
τῶ ὁρίϲοντθ· καὶ ὲν σημϲῖον, ἢ ὁλθ· ὑπϲϲ γῆμ ἀϲλαμβανόμϲθ·.ὲμ ὧ τὰ κϲ-   *circle.*
μϲρα τῶν ἀϲϲων ὅ τε δύϲιν, ὀυ͛· ἀναϲολὴμ ποιϲῖται.ἀλλὰ δ᾿ ὅλϲϲ τὴϲ νυκ τὸϲ
πϲϲὶ τὸμ πύλομ ϲϲϲφόϲκϲϲ θϲωϲϲῖται.

Septentrionalis circulus eft is, qui omnium quos perpetuo cernimus,
planè maximus eft, quiq Horizontem folo puncto contingit, totus
fupra terrã interceptus. intra hunc quaecunq clauduntur aftra, nec ortũ
nec occafum norunt, fed circa polum verti tota nocte cernuntur.

The Arctike cirle is the greatteft of all thofe circles whiche
do alwaies appear, and toucheth the Horizonte in one only
pointe, and is all togither aboue the earthe, and all the ftarres that bee within this circle nother rife nother fette, but are
feene to runne rounde about the Pole all the nyghte.
Thus haue you the fourth parallele, Nowe refteth the fyfte
whiche is defcribed thus of Proclus.

Ἀνταϲκ ϲϲσ δὲ δὴ κύκλθ· ἴσος καὶ παϲάλληλθ· τῷ ἀρκτικῷ,καὶ ἐφαπτόμϲνος   *The Antar*
τῶ ὁϲ ϲοντϲς καθ᾿ὲμ σημϲῖομ,καὶ ὅλθ· ὑπὸ γῆμ ἀϲλαμβανόμϲθ·.ὲμ ᾧ τὰ κϲ.   *ctik circle.*
μϲϲα τῶν ἀϲϲων ϲϲσἀ παντὸς ἡμῖμ ὄϲμ ἀόϲατα.

Antarcticus vero circulus aequalis & aequidiftãs Septentrionali circulo
eft, & Horizontavno puncto contingens. totus praeterea fub terris
merfus, intra quem fita aftra femper nobis occulta manent.

The Antartike circle is equall and equidiftant to the Arctike circle, and toucheth the Horizonte in one only point,
and is all vnder grounde, and all the ftarres that be in it, are
euer more out of our fighte.
Thefe are al the Paralleles which are wont to be fet forthe in
the materiall fphere, and that agreeably of all men, faue that
                              C.ij.   tou-

touchinge the two laſte circles there is a difference, of which
I will inſtruct you at large in the next part of our talke, and
omitting it for this time, will go forward to other thre cirdes whiche yet remaine, and are needfull to oure ſphere. By-

The zodi-
ake.

cauſe our chieffe conſideration conſiſteth aboute marking
of the motions of the Sonne, the Moone and the other
planetes, howe they chaunge their places in the ſkye, and
therfore make diuers apparaunces to vs that beholde them,
and mark their courſes, and yet all they haue (as it were) one
common path or waye, from which they ſwarue not, but
kepe them ſelues ſtill within the limites of it : how think you
is not that path of theirs well to be marked, and worthy to
haue a notable name?

Schollar. Mary that is the principall pointe (as I take it)
of all the reſte : for without knowledge of that, nothing els
can be knowen.

Maſter. That common path of the Planets, wherin all
thei haue their courſe, is called of Aſtronomers the Zodiake : whiche is, as you may engliſhe it, the Circle of the

The .xij.
ſignes.

Signes : whiche ſignes are the greateſt and notableſt partes
of that circle, and were inuented for the more exacte diſtinction of the motion of the Planetes monethlye. For as
there bee but twelue monethes in the yeare, ſo there are
twelue partes of the Zodiake diſtincte by ſeuerall names,
and correſpondent to euery moneth, althoughe they varye
ſomething now from their firſt application, wherof hereafter I will inſtructe you ſufficiently. and now will touch them
briefly as this place doth require. Their order in the zodiak
and their names ar theſe that folow, in greek and latin, which
maye bee engliſhed as I haue vnder written, and are often
tymes mentioned of our engliſh Poetes.

κριὸς.	ταῦρℴϹ.	δίδυμαι.	καρκῖνℴϹ.	λέων.	παρθένℴϹ.
Aries.	Taurus.	Gemini.	Cancer.	Leo.	Virgo.
the Ramme.	the Bull.	the Twinnes.	the Crabbe.	the Lyon.	the Virgin.
♈	♉	♊	♋	♌	♍

χηλαί.	σκορπῖ☉	τοξότης.	αἰγόκερως.	ὑδροχό☉.	ἰχθύσ.
Libra.	Scorpius.	Sagittarius.	Capricornus.	Aquarius.	Pisces.
the Balance.	*the Scorpion.*	*the Archer.*	*the Goate.*	*the Waterman.*	*the Fishes.*
♎	♏	♐	♑	♒	♓

And bicaufe that their names alwaies can not bee placed in
fmall inftrumentes, there ar certain figures deuifed for their
names, whiche I haue alfo fette vnder their names, that
you maye the better knowe them. Thefe Signes are all of *The de-*
one lengthe, eche beynge the iufte twelfte parte of the Zo- *grees of*
diake. And for exacter knowledg of the motion of the pla- *the fignes.*
nettes euerye daye, eche Signe is deuyded into thyrtye
equall partes, which are called Degrees, fo that in the whole
circuite of the zodiake there muft bee 360 degrees, whiche
agree almoft with the dayes of the yeare.

Scholar. And therby I gather, that as the Son doth moue
throughout all the zokiake in a yeare, fo euerye moneth he
moueth, he runneth one figne, & euery daye nere one degree.

Mafter. You gether well, but this mufte you marke alfo, *What a de*
that by this fame nombre of degrees all the cyrcles in the *gree is in*
fphere are deuided, fo that of euery circle greate or leffe, a *meafure.*
degree is the 360 parte and not anye meafure certaine, as a
foote, a yarde, a myle, or fuche lyke.

Schollar. I vnderftande you thus : as a quarter is no meafure certaine, but fometyme is referred to one thinge, and
fometime to an other, and yet ftill it betokeneth the fourth
parte of that wherunto it is referred, for when we fay : a year
and a quarter : an houre and a quarter : a yard and a quarter :
a quarter of a foote : in all thefe fayings, the quarters differ.
fo when wee faye : a quarter of corne : a quarter of clothe : a
quarter of pepper : a quarter of allame : by the accuftumed
meafures all men vnderftande our meanynge, and yet thefe
quarters differ, and be in common meaning, a quarter of a
weye, or eight bufhels, a quarter of a yarde, a quarter of a
pounde, a quarter of a hundreth.

Mafter. So is a degree the thirteth parte of a figne, and a

C.iij.   figne

figne the twelfte parte of any circle. howe be it, commonlye
& chiefly the name of Signes, is attributed to the Zodiak.
(whiche many doo call the Thwarte circle) This Zodiake is
thus defcribed of Proclus.

*The zodi-
ake.*

λοξὸς δὲ ὅτι κύκλ☉ ὁ τῶν.ιβ. ζωδίων, αὐτὸς δὲ ἰκ τριῶν κύκλων παραλλήλων
συνίσηκιν,ῶν οἱ μὲν τὸ πλάϊος ἀφορίζων λίγεται τὸ ζωδιακου κύκλυ,ὁ δὲ δὲ
μίσον τῶν ζωδίων καλᾶται. οὗτος δὲ ἐφάπτεται δύο κύκλων ἴσων καὶ παραλλήλων,τῦ μὸν θερινοῦ τροπικοῦ κατὰ τὴν τῦ καρκίνε πρώτην μοῖραν, τῦ δὲ
χειμρινῆ τροπικῦ κατὰ τὴν τῦ αἰγκέρωσ πρώτην μοῖραν.τὸ δὲ πλάτος τῦ ζω-
διακῦ κύκλυ ὅτι μοῖραι.ιβ.λοξὸς δὲ κίκληται ὁ ζωδιακὸς κύκλ☉,διὰ τὸ τέμνειν
τους παραλλήλους κύκλους .

Obliquus circulus is eft, qui duodecim figna continet, ex tribus aequi-
diftantibus circulis conftans : quorum duo latitudinem figniferi deter-
minant, vnus per media figna ductus vocatur. hic adeo duos pares &
aequidiftantes circulos attingit, Solftitialem in prima Cancriparte, Bru-
malem in Capricorni principio. Latitudo Signiferi continet partes duodecim. Dictus eftautē hic circulus Obliquus, quod æquidiftantes (ad in-
æquales angulos) interfecet.

The thwarte cyrcle (or zodiake) is the cyrcle of the twelue
fignes, and is made of thre circles, wherof two are the boundes of his bredthe, and the thyrd is called the Middle figne
circle, (bicaufe it goeth by the middle of the fignes in the
zodiake) and it toucheth two equal

This whole circle reprefenteth the zodiake, and the
myddle circle fignifieth the
ecliptike lyne.

circles of the parallels: that is to fay,
the Sommer tropike in the firfte
point of the Crabbe called Cancer,
and alfo the Wynter tropike in the
firfte degre of the Goate, called Capricorne. The breadth of the zodiake, containeth twelue degrees. This
zodiak is called a Thwart circle, bycaufe it croffeth the parallele circles,
goynge ouerthwarte them. By thefe
wordes of Proclus you may vnderftande, that the zodiake dooth not

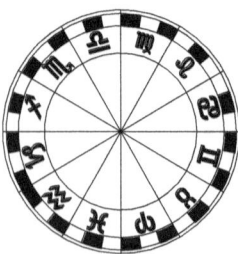

go

go directly betwene the two poles of the worlde, as all the
fiue paralleles doo, but is drawen croſſe the ſphere, ſo that
his middle (in breadthe) doth touche the two tropikes, and
that middle line is called of latin writers the Ecliptike lyne,   *The Ecli-*
bicauſe there can be no eclipſe of Sonne or Moone, onles   *ptike line.*
the Moone be vnder that lyne : as hereafter I wyll declare in
place conuenient. But touching this zodiake (of which wee
ſpake laſte) I ſayde it was diuided into twelue ſignes, according to the twelue monethes of the year. And bicauſe euery
quarter of the yeare maye bee the more exactlye knowen a
ſonder, this zodiake is parted into foure partes principall,
euery part (as it muſt needes folow) containing thre ſignes.

Schollar. This is a very apte agreement of arte vnto nature : for as the whole zodiake agreeth with the whole year,
ſo for the foure quarters of the one, there is foure quarters
in the other : and for the twelue monthes of the yeare, twelue
ſignes in the zodiake : and for the thirtye dayes of the moneth, thirtye degrees in euerye ſigne. But I praye you ſyr,
dooth the beginninge of theſe ſignes anſwere to the begin-
ning of our yeare?   *The yeare*
Maſter. The beginning of the yeare is diuers in dyuers   *when it be*
nations, as I will ſhewe you an other tyme, with the reaſon   *ginneth.*
why we begin our yeare in Ianuary : but for this tyme it ſhal
be ſufficient, to declare the agreement of our yeare with the
Aſtronomers yeare. The Aſtronomers beginne the twelue
ſignes of the zodiake at Aries, and lykewaiſe do they begin
the yeare that daye and hower, that the Sonne entreth into
that ſigne of Aries, whiche is nowe at the eleuenth daye of   *The ſpring*
Marche : and from thence they recken the Springe of the   *of the year*
yeare thre monethes, whyle the Sonne is in the fyrſte three
ſignes. Then at the eleuenth day of Iune, they accompte the
ende of the ſpringe, and the beginning of Sommer, bicauſe   *The Sõmer*
then the Sonne entreth into Cancer, whiche is the fourthe
ſigne, and while the Sonne paſſeth other thre ſignes, (which
maketh the ſeconde quarter of the zodiake) they accompte

C.iiij.   the

the fecond quarter of the yeare, which we call Sōmer, & that
endureth till the 14 day of September, at which time ẙ Son
entreth into Libra, wher the third quarter of ẙ zodiak doth
begin, & fo with it begīneth Harueſt, which is the third quar
ter of the year, and cōtinueth till the twelft day of Decēber,
and then doth the Son entre into Capricorn. & Winter beginneth, being the 4 and laſt quarter, which continueth tyll
the eleuenth daye of Marche, where the olde yeare endeth,
and a newe yeare beginneth.

Harueſt.

Winter.

    Scho. Theſe 4. ſignes, Aries, Cancer, Libra & Capricorn,
feeme to haue a certain prerogatiue, ẙ they begin ẙ 4. quarters of ẙ year, therfore thei wold be well noted in ẙ zodiake.

    Maſter. You ſay well, and yet thei haue other notable qua
lities, for in the beginning of Aries and Libra, ẙ ſon maketh
the daies equall with the nights. & theſe 2. points ar named ẙ
equinoɗtial points. In the firſt part of Cancer, the day is at ẙ
longeſt, and beginneth to ſhorten by the defcending of the
ſon frō our heds, & when the ſon doth enter into Capricorn,
the day is at the ſhorteſt, & then the ſon beginneth to returne
to vs. again, & the day doth thē begin to increaſe and theſe 2.
points ar called the ij. Tropike points : Wherfore as theſe 4.
points are notable, ſo are ther ij. circles appointed for their
lymites, the one going by the beginning of Aries & Libra,
and the other by the beginning of Cancer and Capricorn.
theſe ij. cirdes ar called Colures, wherof the one only which
paſſeth by Cancer and Capricorn, is defcribed of ẙ grekes,
the reafon whereof I will ſhewe you in the fourthe treatiſe.
But this fyrſte colure, whiche is called the Tropike colure,
is thus defcribed by Proclus.

The Colu-
res.

Tropike
Colure.

* λ.ſigni-
ficat. 30.
que ſemif
tis eſt cir-
culi maxi
mi diuiſi
in 60. par
tes, quod
Prodˑ fa-
cit. vr per
pera qui-
da λ.hic, p
λ. ſubſti-
tuerint.

δ̕α τῶ πόλων ἀ̕ ἀσι κύκλοι ὑπὸ τινων πόλυρα πεοσπ χαρινόμδναι, ὡ̕ς ουμ βἀ̕λω
κει ὑπὸ τῶ ἰδ̕ἰων σιρερφραῶρ δ̕υσ τ̕ο κόσμε πόλουσ ἰχαρ.κόλερφι ἀ̕ κίκλω᷈),
δ̕α τ̕ο μιρα τίνα ἀθιώρρτα αὐτῶρ γίνιδ̕ͅ.οἱ μἐ̕ λ γαρ λαιπτοί κύκλαι κα τὸ ᷈χ συ
ριφοφἀρ τ̕ο κόσμε ὅλαι διωρουῶ τοι, τῶ ἀ̕ κολύρφωρ κύκλωρ μιρα τίνα ὅτι ἀθϊαλ-
ρῦντω, ται ἰ̕χ τ̕ο κύταρίζιωσ ὑπὸ ὀρ ὁρίζͅῶντα ἀρλαμβανόμλνη . γράφωντω ἀ̕
ὅρι οἱ κύκλω δ̕α τῶ ＊τροσπκῶρ σημείων,καὶ ἀς μἰρα ＊.λ.ἰσω δ̕ιαρρῶτον τὴρ δ̕α-
μίσωρ τῶ ζωδͅίωρ κύκλωρ.

                                  Sunt

THE CASTLE OF KNOWLEDGE   33

Sunt & per polos ducti circuli quos nonnulli Coluros vocant: in acci-
dit, vt in ambitus fuos mundi polos recipiant. Coluri autem dicti funt,
quod partes aliquas in fe minime confpectas habent. reliqui enim circuli in mundi circumactu integri cernuntur, fed colurorū partes quae-
pia que videlicet ab a Arctico fub Horizonte latent, cerni nō poffunt.
Signantur autem hi circuli per tropica puncta, diuiduntq per ✱ duas
æquas partes circulum qui per media figniferi ducitur.

✱ Antar-
ctico legē
dit, cōtra
exempla-
rium om
nium con
feufum.

✱ duas ad
modū ap-
tè Lina-
cer trāftu
lit loco λ
literæ, q
femiffim
hic figni-
ficare, fu-
prà admo
nui.

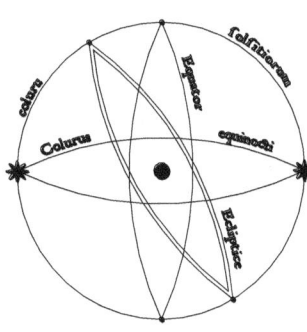

The circles that go by the
poles ar thofe, whiche fome
men call Colures : thei haue
the poles of the worlde in
their circumference. And
ar named Colures in greek,
that is trunked circles, bycaufe fome partes of them
come not into oure fighte.
for the other circles by the
turning of the world are all
feene, but fome parts of the
Colures are not feene, that is, thofe partes whiche are in the
Antartike circle, and remaine vnder our Horizonte. Thefe
cyrcles are deawen by the two tropike pointes of the eclipte
circle, and fo deuide it into two equall partes. The Equinoctiall colure goeth by the poles of the fphere, and by the .ij.
equinoctiall pointes of the Zodiake, in Aries and Libra.
Thus haue you nowe all the cyrcles needfull for a materiall
fphere. let me heare howe you doo remember their names.

*The Equi-
noctiall co-
lure.*

    Schollar. If I fhoulde not remembre theim, I dydde but
leefe my laboure, and occafion you to fpend your tyme in
vaine : for I know that in this fcience and in all other, he that
coueteth to runne ftyll forwarde, and remembreth not that,
that is gone before, fhall neuer attaine that whiche remaineth behynde, but while he deliteth to muche to fee the end,
he deceaueth him felfe of the frutefull ende of knowledge.
muche lyke a man that is delited in hearing a cunning fong
                         of

*A good
leffon.*

of muſyke, but when it is done, doth remembre nothing of
it, ſo is his profite and pleaſure bothe ended, when the ſong
is ended. Therfore (if it pleaſe you) I will repeate the chieffe
pointes that I haue learned ſythe my former repetition.

    Maiſter. Doo ſo then.

    Schollar. This it is as I remmembre,

The ſecond   1.   Fyrſt you taught me what a ſphere is, and howe it is made, alſo what is
repetition.   his Centre, his Axetree, his Diameter, and his Poles, and what the
               Poles are named.

      2.   Nexte you declared two circles, that is the Horizonte, and the Meri-
              diane circle, whiche (I perceaue) ſtand ſtyll, and tourne not with the
              worlde, but keepe their places.

      3.   Then did you deſcribe fiue parallele circles, the Equinoctiall, the twoo
              Tropikes : the Sommer tropike, and the winter tropike, and then the
              other two Paralleles, that is, the Northe circle, and the Southe.

      4.   After that, you ſhewed me what the Zodiake was, and the twelue Si-
              gnes that be in him, and of their diuiſion.

      5.   And laſte of all, you deſcribed the twoo Colures, whiche diuide the
              Zodiake into foure equall and principall partes, accordynge to the
              four tymes of the yeare.

    Maiſter. This good remembraunce declared your good
will to knowledge, whiche I ſhall with as good a will healpe
to further. Now you looke (I think) to be inſtructed in the
vſe of all theſe thinges, and to vnderſtand therby the celeſtiall motions, and the diuers appearances that therby doo
enſue : how be it, bycauſe that a materiall inſtrumet is a great
helpe for them that begin to trauaile in this arte, and dothe
as an image repreſent to the eies thoſe thinges, which by on
ly hearing, were very hard to conceaue, beſides many other
commodities, whiche ſhall be vttered in their place, I think
it moſte conuenient order, fyrſt to teache you the manner
howe to make ſuche a materiall ſphere, as may ſerue both to
learne by, and alſo to worke by, in practiſing the obſeruations needefull to this arte.

THE SECOND TREATISE   35

OF THE CASTLE OF KNOWLEDGE.

Wherein is taughte the makinge of the materiall sphere,
as well in founde or massy forme, as also in
rynge forme with hoopes.

MASTER.

LTHOVGHE THERE BE MANY and wonderfull inſtrumentes wittely deuiſed for practiſe in Aſtronomy, as the Aſtrolabe, the Plaine ſphere, the Saphey, the Quadrante of diuerſe ſortes, the Chylynder, Ptolome his rules, Hipparchus rules, Tunſteedes rules, The Albion, the Torquete, the Aſtronomers ſtaffe, the Aſtronomers ringe, the Aſtronomers ſhippe, and a great numbre more, whiche hereafter in tyme you may knowe, yet all theſe are but parts, or (at the moſt) diuers repreſentations of the Sphere. wherfore as the Sphere is the grounde and beginner of all other inſtruments, ſo is it moſte meete that we begin with it, and the rather bycauſe it dothe more aptlye repreſent the forme of heauen, then anye other inſtrument canne doo. What a Sphere is, you haue learned before : and howe a materiall Sphere or Globe maye bee made rounde, you maye coniecture by the ſame deſcription of Euclide. Therfore muſte you haue an inſtrumente of ſteele made lyke a Semicircle, whiche in the inner circumference muſte haue a ſharpe edge apte to cutte and pare ſmothe, and (as I maye ſaye) by true woorkinge to iuſtifie your Globe, whiche fyrſte maye bee made as rounde, as any Turner can doo it. and then ſhall your inſtrument not only duly examen the Turners work, but correct it exactlye if it be amyſſe.

Inſtrumēts of Aſtronomye.

The tourning of a Globe.

This is the forme of that inſtrumente, and it is thus made iuſtlye. Firſte draw a righte lyne as longe as you wyll haue

the

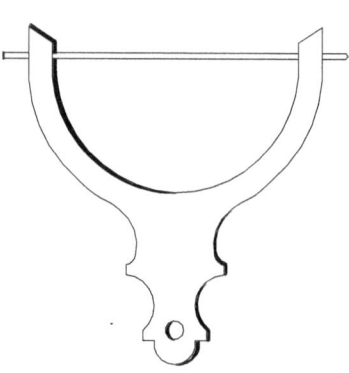

the diameter of youre
fphere, and an yche lon
ger, or more. Then o-
pen youre compas ac-
cordynge to the halfe
diameter of the fphere
that you would make,
and draw halfe a circle,
fo that the fixed foote
of your compas be fet
in the myddle (as you
may nearlye geffe) of
the fayd line, and wyth
the other moueable
foote make the femicircle, but not fullye complete to the
diameter, for there mufte bee twoo holes made as bigge
as a wheate ftrawe or bygger, accordynge to the bygnes of
the Globe, for thoroughe thefe
holes mufte the Turners fpyn-
dles pearfe, that mufte beare the
Globe whyle it is in tournynge :
but you mufte take good heede,
that thofe holes bee fo made,
that the forefayde lyne doo paffe
exactlye thoroughe the verye
myddle of them, for fo muche
as you miffe in makynge thofe
holes, fo muche will your fphere

An other forme of the
fame woorke.

bee falfe in euerye quarter. Againe you mufte take heede
that youre inftrumente doo not bowe inwarde withoute
thofe holes towarde bothe the poyntes, excepte it bee in
true compaffe, but better it is to fyle it fomewhat a flope
outwardelye. What more is to be doone, I leaue it to the ftu-
dioufe deuyfe of your owne practife. for fuche thynges are
better taught by hande, then by mouthe.

                                                      Schol-

THE CASTLE OF KNOWLEDGE 37

Schollar. I wolde I coulde as well vſe it, as I could diuiſe
to make it iuſte rounde.

Maſter. When you haue your globe ſo iuſtified in round *To find the*
nes, marke well the twoo Poles of it, which you may eaſily *Poles in a*
do by the ſame inſtrument, whereby you did iuſtifye it, for *Globe.*
the ſpindles that paſſed through the twoo holes of your inſtrument, doo touche the twoo poles exactly.

Schollar. That can I eaſilye doo.

Maſter. Then muſte you haue a payre of compaſſe
aptelye made for to drawe the circles in youre Globe,
and the poinctes of the ſhankes in that Compaſſe muſte
bowe ſomewhat inwarde (as here you ſee an example)
and the poynctes of it muſte bee verye fine and harde, that
they maye graue deepely, and yet make a fine and ſmall
circle. for the fyner that your circles be, the exactlier will the
diuiſions be made, and the leſſe erroure wyll bee in the ma-

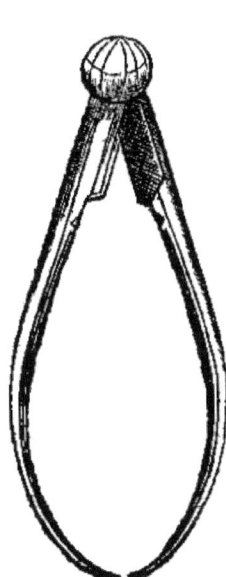

kynge and vſing of the ſame Globe. *A compas*
Then ſette one foote of the com- *for a Globe.*
paſſe in one of the Poles of the
Globe, and open the other ſo wyde,
as you thynke will ſuffiſe to reache
to the myddle of the Globe, towarde the other Pole, and with that
foote make a lyghte marke in the
Globe : and keepynge youre compaſſe vnchaunged, putte one foote
of it in the contrarye Pole, and
tourne the other foote towarde the
foreſayde marke, in the myddle of *To make*
the Globe, and if the foote touche *equinoctial*
it exactelye, then is that myddle *circle.*
duelye founde : but if the compaſſe
reache to farre, or to ſhorte, make
wyth yt an othr lyghte marke,
and the true myddle betweene

D.i.   thoſe

thofe two marks is the iuſt middle of the Globe or Sphere, as by your compaſſe a little opened more or cloſid (as you fee caufe) you maye prooue.

Schollar. That can I do well ynough, by experience learned in often practiſynge the conduſions of youre Pathwaye.

The Path-way.

Maſter. That Pathwaye wyll leade you rightlye to this woorke, if it bee well trauayled as it oughte to bee before you come to this woorke. But to procede with our Sphere : When you haue founde the iuſte myddle of the Globe betwene bothe the Poles, then open youre compaſſe accordynge to the diſtance of that middle marke, and one of the Poles, and fet one foote of the compaſſe in the Pole (whiche you lyſte) and with the other drawe a cyrcle rounde about the Globe. whiche whether it bee truelye doone or not, thus

Proof.

maye you prooue : Remoue the foote of your compas into the other Pole, and with the mouable foote trye the former circle, & if the compaſſe run iuſtly in it, then is that circle truly drawen betwene both the Poles, elfe haue you erred : and therfore graue not ẙ circle to deepe, till you haue examined

The diui-ding of the equinoctial.

it. And when you haue found it true, then without alteringe of the compas, fet bothe feete of it in the fayd circle, & they will take the fourth part of the fame circle, as by remouinge it four tymes, you maye know.

Proof.

Schollar. That haue I learned in the Pathwaye alfo, and if I haue myſſed, it is by the groſſeneſſe of the poyntes of my compaſſe, or elfe by myne owne groſſe negligence, whyche bothe I canne quickly examine and amende, as the cafe requireth.

Maſter. After that you haue marked oute thofe foure partes of that circle, dyuide eche of them into three euen partes, and fo haue you that cyrcle dyuided into twelue equall partes : marke thofe partes with little croſſe lynes, or elfe drawe an other circle wythin a corne breadthe of that other, on which fide you liſt, but let it be fomwhat leſſe

graued

grauid then the fyrſte, that the fyrſte may bee knowen for
the true circle, and this ſeconde cyrcle to ſerue but onlye
for the markes of diuiſion in that other : and ſo drawe a
lyne at euerye twelfte parte, from the one cyrcle to the
other. Then dyuyde euerye one of thoſe partes into
three leſſer partes, and eche of theym agayne into euen
halues, and ſo haue you in all, 72. parts made of that cyrcle.
After this, diuide one of thoſe partes into fiue leſſer portions, equallye, and by the ſame example diuyde all the other
71. partes, and ſo haue you in the whole circle, 360. partes,
whiche you ſhall marke with nombres of figures, from 10.
to 10. beginninge where you lyſte.

Schollar. Thoſe I maye call degrees, as I remembre
by youre former leſſons. and I muſte marke them thus. 10.
20. 30. 40. and ſo vnto 360.

Maſt. So it is : And thys cirde thus drawen in the middle
betwene bothe the Poles, is the Equinoctiall cyrcle in that
ſphere. Now to make the two Tropiks, open your compas _To drawe_
ſo : that they maye extend to 66. degrees and an halfe of the _the twoo_
ſaid Equinoctiall cyrcle. and then ſet one foot of the com- _Tropikes._
paſſe in which Pole you will, and with the other foot draw a
circle on the Globe, which ſhal ſtand for one of the tropiks,
and ſetting the foote of the ſame compaſſe vnaltered, in the
other Pole, draw about it an other circle, for the other tro- _The Poles._
pyke. Now appointe names for the Poles, callynge one
the South pole or Antartike pole, and the other the North
pole or Arctik pole : and then the tropikes of neceſſity will
take their names : for that Tropike which is next the North
Pole, muſt be the tropike of Cancer, that is, the Sōmer tro- _The Tro-_
pike, and the other that is nexte to the Southe Pole, muſt _pikes._
needes bee the Tropyke of Capricorne, or the Wynter
Tropyke. Then marke where you beganne the noum-
brynge of the degrees in the Equinoctiall (whiche maye
well be called the begynninge of the Equinoctiall) and ſet _The tropik_
one foot of your compas in that beginning, openyng the _Colures._

D.ii.   other

40   THE SECOND TREATISE

other foote tyll it will reache vnto 90. degrees iuftlye, and
fyrſte holde the one foote fteddye in the begynninge of
the Equinoctiall, and drawe a circle with the other foote,

Proof.
and if that circle touche bothe the Poles of the Globe,
then is it trulye drawen. but it fhould go alfo by the ende of
the 270 degree of the Equinoctiall, and if it miffe anye
whitte, examine it well, and amende the faulte, before

A generall
rule.
you woorke anye farther. whiche rule you fhall obferue
ftyll, for els of one faulte neglected, many other may enfue.

The Equi-
noctial Co-
lure.
        This doone keepe youre compaffe at the fame wyde-
neffe, and fette one foote in the Equinoctiall circle, at the
ende of 90 degrees, and holdynge it fteddye, with the other

Proof.
foote defcribe a circle, whiche fhall paffe by both the Poles of the Globe, and by twoo pointes of the Equinoctiall,
that is the beginninge of it, and the ende of 180 degrees.
and if you haue miffed, amende it by and by. This lafte circle is the Colure Equinoctiall, and the other laft before
drawen is the Colure Tropikall, or Solftitiall, or the Tropike

The diuifi-
on of the
Colures.
Colure. Thefe twoo circles fhall you diuide into 360 parts
eche of them, beginninge your numbrynge at the Equinoctiall, and rekeninge towarde the Pole, in euery quarter of
them feuerallye, fo fhall you neuer recken aboue 90. But
it is eafilye knowen, that foure tymes nynetye doothe
make .360.

        Scho. But in this ordre of numbrynge, the cōmon forme
of accompte is not kepte, as it was in the Equinoctiall :
for when I haue reckened in one quarter 90. degrees from
the Equinoctiall to the Pole, then if I go forwarde in
the fame circle, the nexte numbre beyonde the Pole is
nynetye againe, and fo that feconde quarter decreafeth from
90 to 10, goynge backwarde, and then the thyrde quarter
increafeth from 10 to 90, and the fourth quarter decreafeth
againe from 90 to 10.

Proof.
        Mafter. So muft it be in thefe circles for mofte apteneffe
in accompte, as you fhall perceaue hereafer. Nowe fhall

it

it be conuenient to mark in what degrees the two Tropikes
do cut thofe Colures, for if you haue not erred, they touch
the myddle of the four and twentith degre in euery quarter
of the Colures. And if you haue doone well, then procede
to the making of the Zodiake, whiche you fhall draw thus.
Open your compaffe to the fame wydeneffe that you dydde
for makyng the Colures, or the equinoctiall, & then recken
from one of the poles (whiche you will) 23 degrees and an *Pole Cir-*
halfe, in any one of the Colures, and it will lighte in 66 de- *cles 2.*
grees and an halfe, bycaufe the numbres from the poleward
go backward. (as you confeffed before) then with a leffer cō-
paffe (for it fhall bee meete that you haue diuers forts) draw
a circle of that circuit about eche Pole, fetting the fixed foot
of the compas in the Pole, and ftretching the other foot vnto 66 degrees & a half. After this looke whether thefe circles
do cut lyke degrees in euery quarter of the Colures : and if
it do, your woorke is righte, els it muft be redreffed. Thefe
circles maye well bee called Pole circles, or Polar cyrcles.
Then take your greater compaffe opened (as is before *The dra-*
declared) to the wydeneffe of a quarter of the Equinocti- *wing of the*
all, and fette one foote of them in that poyncte where the *zodiak.*
Polare circle that is aboute the Northe pole, dooth croffe
the tropyke Colure in that quarter, whyche goeth from
that fame Pole, to the .270. degree of the Equinoctiall,
and holdynge that foote fteddye, with the other drawe a
circle aboute the Globe. This circle will touche the *Proof.*
twoo Tropikes in twoo of thofe places, where they croffe
the Tropike Colures : and alfo it wyll croffe the Equinoctiall in twoo pointes, that is, in hys very begynnynge, and
in the ende of the 180. degree. Nowe to proue whether it be *An other*
truely drawen or not, by an other meanes, fet one foote of *proof.*
that compaffe (with whiche you drew the Zodiake) in that
pointe whiche is directly contrarye to the firfte place, where
you ftayed hit : that is to faye, in the croffynge of the
fourthe Polare circle, and that qnarter of the tropike Colure, whiche goeth from the South pole to the 90. degree of
D.iij.   the

42 THE SECOND TREATISE OF

the equinoctiall, and on that point proue whether the mouable foot of the compafe will exactly agree with the fore-
fayd circle, whiche yf he doo, it is well drawen, els is there
fome erroure, which mufte bee amended. This circle thus
drawen, is the Ecliptike circle, whiche goeth by the myddle of the Sygnes and of the Zodiake. and thefe twoo
poyntes wherein the fyxed foote of the compaffe was
ftayed, are the Poles of the Zodiake. But confidering that
the Zodiake (as you hearde before) hath in it twelue degres
of bredthe, that is, on eche fyde of the Ecliptike lyne fixe,
therefore open your compaffe to 84. degrees only, that is
fixe degrees leffe then a quarter of the Equinoctiall, and fet
one foote of it fixedly in the one Pole and the Zodiake, and
with the other moueable foote drawe a circle, which wyll
be a Parallele to the Ecliptike circle, diftaunte from it in all
partes by 6 degrees, and with the fame compaffe vnaltered,
draw a lyke circle on the other Pole of the Zodiake, whiche
fhall bee a Parallele to the other twoo, and they three do
make the full Zodiake in length and breadth.

The Poles of the zo-diake.

Schollar. I vnderftande all this verye well, but I mufe
what thofe Polare circles meane, of whiche I hearde no
woord before in the firfte treatife.

The Polare circles, and their vfe.

Mafter. I dyd of purpofe omytte them before, bicaufe
they ar named of diuers men, as of Ioannes de Sacro Bofco
and other later writers, for the circles Arctike and Antarctike, contrarye to Proclus, and all the greeke writers, and
I pourpofed (and fo doo I ftill) to referue the difcuffing of
that repugnance, to the fourthe treatife, yet here was fuche
iufte occafion miniftred to vfe theire helpe in fyndynge the
poles of the Zodiake, by whiche poles they are defcribed
euery day, by the reuolution of the heauens, that I coulde
not willyngly neglecte them : for although I myghte fynde
the poles of the Zodiake without them, yet they bringe a
proof of the woorke with them, as before I haue fhewed,
and alfo they enclofe all fuche ftarres as are within 23. degrees

grees and a halfe of the Pole, and are the lymites of the
motion that the Poles of the Zodiake doo make about the
poles of the worlde, as you ſhall better perceaue hereafter.
And bycauſe their names ſhoulde not bee confounded with
the circles, Arctike and Antarctike, I thinke it moſte meete
to cal them only Polare circles, or Pole circles, which name
the other circles may not iuſtly chalenge, eſpecially bycauſe
they are not fixed (as the Pole circles are) but be chaungea-
ble as the regions chaunge. which thing I will declare more
largely hereafter, but nowe for the drawinge of the circles
Arctike & Antarctike, that is (as I named them) the Northe   *Circles ar-*
circle, and the Southe circle, you muſte learne the eleuation   *ctike and*
of the region for whiche the Globe is made, and according   *Antarctike.*
to it muſt you draw thoſe circles, whiche thinge bicauſe as
yet it is not eaſye for you to doo, I will in example of oure
owne cuntrye ſhew their deſcription, namely for the vniuer-
ſitye of Cambridge, whiche ſtandeth in euen degrees of 52.
Therfore recken from one of the Poles 52. degrees in anye
Colure, and it will lyghte on 38. degrees (bicauſe the numbres go backward) and there ſet one foote of your compas,
extending the other foote to the next Pole, where you ſhall
ſtaye it, and with the other foote deſcribe a circle fyrſt about
the one Pole, and then about the other : and thoſe two circles ſhall ſtand for our circles Arctike & Antarctike. And
thus hath the Globe all thoſe circles whiche were accompted needfull vnto it, excepte the Horizonte and the Meridiane circle, whiche are not ſo well placed in the Globe as
without it, bicauſe they ought not to moue with the Globe.

    Schollar. Where ſhall they be made then?

    Maſter. That will I ſhewe you, as ſoone as I haue ended
the Globe, whiche yet is not doone, for the Signes in the
Zodiake are yet vndrawen. Firſt therefore ye ſhall drawe by
the Ecliptike line within a corne bredth of it, an other circle   *The diuiſi-*
as you did by the Equinoctiall, it forceth not on whyche   *on of the*
ſide, but let the Ecliptike line be more notable then it. Then   *zodiake.*

                  D.iiij.   conſider

confider that the Zodiake is all ready diuided into foure e-
quall quarters by the two Colures, now it is meet to diuide
euerye quarter into three equall partes, and fo haue you
twelue partes in the whole Zodiake, whiche ftande for the
twelue Signes, which fhall be diftinct by lynes drawen ouer
thwarte all the breadth of the Zodiake.

 Schollar. Thofe are not eafye to drawe, but errour may
quickly be committed, in making them wyder in one place
then in an other.

 Mafter. Therfore to auoyde that errour, thus fhall you
do. Open your compas equally with a quarter of the Zodiake. then keepe one foote of it fteddy in eche diuifion, one
after an other, and with the other drawe a portion of a circle croffe ouerthwart all the breadth of the Zodiake, & thus
fhall you do it exactly, and in fo doing, your compaffe doth
trye and examine the former diuifion : for if at anye fetting of your compaffe it reache to fhorte, or to far, and not
iuftly on the thyrde figne, then muft you correct your fyrft
diuifiō. When you haue drawen thefe twelue fignes, thē muft
you diuide euery one of them fyrfte into two parts equally,
and eche of them againe into three euen partes, and laftlye,
euery one of them into fiue iufte portions, and fo haue you
in euery Signe, thirtye partes or degrees.

Proof.

 Schollar. This diuifion is like the diuiding of the Equinoctiall and the Colures, fo that I maye conceaue the one
by the other.

 Maft. Indeed they ar all thre lyke in their general diuifiō,
but yet in placinge of their numbres, they differ eche from
other, for the Equinoctiall had his numbres continuallye
proceding from 1. to 360. The Colures, ftay their numbres
at euery quarter, neuer procedinge aboue 90, but the Zodiake ftayeth in a leffer numbre, for at euery figne, his numbres chaunge : fo that from the beginning of eche Signe to
the ende of the fame, you fhall marke them from 10. to 10.
thus: 10. 20. 30. and fo lyke in all the Zodiake no numbre is
               greatter

greater then 30.

Schollar. I perceaue that, sith you tolde me before, that
euery Signe seuerally hath 30 degrees.

Master. Those diuisions shall you marke with a little line
drawen from the Ecliptike circle to that other which is drawen within a corne bredth of it : yet at euery tenne degrees it
will do well to draw the line somwhat longer from the Edi-
ptike, that those degrees maye be the easier to see and to re-
ken, and so maye you doo at euery fiue degrees, but some-
what shorter then that other, and so shall you haue the degrees more notablye distincte in sonder. Nowe resteth no
more but to geue euery Signe his name, which you may do
other by writinge it at lengthe, or els by settinge their Cha-
racters and figures for their names, which I before haue set
forthe vnto you in bothe formes.

Schollar. That is easye inough to vnderstande, but how
shall I knowe their places?

Master. That is as easye also, if you marke the ordre of
the circles. but for a full plainesse you maye beginne at the
Tropike of Cancer, where the signe of Cancer doth begin,
and in that quarter of the Zodiake, which is on your right
hande, and descendeth toward the Equinoctiall, sette these
three signes, Cancer, Leo, Virgo, and so procede forward
as the signes succede in ordre : then will the seconde quarter
haue Libra, Scorpius, and Sagittarius : and the third quarter, Capricornus, Aquarius and Pisces : and to make vp
the fourth quarter, ther resteth Aries, Taurus and Gemini.

Schollar. You name the seconde quarter of the Zodiake
to the the fyrste, and so commeth it to passe, that you call the
fyrste quarter the fourthe, as I remembre youre former
doctrine.

Master. You maye perceaue, that I named them nowe
not in their custumable ordre of quarters, but accordynge
to the ordre of this woorke, els if you can discerne the place
of Aries from the place of Libra, you may best begin with

Aries

46   THE SECOND TREATISE

quarters of the zodiake.	The quarters of the yeare.	The Signes in euerye quarter of the zodiake, aunſweryng to eche quarter of the year.		
1.	Springe.	Aries,	Taurus,	Gemini.
2.	Sommmer.	Cancer,	Leo,	Virgo.
3.	Harueſt.	Libra,	Scorpius,	Sagittarius.
4.	Winter.	Capricornus,	Aquarius,	Piſces.

Aries, & thē not only the ſignes, but ẙ quarters wil keep their
accuſtomed ordre, as here in a table it doth apear : wher I haue
alſo annexed the quarters of the year for readines of remem
brance, & for the better occaſion to marke the motion of the
ſon in eche of thoſe quarters. And thus haue we ended the
globe or ſphere, with al ẙ cirdes in it cuſtomably vſed, whoſe
picture here you may ſe, as it will be drawen in flatte forme.

A, C. is the Equinoctial cirde.
E, K. the tropik of Cancer.
Q.L. the tropik of Capricorn
Q.K. The Zodiake.
B, and D, The ij. Poles of the
    worlde.
F, I. The Arctike cirde.
P, M. The Antarctike cirde.
G, H, and O, N. The two Po-
    lare cirdes.
G, and N, The ij. Poles of
    the Zodiake.

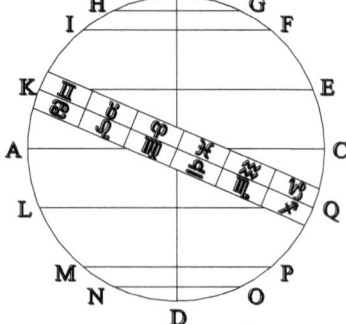

*The making
of the Ho-
rizonte.*

Now for the Horizōt
& the Meridiā thus ſhal
you do. Take 2. ſquare
bords of a quarter of an inch thick, & let ẙ one be in bredth 3,
inches, & the other one inch & a half more then ẙ diameter of
your globe, in ẙ middle of the broder borde take a centre, &
on ẙ cētre make a cirde, ſcarſly a corn bredth wider thē your
globe is, which you ſhal thus find out. Open your cōpas as
wide as ij. ſigns in ẙ Zodiak, or 60. degres in ẙ Equinoctial,
                                                        or any

THE CASTLE OF KNOWLEDGE   47

any other of his greate circles, and that compaſſe wyll make a circle iuſt in bigneſſe with any great circle of your Globe, therfore make you the circle in the ſquare borde, almoſte a corne bredthe wyder then that circle of youre Globe. And without alterynge of the compaſſe, make the lyke circle on the myddle pointe of the narrower borde. Then haue you taken the iuſt meaſure for the inner part of your Horizont, and alſo of your Meridian.

Schollar. I doubt not but I canne doo that with a lyttle labour by often triall where the myddle of the bord is. but is there no waye to fynde the place of the centre quickly?

Maſter. Yes truly, and that maye you doo diuerſly, but one redye way is this. *To find the middle in any ſquare.*

Drawe with your ruler a right line from corner to corner, or if you lyſt, make it onlye about the myddle of the bord, as you can ayme with your eye, but be ſure that you drawe it longe ynough, then turne your ruler to the other two cor-ners, and make a lyne croſſe that other, and where they doo croſſe, there is the myddle of the borde, on whiche, as on a

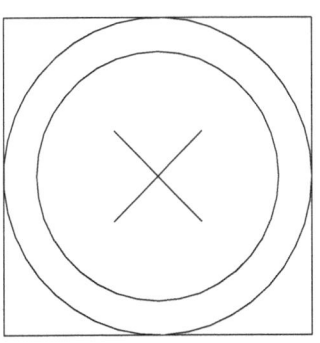

cētre you may make your circles. This work might you eaſilye gather out of the 35 concluſion of the Pathway. *The Path-way of G-ometrye.*

Schollar. I ſee now cō-tinually more and more, that the Pathwaye ſer-ueth to other vſes, then I toke it.

Maſter. It is a commō inſtrument to many arts, and infinite concluſions : and if you procede to farther knowledge of higher artes, without good exerciſe in it before, you do as a carpēter that goeth to woorke without his tooles. But nowe to proceede,

when

48   THE SECOND TREATISE OF

When you haue drawen this circle on bothe thofe bordes,
on the fame centre make an other circle in eche borde, a corn
bredth wider then that other : and after that an other fomewhat wider, as you may ayme two corne bredthes : and then
the fourth wider then the thyrde by a quarter of an ynche :
and yet againe one other a quarter of an ynche wyder then
the fourth. and thefe fiue circles fhall you make in bothe the
bordes, and you fhall diuide them bothe in one manner, after this forte.

Diuide the innermoft circle faue one, into 4. quarters firft,
and after that, euerye quarter into three partes, and eche
of thofe partes into 30. as you dyd before in dyuers cyrdes of
the Globe. then fet your rules to the centre, and to euery di-

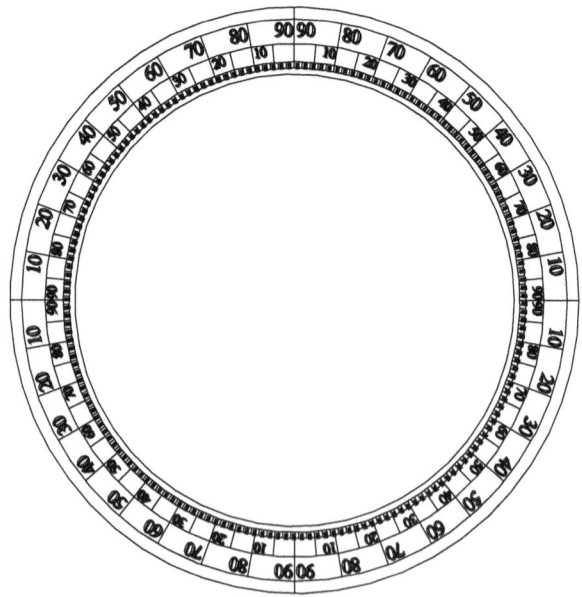

vifion, and make a lyne from that fecond circle to the third :
but at euery 10. degree you fhall drawe the line longer, that
                                                                   is,

THE CASTLE OF KNOWLEDGE   49

is vnto the fifte circle, and at euery fift degree, you fhall draw
the lyne to the fowerth circle, fo fhall you both place your
numbres beft, and alfo recken them mofte furely and moft
fpeedily in all vfes of them.

Schollar. All this I can do by the former examples, if I
knewe how hyghe the numbres fhall proceede. for in them
I remembre ther was 3. varieties before, eche vnlike to other.

Mafter. And in thefe fhall be fomwhat diuers from them
all. for here fhall be fet double numbres, but yet the fyrfte
placynge of the numbres fhalbe lyke as it was in the Colures, I meane in eche quarter 90. and thofe numbres fhalbe
fet in the fpace, betweene the thyrde circle and the fourthe.
Then fhall you fet the lyke numbres betweene the fourthe
circle and the fyft, but not in lyke ordre, for their ordre fhal
be contrary to the other, fo that where 10. ftoode in the fyrft
ordre, & then 20, and fo increafyng to 90. in this 2. ordre you
fhall fet 90, and thē 80, & fo decreafe vnto 10. as here in exam
ple you may fe, wher I haue drawen ỹ Meridian lyne fufficiently diuided, for ỹ vfe of ỹ fphere : but thē ỹ horizont muft
haue other things drawē in it, as in this figure folowing you
may fe. for in ỹ inner part it is deuided like vnto ỹ meridian,
but then without thofe diuifions it hath a certain fmal fpace
all black, left for a partition, without which ther are drawē 3.
other circles, eche one a lyttle wider then other, & the wideft
is vttermoft, and ỹ laft circle is as large as the borde will per
mit, fo that ỹ whole bredth of ỹ Horizont is an inch & a half,
for bicaufe the whole bord was 3. inches wider thē the globe.
And ỹ Meridian fhalbe but 3. quarters of an inche brode, fe
ing his bord was but 1. inch & an half wider thē ỹ globe. Now
for the diuifion of the vtter part of the Horizont, you fhall
dyuide the vttermofte of the three circles into eyghte
partes only : The feconde circle fhalbe diuided into 16. parts
And the third or innermoft of thofe 3. fhall be parted into
32. partes, whiche do betoken the points of the Shypmans
compas. or the 32. winds notable in failyng, as fome mē lyft
E.i.   to

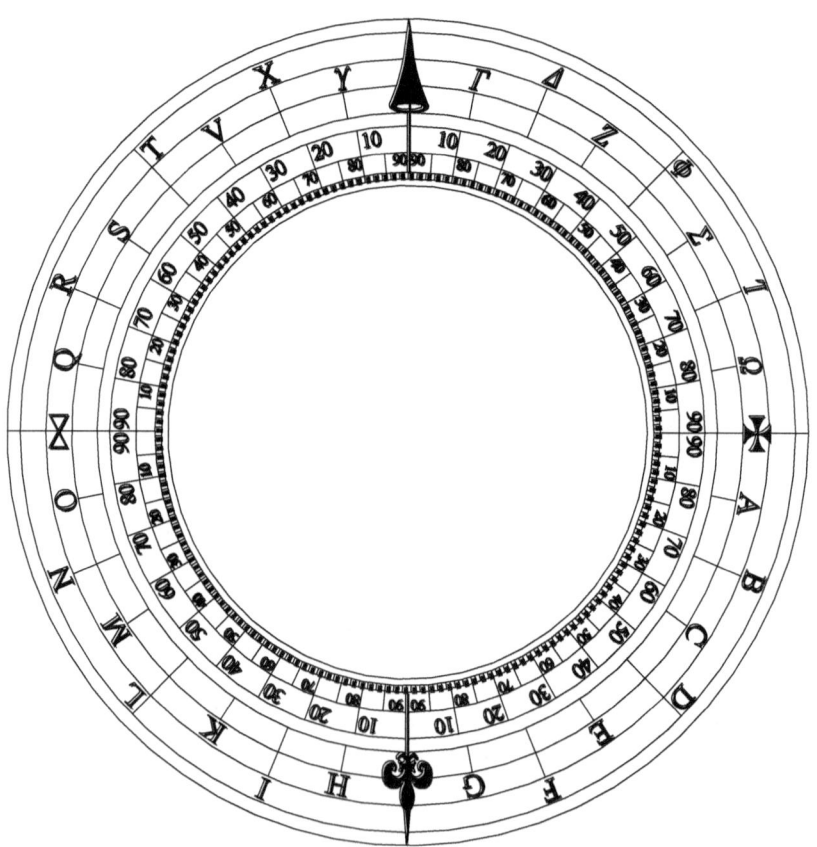

to call them. If your Horizonte bee large inoughe to receaue their names, you fhall write them at lengthe, els maye
you write letters for theym, as youre owne phantafye lyketh.

Their names are thefe folowinge, agreable to thofe places
and letters, whiche I haue drawen in the Horizont.

                                                              The

THE NAMES OF THE

THIRTY AND TWO POINTES IN THE SHIPPE
compaſſe, whiche bee the Windes names that Mariners ſayle by.

♇	Northe.	N.	Eaſte northeaſte.
♈	Southe.	O.	Eaſte and by northe.
X	Eaſte.	Q̣.	Eaſte and by ſouthe.
♇	Weſte.	R.	Eaſte ſoutheaſte.
A.	Weſte and by northe.	S.	Southeaſte and by eaſte.
B.	Weſt northweſte.	T.	Southeaſte
C.	Northweſte and by weſte.	V.	Southeaſte and by ſouthe.
D.	Northeweſte.	X.	Southe ſoutheaſte.
E.	Northweſte and by Northe.	Y.	Southe and by eaſte.
F.	Northe northweſte.	Γ.	Southe and by weſte.
G.	Northe and by weſte.	Δ.	Southe Southeweſte.
H.	Northe and by eaſte.	Z.	Southeweſte and by Southe
I.	Northe northeaſte.	Φ.	Southe weſte.
K.	Northeaſte and by northe.	Σ.	Southweſte and by weſte
L.	Northeaſte.	Τ.	Weſte ſouthe weſte.
M.	Northeaſte and by eaſte.	Ω.	Weſte and by ſouthe.

And thus nowe is the horizonte fully drawen. That Hori- *The foote* zonte muſte you ſet vpon a foote, that it may ſtande lyke a *of the Ho-* rounde table : and that foote muſte be made of twoo halfe *rizonte.* circles of woode, ſomewhat thycker then the Horizonte, but of the ſame compaſſe in the innermoſte parte, and they muſt be ioyned ſo, that the one maye croſſe the other, wyth ryghte corners, and them ſelues bee faſtened on a ſtronge foote, that may beare all the whole frame, wyth the Globe. The ioyninge of them vnto the Horizont is diuerſly to be ymagined, for if their headdes be flat, then muſte you haue nailes or els pinnes, that muſt perſe the Horizont and enter into their heddes, otherwaies there maye be left certaine te-nauntes on their heddes, and then muſt you make lyke mor teyſes agreable to them, in the Horizonte, to receaue thoſe tenauntes. & ſo may there be ymagined diuers other formes, whiche I leaue to your owne deuiſe.

Schollar. If I myght ſee their forme I ſhoulde be muche eaſyed in framynge it.

E.iJ.   Maſter

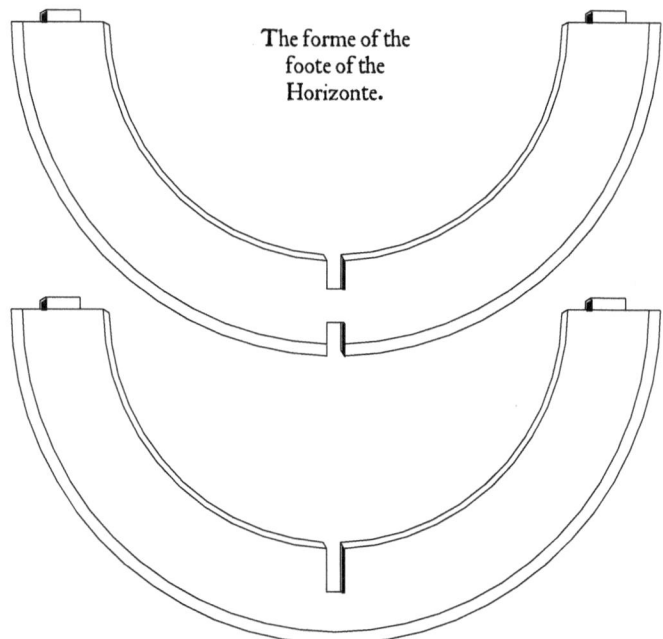

The forme of the
foote of the
Horizonte.

Maſter. Here is the form, with their ſockets, & one namely
for the Meridiane, in that arme alſo that goeth from Eaſt to
weſte. Howe be it, it ſhall be beſt, to faſten thoſe armes
vnder the Horizonte in the Southe eaſte, Southe weſte,
Northe eaſte, and Northe weſte, and ſo ſhall the Meridiane ſynke beſte into the Horizonte, with an eaſye ſocket
in the meetinge of thoſe armes, ſo that the iuſte halfe
of the Meridiane onlye maye appeare aboue the oouer
edge of the Horizonte : in whyche thynge praĉtiſe ſhall
inſtruĉte you farther. As for the foote, make it as you
thinke beſte. But nowe muſte you cutte out of bothe, the
Meridiane and the Horizonte all that is within the innermoſte

THE CASTLE OF KNOWLEDGE   53

moſte cirde, and ſo muſte you pare awaye all that is without the vttermoſte cirde, to make them bothe lyke iuſte
cirdes. Alſo you muſte make in the Horizonte twoo ſockets, one by the Southe lyne, and the other by the Northe
lyne, ſo that the one ſyde of thoſe ſockets whiche is toward
the eaſte, ſhall touche the Southe and Northe lynes, and
the other ſide ſhall go weſtwarde from bothe thoſe lynes,
as muche as the thicknes of the Meridian is : and the length
of eche of thoſe ſockettes ſhall bee agreable to the iuſte
breadthe of the Meridiane, ſo that the Meridiane maye
entre iuſtlye into thoſe ſocketts, and turne in them without
ſtreſſynge.

The forme of the foote vnto whiche the armes are faſtened that beare the Horizonte.

which therfore wolde be made large, that it may beare the Globe with all his circles ſted-dilye.

Schol. This trobleth me ſomwhat, bi-cauſe the ſoc kettes be not iuſtelye one agaynſte the other, but bothe ſtande towarde the Weſte halfe of the Hori-zonte.

Maſter. It wolde trou-ble you worſe to re-membre that the Globe muſte be faſtened to the Meridiane
on the two poles, & both they placed within the Horizonte.

Schollar. That is ſtraunge in deade, for ſo ſhold the globe
beare more toward the weſt, then toward the eaſte : and ſo all
were miſframed.

Maſter. To auoide all that, you ſhall make twoo ſmall
E.iij.   clampes

34   THE SECOND TREATISE

clampes of thinne braſſe plate, and bow them ſo in the middle, that when they are tacked to the ſide of the Meridiane
in twoo contrarye pointes, iuſte ouer that line where 90. is
ſet, thei may receaue in their bought the poles of the globe.
I meane here by the poles two ſhorte pinnes, which ſhall go
through thoſe clampes of braſſe, and be faſtened or driuen
into the twoo Poles of the Globe, excepte you will take the
paine to pearſe a hole through the globe, from one Pole to
the other, for ſo maye you make an axetree to run thoroughe bothe the clampes and the whole Globe, whiche is
all to one effecte. And by this meanes ſhall the Globe not
onlye hange in the iuſte middle of the Horizonte, but alſo
the one ſide of the Meridian (whiche hathe the diuiſions in
it) ſhall pointe exactly the ſouthe and north partes of your
Globe, whiche will be moſte exactly ſeene, if you conſyder
the thicknes of your axetree, and frame youre clampes ſo,
that the one halfe of the thicknes of the axetree, may be let
into the ſyde of the Meridian.

Schollar. I thynke I doo conceaue the true meanynge of
your woordes, howe be it to bee oute of all doubte, I wyll
be bolde to ſee your Globe, at ſome conuenient tyme.

Maſter. So ſhall you doo well, for manye thynges in the
makinge, and in the vſe alſo of inſtrumentes, are better perceaued by a lyttle ſighte, then by many woordes. and thus
haue I ended the making of this Sphere.

Schollar. Yet is this ſphere vnlyke to that, whiche is cōmonly vſed, by the name of the Sphere, and is made all togither of hoopes.

Maſter. You ſhall vnderſtand that this is the true ſphere,
whiche I haue deſcribed, and that other (which you meane)
ought rather to be called an Armylle or Ringe ſphere, then
abſolutely a ſphere, for it is but a part of this other Sphere :
I meane, that it doth contayne only the circles of the ſphere
and not the ſubſtaunce of it. And therfore dothe many men
cal that a Perſed ſphere, and is named in Latin Sphaera per-
tuſa,

THE CASTLE OF KNOWLEDGE   55

tuſa, where as they call the other ſphere, a Sound or Maſſye
Sphere, that is in latine, Sphaera ſolida. but ſeynge that it
is not only commonly receaued by the name of the Sphere,
but the vſe of it is very apte in teaching, and it is more eaſy
to bee made in ſlyghte forme for yong learners then is the
ſoonde ſphere, and for other conſiderations, whiche nowe I
omyt, I wyll alſo deſcribe the compoſition of that Armylle   *The makig*
ſphere. Fyrſt you ſhall make of woode or of braſſe (as you   *of the Ring*
lyſte to beſtow the coſte) four hoopes of one bignes in com   *ſphere.*
pas, the one of them beyng three times ſo broade as any of
the other, as your eye may ayme. Then diuide eche of thoſe
circles into 360. partes, one of them accordynge as you did   *The equi-*
diuide the Equinoctiall in the former ſphere, and the other   *noctiall.*
two lyke vnto the two Colures, and the fourthe which muſt   *ij. Colures.*
be the brodeſt of them, you ſhall diuide, as you learned to
diuide the Zodiake in the other ſphere. And when they are   *The zodi-*
thus diuided, you ſhall call them by the names of thoſe cir-   *ake.*
cles whoſe diuiſion they folowe, wherefore if the Zodiake
haue more breadth then twelue degrees are in lengthe, you
ſhall abate the ouerplus, allowing it but 6. degrees in bredth
on eche ſyde of the Ecliptike line, whiche as you remembre
before, did run by the mydle of the Zodiake.

    Schollar. Then I perceaue I muſte make in this Zodiake
an Ecliptike line, and all the ſignes with their diuiſions, as
I learned in the other Zodiake.

    Maſter. You ſhall make them as like as you can deuiſe.
Then ſhall you ioyne the two Colures ſo togither, that the
one of them may croſſe the other, (as thei do in the Globe)
with righte and equall corners, obſeruing well that the places of their croſſyng be in the iuſte pointes where 90. is ſet,
in eche of them : and thoſe places muſte be called the Poles
of the ſphere. Then put on them bothe croſſewaies (like a
girdle) the Equinoctiall circle (ſo that it do croſſe them ex-   *The Poles.*
actly with his middle, in thoſe pointes where the numbre of
eche quarter dooth beginne, and that the beginning of the
                              E.iiij.   Equi

Equinoctiall, in numbre be againſte the iuſte middle of one
of them, that is, of it that ſtandeth for the equinoctiall colure, and then ſhall the 180. degree of the ſame Equinoctiall
ſtand iuſtly on the middle of the ſame Colure, in the contra
rye pointe : and the other Colure whiche is the Tropike Co
lure, ſhalbe ioyned with the 90. degree, & the 270. of the equi-
The .ij. tro- noctial, in ij. cõtrary points. Then ſhal the 2. tropike circles
pikes. be ſet on ẙ Colures equidiſtantly to the equinoctial, ſo that
thei be faſtened on the 23. degree & a half from ẙ Equinoctial,
wherby you may eaſilye conceaue, that they muſte be ſomewhat leſſer then the equinoctiall, that they may ioyne cloſe-
ly to the foure Colures. Then muſte you haue twoo other
circles of one bygneſſe, that may ioyne iuſtly with the Colures, 52. degrees from the Equinoctiall, on eche part equallye diſtaunte : and thoſe muſte be called the Arctike, and
The Arctik Antarctike circles, or the South circle, & the Northe circle.
and Antar- Beſide theſe you ſhall make two other leſſer circles of equall
ctik circles. bygnes, whiche ſhall be ſet on the Colures alſo equidiſtante
frõ the other paralleles : and they muſt be faſtened with their
middle on the 66. degre & and half frõ the equinoctial on both
ſides, that is 23. degrees & a half from eche pole, and therfore
The Pole I thinke meeteſt to call theſe circles peculiarly, Pole circles.
circles. This beinge doone, you haue 2. Colures and 7. Paralleles
fixed on them. Nowe muſte you ſette the Zodiake a ſlope
waies croſſe the Equinoctiall, ſo that his myddle lyne, named the Ecliptyke lyne, maye touche the myddle of eche
Tropyke, and that maye you trye by the vtter edges of the
The zodi- breadthe of the Zodiake, for the one muſte touche the
ake. 29. degree and an halfe, and the other the 17. degree and an
halfe from the Equinoctiall. And thus is this ſphere plaine-
lye made, whoſe picture I haue here ſette, as it will bee drawen in a flatte forme. Then if you make twoo ſmall holes
The Axtre thoroughe bothe the Colures, in the places of theyr
The Meri- croſſynge, where the Poles of this Sphere are, and putte
diane and
Horizonte. a ſmall axe tree thoroughe theym, you maye thereby
                                                          ioyne

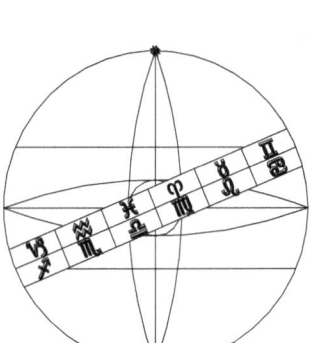

ioyne this Sphere to his
Meridiane fyrſte, and then
place it in the Horizonte,
as you didde place the
Globe : for thoſe twoo circles, are like in both theſe
Spheres.

The Pro-
Schollar. I vnderſtand al *portion of*
thinges herein wel inough *the circles*
as I thinke, ſaue y̆ I doubte *in a ſphere.*
ſomwhat of the quantitye
of the parallele circles. for
although I know by triall
I maye att lengthe make
them meete, yet woulde I gladlye knowe their meaſure before hande, if I myght, for ſo ſhall I be ſure to woorke moſte
certenly.

Maſter. Your deſire is good. and all be it that the writers
of the Sphere haue omitted it, as they haue doone manye
thinges els, yet will I geue you a rate of proportion drawen
out of the tables of Cordes and Arkes, called commonly in
latine Tabulae Sinuum.

Fyrſte you vnderſtand, that the Equinoctiall, the Zodiake and the two Colures muſt be of one compaſſe, that is
of one bygnes, althoughe not of one bredthe, for the Zodiake muſt be in bredthe twelue degrees, and the other circles as ſmall as they maye be, and beare any ſtreſſe, for the
ſmaller they be, the better they are, and moſte apte for the
vſe of the ſphere. The other ſyxe paralleles wold be made
as ſmalle as they maye beare conuenientlye, and in lengthe
they muſte haue three dyuers rates, whyche I wyll ſette
forthe, bothe in meaſure, and alſo in numbre, to the intent
that you may alter the meaſure to what bignes that you liſt,
by the helpe of the numbre.
And loe here is there formes.

The

1.   The Equinoctiall with his diuiſion.

2.   The Colures both of one forme.

3.   The Zodiake with the 12. ſignes, and his bredth of 12. degrees.

♈ ♉ ♊ ♋ ♌ ♍ ♎ ♏ ♐ ♑ ♒ ♓

4.   The length of the twoo Tropikes.

5.   The proportion of the Arctike and Antarctike circles.

6.   The proportiō in length of the two Polare circles.

THE CASTLE OF KNOWLEDGE 59

Here you fee fixe feuerall formes.

The fyrfte reprefenteth the iufte lengthe of that plate or hoope, that fhalbe the Equinoctial, and in it is the diuifions fett forth as they ought to fuccede in ordre, with their numbres agreeablye.

The fecond is the forme, that ferueth for the two Colures with their numbres and diuifions, as thei fhould be fet.

The thirde is the draughte of the Zodiake with his iufte bredth of fixe degrees, and the twelue Signes fett forth with their degrees ordrely. And thefe three circles be all of one lengthe.

The fourth circle dothe reprefent the due lengthe of the two Tropikes, whiche muft be fhorter then the Equinoctiall by 30 degrees, for it is equall to 330 partes of the fame : fo that the lengthe of the Tropike dothe beare the fame proportion to the Equinoctiall, as 11 doth to 12.

The fyfte plate, refembleth the meafure of the circles Arctike and Antarctike, and is in lengthe equall with 222. degrees of the Equinoctiall, which proportion is as 37. to 60.

The fixte plate fetteth forth the iufte meafure of the twoo Pole circles, whiche is equall to 144. degrees of the Equinoctiall, and fo it beareth to him the fame proportion that 2. dothe beare to 5. and eche of thofe circles Paralleles are diuided lyke vnto the Equinoctiall, into their 360. degrees.

Schollar. This is fo plainly fett forthe, and fo certenlye, that I fee no doubtfulnes nowe in the whole worke, for the makinge of it : for thefe plates are fo made, as if they were of metalle, and fhoulde haue bothe the endes foudred togither. fo that if any man wil make them of woodden hoopes, he muft allow fo muche more in the length of eche of them, as will fuffice for to bynde them fafte in compaffed forme. But thefe hoopes of this lengthe will make but a very fmall Sphere, yet by the fame forme of the numbres, and their proportion, I may make a fphere of what bignes that I will.

Mafter. So may you do certenly. and if you will haue a
fphere

Sphere twiſe ſo much in cōpas as theſe hoopes wold make, or thriſe, or 4. tymes, and ſo forth, this meaſure alſo may ſerue you, taking for eche circle ſo often tymes the length of the lyke here in this patron, as you wil haue your Sphere greater then this in numbre of tymes.

Schollar. And ſo I perceaue, if I woulde make an other three tymes and an halfe ſo bigge as this, I ought to take the meaſure of eche circle thre time and an halfe. and ſo for all other proportions.

Maſter. Truthe it is, ſaue that you muſt augment the breadth of the Zodiake only in like numbre of times : But as for the other circles, they are brod inoughe if they be not to weake, for the ſmaller they be, the better is the Sphere, ſyth their breadth dothe ſerue only for ſtrength, and for to receaue the diuiſions as here you ſee.

And thus haue I deſcribed vnto you both ſorts of Spheres, that is the Globe or Maſſye ſphere, and the Perſed ſphere or Armille. One other forme of Sphere there is, whiche excel-leth both theſe formes, and is wonderful apt for the teaching and expreſſinge of the Theorikes of Planetes, therfore I wyll reſerue it to that place.

Here needeth no repetition, bycauſe all ſtandeth in woorkynge of the former leſſons before re-peted, and therfore this ſeconde treatiſe ſhall ende here.

THE THIRDE TREATISE   61

WHERIN IS BRIEFLY TAVGHT

the vſe of the Sphere, for certaine concluſions of daily
appearaunces and other lyke matters.

MASTER.

OW YOU LOOKE TO HEARE SOM
what of the vſe of the Sphere, as you ſhall
do anon. And for an induction therunto,
you muſt diligentlye knowe the plages of *The plages*
the world, amongeſt whiche there are four *of the*
principall, that is, the Eaſte, the Weſte, the *worlde.*
Northe, and the Southe : and betwene theſe
are there other diuers, which are ſufficiently ſet forth in the
Horizont of the Globe, as muche as ſhall at this time bee
needefull.

  You muſt knowe alſo, that euery one of the Paralleles in *The Paral-*
the heauen hath a lyke circle in the earthe proportionably *leles in the*
drawen, and anſweringe to thoſe that are in heauen, in iuſte *earthe.*
rate of diſtance. So is ther fyrſt an equinotiall in the earthe *The earth-*
exactlye drawen vnder that Equinoctiall in heauen, and it *ly equino-*
diuideth the whole earthe into twoo equall partes, betwene *ctiall.*
the ſouthe and the northe, ſo that it poynteth preciſely the
myddle of the earthe, in that reſpecte : and all the partes of
the earthe from that earthly Equinoctiall toward the north, *The middle*
is called the Northe parte of the earthe : and of the world *of the earth*
lykewayes all that is beyond that cyrcle towardes the ſouth, *The northe*
is called the Southe partes of the earthe. *part of the*
 earthe.
  Schollar. Yet wee doo call that parte only Northe, that *The ſouthe*
is northe from vs : and that wee call Southe, that is ſouthe *parte of the*
from vs. *earthe.*

  Maſter. You muſte conſider that there is two formes of
ſpeakinge in ſuche talke, the one vulgare, and commonly
vſed, as well of the vnlearned as of the learned, and that ma
keth not the compariſon to the whole world, which few men
F.i.   doth

doth know, but it regardeth principally their owne cuntry,
which they do beſt know. The other talk is general in forme
of ſpeakinge, bycauſe it hathe reſpecte to the whole earthe,
and yet is it not generall in knowledge, for fewe men canne
aptlye ſkyll of it : ſo that bothe are true in their due vſe, but
the one is leſſe knowen then the other.

    Schollar. So I perceaue then, that although in common
talke we do call Spaine ſouthe, and likewaies other cuntries,
yet is not that true in compariſon to the partes of the
whole worlde, but in compariſon to vs, for our common
talke hath chiefe relation in ſuche thinges to our owne cuntrye. But I pray you then, where is the myddle of the earthe,
from which we muſt make our accompt, and vnto whiche
we muſte haue regarde in all ſuche generall talke?

    Maſter. That wyll I tell you anone, but firſte we muſte
ende that matter that we beganne withall, touchynge the
Paralleles on the earthe, whereof I haue named yet onlye
the Equinoctial, but
nowe muſt you ima-
gin other 2. parallels
next vnto it, the one
toward the Southe,
& the other towarde
ẙ north, which maye
anſwer to the 2. Tro
piks. And for a gene
ral knowledge fyrſt,
vnderſtand this, ẙ al
nations ouer whoſe
heds ẙ ſon doth run
directly, whē he is in
ẙ hyeſt point toward
ẙ north ẙ is in ẙ begi
ning of Cācer, wher

An example of the Paralleles in earth agreably
to the Paralleles in the ſkye.

The Tro-
pikes on
the earthe.

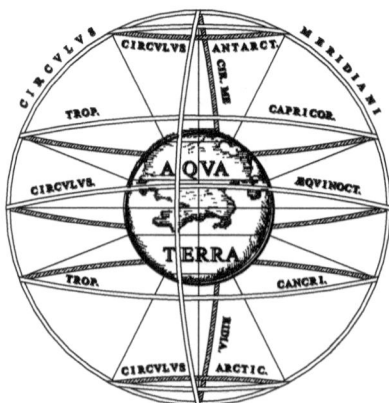

he deſcribeth ẙ tropik of Cancer in the ſkie, al thoſe people I
                                 ſaye

THE CASTLE OF KNOWLEDGE   63

faye dwell iuſt in the courſe of the like tropike in earth : And
contrary waies, all thoſe people ouer whoſe heddes the Son
paſſeth directly, when he is in the Winter tropike, they dwell
in the courſe of that ſouth Tropike in earthe, and haue the
ſonne right ouer their heddes that daye that he entreth into
the firſte degree of Capricorne.

Schollar. By theſe examples I can imagine the ſouthe and
north circles in y̆ earth to be vnder the Antarctike and Ar- *The other*
ctike circles in heauen, and ſo two Polare circles in earthe *Parallels.*
vnder the two Pole circles in heauen. Then are there ſeuen
Paralleles in earthe, anſwering to ſeuen other in the ſkye.

Maſt. That is ſufficient. howbeit for this time I will omit
the circles Arctike & Antarctike, bicauſe in mine opinion,
they make no Zone in earth, though all the Grekes in appe-
rance do ſay the contrary, but I will bringe inuincible reaſõ
for my purpoſe, when we come to the ſcanning of repugnãt
ſentences, eſpecially whẽ I do diſagree with the grekes, which
are the fathers of witte. but in this pointe of the fiue Zones, *Ioan. de*
I like much better our own cuntry man Iohn de Sacro boſco *S. Boſco*
as I will now only affirme, & in the fourth treatiſe wil proue *zonarure*
it ſubſtantially. Therefore to continew our matter as we be- *ſtaurator*
gan : there are made by theſe v. paralleles, v. large roomes in *The fyve*
the heauen, and other v. in the earthe, agreable to them in *zones.*
heauen, whiche ſpaces are called Zones.

Scholl. By your fauour, ther are ſixe Zones, if euery ſpace *Example of*
betwene the Paralleles be accompted for one zone, and that *the zones.*
doth not only the accompt of thẽ by memorye declare vnto
me, but alſo the ſighte of them in this figure, which is commonly named the figure of the Zones.

Maſter. Nother doth the accompte deceaue you, nother
yet the ſight of the figure, but wante of knowledge of their
naturall qualities, whiche therefore I will tell you by and by,
though theſe parallele circles do ſufficiently diſtincte them,
as their notable boundes, yet by the qualities bee they di- *The quali-*
ſtincte alſo. for as reaſon doth leade you, all the ſpace be- *ties of the*
F.ij.   twene *fiue zones.*

tweene the 2. Tro-
pikes, muſt needes
bee eſteemed verye
hotte, bycauſe the
Sonne runneth al-
waies betwene thē,
ſo that in the myd-
dle betwene the two
Tropiks is ẙ equi-
noctial line, frō the
which the Son is ne
uer fully 24. degres
ſo muſt it ſeem to be as hotte there in the myddle of winter,
as it is in Spaine in the myddle of Sommer, and for this
cauſe all the olde Coſmographers dydde thynke that that
countrye myghte not be inhabited for heate : and therefore
The Bur- called all that ſpace betweene the twoo Tropykes, the Bur-
ning zone. nynge Zone, called in latine Zona torrida. And of eche
ſyde of it, they noted twoo Zones, one vnder eche Pole,
whiche they called the Frozen zones, (and are named in la-
The Froſen tine, Zonæ Frigidæ) where for extreme could, they thought
zones. that no man might dwell. and betweene thoſe Froſen zones,
& the Burning zone, they appointed two Temperat zones
(called Zonæ temperatæ of latine men) which were parta-
The Tempe kers of the heat on the one ſide, and of the cold on the other
rate zones. ſide, ſo that of bothe, there was made a temperate mixture.
Now ſe you that betwene the Equinoctiall and the one tropike, there is no other qualitie, then is betwene the ſame equi
noctiall and the other tropike, wherfore all men (except only Polybius) did accompt the ſpace betwene the Tropikes
but as one Zone : ſo that the Equinoctiall is the bounde of
no Zone, but paſſeth by the middle of the Burning zone.

Schollar. Nowe I ſee (as I haue had at other times often
occaſion) ẙ we learn many things when we be childrē, which
we vnderſtande not all when we bee menne, for by this talke
I remem-

THE CASTLE OF KNOWLEDGE   65

I remember that both in Ouide & Vergile I learned ỹ diſtinction of thoſe 5. Zones, but what was to be vnderſtande by
them, I neuer knew till now. And nowe I ſee reaſon that betwene the 2. Tropikes, all may well be accompted the Burninge Zone, where no man can dwell, as bothe my authors
affirme.

Maſter. They had ſpoken more modeſtly, yf they had ſaid
that ther had been painful dwelling for heat. & likwaies of the
cold Zones, ỹ ther is hard dwelling for cold : but of this wil
I more exactly reaſõ in an other place, and for this time (as ỹ
truth by experience is knowen) I ſuppoſe that all ỹ 5. Zones
haue their inhabitants, though non ſo plentifully as the two
Temperat zones now haue, eſpecially this tẽperat zone that
we dwell in. Who is it that hathe not hearde of the iſles of
Molucca, and of Samatra, where the Portingales gette the
greate plentye of rich drugges and fine ſpices? and all that
haue been there, confeſſe that thoſe places ar right vnder the
Equinoctiall line : and Calecut is but little from it, for it is

A. C. The Horizonte.
B. The pointe ouer the heade.
＊ The Poles of the worlde.
The zodiac and the other circles doth ap-
peare of them ſelfe.

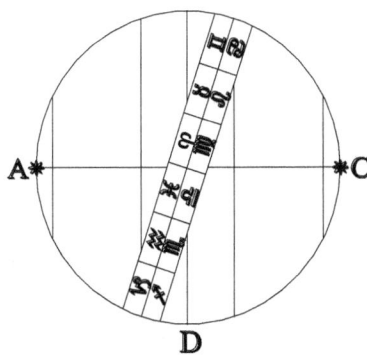

D

more thẽ 19. degres be
yond the Tropike of
Cãcer toward ỹ ſouth
ſo ỹ it is within 5. de-
grees of the very equi
noctial line. Now ther
fore I thinke it moſte
apt place for my pur-
poſe to begin at theſe
cũtries, ouer whoſe hed
the equinoctiall dothe
rightly paſſe, ſo ỹ they
muſte nedes ſee both ỹ
Poles in their Hori-
zonte.
Sc. That doth reaſona
bly folow, bicauſe half
F.iiij.   the

the heauen iuſtly appeareth aboue the Horizont, and the o-
ther halfe is vnder the Horizont. And alſo I perceaue that if
I ſet the ſphere ſo that the Equinoctiall ſtand full vprighte,
then will both the Poles be in the very Horizonte. as this
poſition of the Sphere doth ſhewe.

Maſter. You conſider it righte. And bicauſe the Equinoctiall doth croſſe the Horizont with right angles (for all 4.
angles are equall) therfore is this placing of the ſphere cal-
A ryghte led a Righte ſphere : ſo that all other nations, whiche haue
Sphere. the one Pole aboue their Horizonte, muſt needes haue the
other Pole vnder their Horizonte, and the Equinoctiall de
clinith from the point right ouer their heddes, that waye as
the hidden Pole is, whether it be toward the South, or els
toward the Northe.

The vſe of Schollar. All this ſeemeth eaſye to me, as longe as I be-
the materi holde this materiall ſphere : but when I doo not conferre it
all ſphere. wyth your woordes, then your ſaynges appeare the more
doubtefull.

Maſter. For that cauſe did I teache you the making of it,
before I inſtructed you in the vſe of it, knowing how greate
a helpe the ſighte of the eye doth miniſter to the righte and
ſpeedye vnderſtandyng of that, whiche the eare doth heare.
But againe to our matter : in all places where the equinoctial
doth decline from the pointe ouer the heddes of any inha-
The Zenith bitauntes (whiche pointe is commonly called the Zenith)
there the Equinoctiall maketh vnequall corners with the
A bowing Horizont, and therfore is that called a Bowyng ſphere, or a
Sphere. Leanynge ſphere, bycauſe the Equinoctiall boweth or leaneth toward one ſyde of the Horizont, more then towarde
the other ſide.

Schollar. I haue hearde it called a Crooked ſphere alſo.

Maſter. That name is vnapte for this arte, for there can
bee no crooked corner betweene the Equinoctiall and the
Horizonte, which myght make that name meet for the mat
ter : and (as I haue ſayde) the Sphere taketh thoſe ſeuerall
                                                         names

THE CASTLE OF KNOWLEDGE   67

A. C. The Horizonte.
B. The Zenith.
* The Poles.
The Zodiake, the Equinoctiall and the o-
ther circles do appear of them felfe.

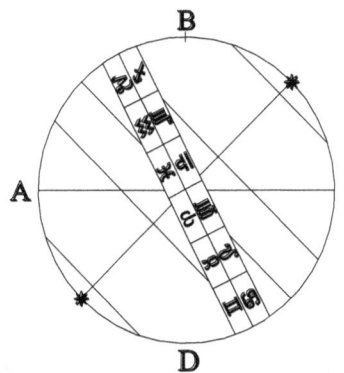

names of his diuers po
fition, and accordyng
to the corners that the
equinoctial doth make
with the Horizonte.

And this may you cō
fider herein, that there
is no Zone but one
that canne haue a right
Sphere : and to fpeake
precifely, but one tracte
in that zone, whiche is
the very middle of the
Burninge zone, righte
vnder the Equinoctiall
where as there be innu-
merable places ý haue
Leaninge fpheres, whi-
che you may call Oblique fpheres or Declininge fpheres, if
you delite more in latine lyke names then englifhe.

Schollar. So I perceaue that bothe we and all other nations whiche dwell not righte vnder the Equinoctiall lyne,
mufte be named to haue a Leaning fphere. And this I confider refonably, that in fome cuntries the fphere dothe leane
and bowe more then it dothe in other, whiche difference I
wolde gladly vnderftande.

Mafter. The diuerfitye in leaning of any fphere, is agreable to the eleuation of the Pole in euerye cuntrye, fo that
where the Pole is hygheft aboue the Horizonte, there the
fphere leaneth moft and where the Pole is lower and nearer
to the grounde, there the fphere leaneth leffer.   *The height*

Schollar. Howe fhall I iudge truly the height of the Pole? *of the Pole.*

Mafter. That true and exacte iudgement will I not treate
of as now, to auoide interruption in teaching : it fhall be fuf-
fcient for this place to fhewe you a plaine and eafye forme,

                F.iij.   with

with the vſe of an inſtrument that may helpe you ſumwhat
in markinge the height of the Sonne and Moone and anye
other ſtarres that you lyſte. and the manner of it is thus.

You ſhall take a Quadrate
(whoſe compoſition I haue
taught amõgſt other inſtru
ments in the Gate of know
ledge, but this which you ſe
here, is the forme of the
moſte playneſt ſorte) and
by the twoo ſyghtes of it,
you ſhall marke the height
of the Northe ſtarre com‐
monly called the Pole, and

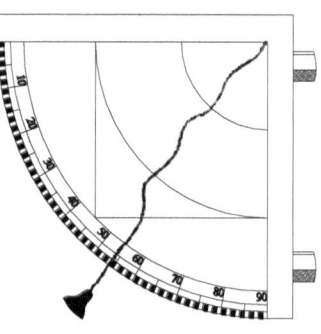

when you ſe it through both ỹ ſights, thē mark what degree
the lyne of the plõmet doth touch in the margent, and, that
may you call Latitude of that region, or the heighte of the
Pole, for this tyme and place where no preciſenes is requi‐
red, for nowe it is ſufficiente for you to vnderſtande ge‐
nerallye, that there are ſuche diuerſyties of eleuation of
the Pole in diuers countries : and thereby maye you vn‐
derſtande, that all Spheres be not alyke in theyr poſi‐
tion. As for example. In the ſouthe partes of Englande
Southham aboute Sowthehampton, the Pole is not fullye 51. degrees
pton. hyghe, and in the iſles of Orkenaye, beyonde Scotlande,
the Pole is aboue 62. degrees highe : this maye eaſilye bee
tryed by them that liſt to trauayle, but if you lyſte to go no
Yorke. farther then Yorke, you ſhall fynde the eleuation aboue
Edynburgh. 54. degrees, and ſo at Edynburghe ſhall you fynde the ele‐
uation aboute 57. degrees. And thus within your owne
cuntrye maye you vnderſtande a greate diuerſitye, wherby
you may coniecture the diuerſities that bee in other partes
of the worlde.

    Schollar. This is ſo appearaunte to them that will tra‐
uel any thing for knowledges ſake, that they cã not pretend
                                                            any

THE CASTLE OF KNOWLEDGE   60

any ignoraunce, but wilfull ignorance : but herin I fynde
one doubte, that maketh me to mufe, for in trauelyng thus *The altera*
from one place to an other, whereby the Pole is diuerflye *tion of the*
chaunged in his eleuation, I can not thinke that the Pole it *Horizonte.*
felfe dothe chaunge his place, but that rather the Horizont
doth alter, from which we mufte take the meafure of height
of the pole.

Mafter. You fay well, for indeed there is no fuche motion
in heauen, that maye make the Pole fo notably to chaunge
his place : but as we doo chaunge our ftandinge, fo dooth
there appear a newe Horizonte, whiche caufeth the Pole to
feeme higher, if we go towarde the northe, for then wee fee
more of the fkye (that waies) aboue our Horizont, then we
did fee before : but if we go toward the South, then will the
Pole feeme lower and lower, ftill as we go Southward : not
bycaufe the Pole chaungeth, but our Horizont chaungeth :
for nowe wee fee more of the fkye towarde the Southe,
and leffe towarde the Northe : but yet generally as much as
wee leefe in the one parte, fo muche wee wynne in the other
coafte, fo that euermore we may fee halfe the fkye.

Schollar. Then this is my doubte, how I fhal vnderftand *Whether*
your former woordes : for I remember you fayd that the Ho *the Hori-*
rizont was a circle immouable, and did not turne as the cir *zonte doo*
cles in heauen do : & now you haue plainly declared that the *moue or*
Horizonte dothe chaunge, whiche can not be without mo- *not.*
uinge of it.

Mafter. You haue anfwered your owne queftion, if you
marke it well : for the Horizonte moueth not as the circles
in heauen do moue : that is to fay, it goeth not round about
the earth by a daily courfe, but it ftandeth fteddye whyle the
heauen moueth, fo that if you neuer chaunge your place,
your Horizont will neuer moue. And to fpeake more exactly : the Horizont moueth not, thoughe you moue neuer
fo farre : but rather fhould we faye, that you are come into
an other Horizonte, when you are come into an other
                                                      countrye,

cuntrye.

Schollar. It mufte needes appeare fo, nowe that I do con
fider the matter more earneftly : for when I am at London, I
fee the fame Horizonte that all other men there do fee : then
if I go to Yorke, I fee the Horizont of Yorke, and not of
London, fo that the Horizont of London remaineth as it
was, and fo doth the Horizont of Yorke, whether I tarry or
go. And thus I perceaue great alteration in the Horizonts
betwene fouthe and northe, wherby the pole is diuerfly altered in height aboue the Horizont. What if I go eaftward or
weftward, fhall I not fynde the lyke alteration?

Mafter. It muft needes appeare yes. for the fame reafon
that caufeth you to chaunge your Horizont betwene south
and north, the fame will caufe it to chaunge betwene eaft and
wefte. And for declaration thereof, anfwere me to this que-
Example of ftion : Do you think that there is any fuche cuntry farre eaft
Calecut. from vs, as the Portingales reporte Calcut to be?

Schollar. It were as muche folly to make a doubte of it, as
it were to make a doubte of Babylon, or Hierufalem.

Mafter. And do you thinke that the fonne doth rife to vs
and to them at one tyme?

Schollar. It can not be. for this muche I maye gether by
that I haue learned already, that the rifing of the fonne and
of all other ftarres, is the apearing of them aboue the Hori
zonte, fo that they rife to vs, when they beginne to appeare
aboue our Horizont : and they rife to them in Calecut, whe
they appeare aboue their Horizonte. And further I gether
now by your briefe admonition of the chaunge of the Horizontes, that as betwene fouthe & northe in our owne cuntry, we maye perceaue notable diuerfitie, fo maye wee confy-
der ỹ fame much more in fo greate a diftaunce, as Calecut is
noted to be from vs, which I haue heard to be named aboue
15000. myles, and that is farre greater (yea 20. tymes) then
all the lengthe of Englande and Scotlande togither. wherefore I gather that the diuerfities of the Horizontes, muft be
                                                                    twenty

THE CASTLE OF KNOWLEDGE   71

twenty times fo muche, as was betwene Southhampton and
the northe parte of England.

Mafter. The diftaunce is not fo muche, nor the difference
fo great, but by meanes that the Portingales do faile a meruailous compaffe in goynge thether, they accompte the diftaunce by that compaffed courfe, whiche is farre from oure
talke now, for we muft euer take right diftaunce by a ftraight
line, as often as we do fpeake of any fuche matter. how be it
for examples fake, fuppofe it to be. *6000.* miles eaft from vs, *The diuer-*
it feemeth to be more then a quarter of the whole compaffe *fities of the*
of all the earthe, (as I will proue it in the nexte treatife) and *uers Regi-*
therfore muft the Sonne at the leafte rife *6.* houres to them *ons.*
foner then it dothe to vs. do you perceaue that?

A. C. The Horizonte of London.
B. The Meridian of it.
A. The eafte to London, and the noonefteede
    to Calcut.
D. B. The eafte to Calcut, and the line of mid-
    nyghte to London.
C. The wefte to London, and the lyne of
    mydnighte to Calcut.

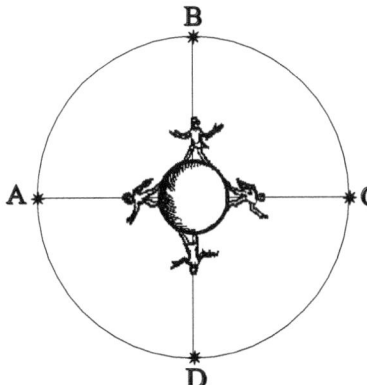

Schollar. The Son
(as all men knoweth)
doth compaffe all the
earthe in 24. houres,
then mufte it compas
halfe the earthe in 12.
hours, and a quarter
of the earthe in *6.* ho-
wers. this is as plaine
as can be : & thē it muft
needes folow, that if
they bee a quarter of
the earthe more to-
ward the eaft then we,
they muft fee the Son
6. houres fooner then
wee.

Mafter. And like-
waies they that dwell
farther eafte then thei,
as the inhabitantes of
Molucca doo, muft needes fee the fonne before them : and
thofe

thofe that dwell more wefterly then they do, as at Hierufa-
lem, or at Conftantinople, muft haue the daye fpringe later
then they that be at Calecut. And thus you maye confider,
that the Horizontes doo chaunge as well betweene eaft and
wefte, as it dothe betwene fouthe and northe : As this figure
fheweth for London and Calecut.

Schollar. That is plaine. for if all thofe places had one
Horizonte, then fhould the fonne rife to them all at ones.

Mafter. And as their morninges do differ, fo muft their
noonetyde differ alfo.

Schollar. No man that hathe reafon can denye that.

Mafter. Then mufte their Meridian circles differ in lyke
forte, feeynge they be the limites of the nonetide.

Schollar. So I perceaue that betweene eafte and wefte,
the Meridianes do chaunge, as well as the Horizontes : and
hereby I vnderftande, that when it is fonne rifinge at Calecut, it is not day with vs, by 6. houres : and when it is noone
with them, it is 6. of clocke in the mornynge with vs.
and fo of all other houres, whiche all appeareth by the former figure.

The diuer-
fities of
daies in one
Region.

Mafter. This ftandeth for the declaration of diuerfities
of dayes in diuers regions : but yet you haue not heard what
caufeth the diuerfities of dayes in one region.

Schollar. Yes for foothe. I remember that you reproued
me for faying that the longe daies caufed the Sonne to fhine
longe : and you tourned that fentence, affirminge, that the
longe fhinynge of the fonne dothe make the daies long, and
the fhorte fhinynge of the fonne, doth make fhorte dayes.

Mafter. And are you fatisfied with that reafon?

Schollar. I thinke it reafon good ynoughe.

Mafter. The reafon is good, but not inough, fyth farther
reafon is to be giuen. What maketh the fon to fhyne longe?
can you tell?

Schollar. By your helpe I trufte to know it.

Maft. Set your Sphere before you, and firft turn it fo that
both

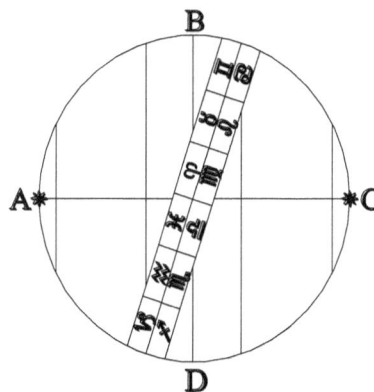

bothe the Poles may touch the Horizōt, which is the situation of the right Sphere. Then do you se y̅ the horizōt doth cut not only the equinoctiall circle in 2. equall halues, but lykewayes doth it cut bothe the tropikes, equally into 2. euen partes, so that there is as much of eche of them aboue grounde, as there is beneth the Horizonte : and contrarye waies. Wherfore it muste needes appeare, that the son when he runneth in anye of those three circles, is lyke tyme aboue the Horizont, as he is vnder it, so must the daies and the nyghts be equall, not only when the son is in the equinoctiall circle, but also when he is in any of the both tropykes : but this equalitye of dayes and nyghtes, when the sonne is in any tropike, is priuately appropried to the ryght sphere : for in all other varieties of the Bowinge spheres, then is the greateste difference in all the yeare, betweene the day and the nyghte, when the sonne is in any of the tropikes. as for example : Set the sphere to what eleuation that you lyst. that is to saye : Raise the Pole as many degrees aboue the Horizonte as you will.

Schollar. I haue sette it nowe (as heere you see) to the eleuation of 52. degrees, whiche you saye is the eleuation at Cambridge.

Master. And nowe maye you see that the Equinoctiall onlye is equallye dyuided by the Horizonte, and that the twoo Tropikes are verye vnequallye diuyded, so that the tropike of Cancer hath almoste thre quarters aboue the Ho

G.i.   rizont

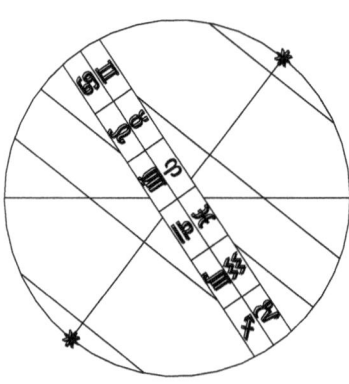

rizonte, and litle more then a quarter vnder the Horizont, wher cō-trarye wayes the Tro-pyke of Capricorne, hath almoſt thre quar-ters vnder the ground, and litle more then one quarter aboue the Ho-rizont : wherof it muſt nedes folow, that when the ſonne is in the Som mer Tropyke, he is al-moſte thre quarters of the Naturall daye aboue grounde, and lyttle more then one quarter of the ſame daye vnder grounde.

Schollar. I knowe your mynde very well, and I doo ga-ther thereby, that when the daye is at the longeſt, it is al-moſt 18. howers daye, and but lytle more then ſyx howers nyghte. And contrarye waies in the ſhorteſt of winter, the daye is lyttle more then ſixe howers longe, and the nyghte almoſte 18. howers. And farther I heare you call the whole *A Naturall Daye.* ſpace of 24. howers a Naturall daye : But I know not yet the reaſon of that name.

Maſter. By that name of addition, the whole daye of 24. *An Artifi-ciall Daye.* howers is diſtinċte from the Artificiall daye, which is from ſonne ryſinge to ſonne ſettinge : and that Artificiall daye is moſte commonlye vnderſtande, when men ſpeake of the daye. therfore for a difference it is good to vſe ſuche an ad-dition. But nowe for the better praċtiſe, ſet your globe to ſome other eleuation.

Schollar. I trow I haue ſet the pole highe ynoughe.

Maſter. Let it ſtande. What is the numbre of the ele-uation?

Schollar. I do ſee betwene the Pole and the Horizont in ỹ Meri-

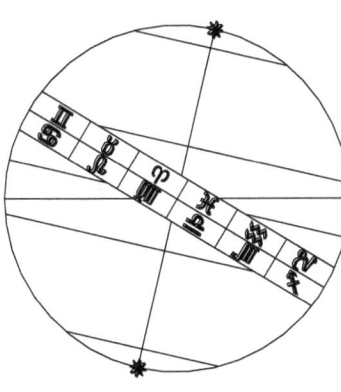

Meridian dyuers num-
bres, but I take that num
bre onli, which touchith
the horizont, and I take
that alſo of the twoo or-
ders of numbres, which
deſcendeth from ẙ Pole,
and that is here now 71.

Maſter. That is the
latitude or eleuation of
the Pole at Wardhouſe,
where our newe vente-
terers into Moſcouia
do touch in theyr viage :
but now mark the varietie of the tropiks to the Horizont :
The Tropike of Cancer is (as you ſee) more then foure degrees aboue the Horizont cleare, ſo that the whole 2. ſignes
of Gemini and Cancer, with 5. degrees of Taurus, and as
muche of Leo, doth neuer ſette vnder the Horizont.

Schollar. Then while the ſonne is goyng through thoſe
ſignes, from the 25. degree of Taurus, to the 6. degree of
Leo, it is continuall daye, bicauſe the ſonne doth not ſet vnder their Horizont. but I pray you how long tyme is that?

Maſter. It is from the 7. day of May vntill the 19. daye of
Iuly, ſo that it is continuall day with them by the ſpace of 73
of our dayes, whiche is almoſt two monethes and an halfe.

*The lōgeſt
Daye at
Wardhous
in 73. daies
continuall.*

Schollar. This is meruailous ſtraunge to me.

Maſter. Yet ſhall you hear more ſtrang matter then that :
Sette your Sphere ſo, that the Equinoctiall maye be iuſtlye
in the Horizont, and the north Pole righte vp in the place
of the Zenith.

Schollar. That haue I doone, as here you maye ſee.

Maſter. Nowe marke how muche of the Zodiake dothe
neuer go vnder that Horizont.

Schollar. Howe ſhall I perceaue that?

G.ij.   Maſter.

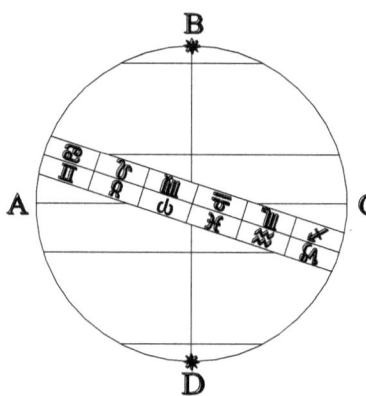

Mafter. Turne the Sphere rounde, as it fhulde moue naturally on his owne poles, but fturre not the Horizonte.

Schollar. Hereby I perceaue that 6. fignes, Aries, Taurus, Gemini, Cācer, Leo, Virgo, doo neuer fette vnder the Horizont, but con tinewe alwayes aboue it.

Mafter. Then while the fonne is in thofe fixe fignes, he can not bee out of theyr fyghte, that dwell within that Horizonte.

Schollar. It is truthe, yf any body doo dwell directly vnder the Pole.

Mafter. It is not now my purpofe, to prooue what partes of the earthe be inhabited, (for that appertaineth to Geographye) but to declare howe the fonne doth fhewe in all partes of the worlde, as well on the fea, as on the londe : and as well in wyldrenes, as in populous countryes. Whereby it doothe appeare fufficientlye, that vnder the Poles of the worlde, it is halfe a yeare continuall daye, and *The length of the daye vunder the Poles of the Worlde.* the other halfe yeare, contynuall nyghte, bicaufe fo longe againe the Sonne is not feene aboue that Horizonte.

Schollar. This is as true as canne bee. the reafon of it is fo certayne and manifefte, that I coulde not better vnderftande the ftate of that place, if I were there to fee it, then I doo by thys beholdynge of the Sphere, and the motion of it. And thys (as I take it) is a meruaylous *The excellencye of knowledg.* excellencye in knowledge, to bee able fo certaynly to iudge of thinges abfente, as if they were prefent : to bee able to tell
what

what houre of the daye it is in all the partes of the earthe,
and when the Sonne ryfeth and fetteth in all nations vnder
heauen.

Mafter. Yow wolde accompt this knowledge more mer-
uelous, if you vnderftoode other more wonderfull conclufions in it, whiche hereafter I will vtter as I fhall haue occafion conuenient : but in the meane feafon, I will fhewe you two
or three conclufions, appertaining to our prefente matter
whiche we haue in hande.

As the houres of the daye are dyuers in dyuers regions,
fo the fhadowes that the fonne caufeth in their dialles, and all
other fhadows, doth difagree many waies, not only from our
fhadowes, but alfo one of them from an other. Againe the
times of the yeare are not alyke through all the worlde, but
when it is Sommer to vs, it is winter to fom other : and when
it is Springe time with vs, it is fommer in an other cuntrye :
and when it is Harueft with vs, other people haue fommer :
fo when it is Winter with vs, fom nations haue fommer : yea
when the fpring time beginneth with vs, it is harueft in fome
cuntries, and in other cuntries it is midfommer at the fame
time : but when it is midfomer with vs, it is harueft no where
in the worlde, but midde winter it is then in two diuers partes of the worlde.

Schollar. This talke is meruailous, and in mine opinion
the greateft meruaile is, ў you can vnderftand the fhadowes
of their dials or any other thinges, in all partes of the world.

Mafter. Peraduenture it wold feem more merueilous if I
fhoulde fay, that by the knowledge of the fhadow of a ftaffe,
or any thing els that ftandeth vpright, (if I heare it trulye
reported) I will tell you in what part of the worlde that fhadowe was marked. And thinke you this no meruell, to tarry
within Englande, and yet to meafure all the compaffe of the
earthe, as certenly, as any man can do it, by going rounde
about the earthe?

Schollar. Thefe thinges do exceede credit, faue that other
G.iij.   thinges

thinges, whiche before I iudged impoſſible, and now I know
them certenly, do perſwade me to thinke many thinges poſſible by learning, that ſeeme vnpoſſible to the ignoraunte,
thoughe their wittes be neuer ſo good. I heare ſuche men ſay
ſometimes, that learned men and farre trauelers may be per
mitted to talke at their pleaſure, ſyth no man canne comptroll them.

Maſter. By thoſe woordes they ſignifie, that they do not
credite all that learned men do write or ſaye : wherfore I will
conſtantly ſaye to them, that if they wolde vouchſafe to imploye ſomtyme in learninge, they ſhoulde be eaſilye perſwa-
ded, not onlye to beleue ſuche thinges as nowe they thynke
impoſſible, but alſo to know them ſo certenly, as they know
howe many fingers they haue. But to perſwade you in the
meane ceaſon, I will preſently ſhew you ſome of theſe thre
concluſions before named, I meane for the generall knowledge of the times of the year : for the declining of ſhadowes
in diuers nations : and for the ordre to meaſure the whole
earthe, and yet go not out of England.

The con-
cluſions.

Schollar. If I maye vnderſtande but the generall forme
of thoſe three, I will truſt hereafter to attayne all the reſte
more certenly.

Maſter. I will begin with the laſte, whiche ſeemeth moſte
hardeſt, and I wyll alleage nothinge, but that whiche you
ſhall gaunt vnto.

The decla-
ratiõ of the
fyrſte con-
cluſsion for
meaſuringe
of the
whole
earthe.

Schollar. Then ſhall your proofe bee as certaine as I can
wiſhe.

Maſter. Can you with a Quadrante marke the eleuation
of the Pole aboue the Horizonte?

Schollar. This is eaſye inoughe.

Maſter. Then marke it fyrſte at Southehampton, or in
ſome other more eaſterlye place, on the ſouth ſhore of England. after that go to Newcaſtell beyond Yorke, and there
take the eleuation with your Quadrante againe, and marke
it well, and the difference of thoſe two eleuations ſhall you
                                                                    ſet

fet in your tables, and by it you fhall write the numbre of
myles diligently and truly taken betwene thofe two places,
where you toke thofe two eleuations.

Schollar. This can I doo with diligence, although it bee
as harde to marke the myles truly (the reportes of them being fo diuers) as it is to woorke truly with the Quadrante,
but diligence will auoide errour in them bothe.

Mafter. Then go forwarde to Edynburghe in Scotland,
and marke the eleuation there : lykewayes go to the mofte
northerlye pointe of Catneffe, and take the eleuation there
alfo, alwaies markinge the difference of euerye twoo places
in myles of equall quantitie, and alfo the difference of the
degrees of the Pole in eche of thofe places from other, and
fet them in your tables in ordre the one by the other, as here
for examples fake only, I haue fet them.

The places.	The Eleuation of the Pole.		The difference in degrees.		The diftaunce in myles.
Southehampton.	51.	o.	o.	o.	ooo.
Newecaftell.	55.	o.	4.	o.	240.
Edynburghe.	57.	o.	2.	o.	120.
Catneffe pointe.	62.	o.	5.	o.	300.
The fumme of all			11.	o.	660.

Here you fee for Southehampton, where the fyrfte eleuation was taken, no myles fette, bicaufe it is the beginning of
your iourneye, but the eleuation of the Pole there is 51. degrees : then at Newecaftell the heighte of the Pole is 55. degrees, and that is more then the other by foure degrees, fo
that foure degrees mufte be fet downe for their difference in
degrees, and their diftaunce in equall myles, is 240. Nowe
to fee how many myles dothe anfwere to a degree, I do diuide 240. by 4. and the quotient will be 60. wherfore I faye,

                          G.iiij.   that

that 60. miles in earthe (by this triall) doth answere to one degree in heauen. Then at Edynburghe I find the eleuatiō of the Pole to be 57, that is twoo degrees more then it was at Newcastell, and the distaunce betweene them in myles, is 120, whiche if I dyuide by 2, the quotient will be 60. as it was before : so that one of these workes doth confirme the other, bicause they agre so iustly.

Schollar. I vnderstande all this, as by declaringe of the thirde woorke it shall appeare to you. At Catnesse pointe, the Pole is 62. degrees aboue the Horizont, whiche maketh 5. degrees more then it was at Edenburghe, and the space betwene those two places is 300. myles : now if I diuide 300. by fiue, there will amounte 60, whiche quotient doth agree with the other twoo before found : so it appeareth that in all Englande, 60. mile in earthe, answereth to a degree of latitude in the skye.

Master. Prooue you also the whole difference in degrees with the whole distaunce in myles.

Schollar. The whole difference in degrees betwene South-hampton (where the Pole is 51. degree highe) and Catnesse pointe, (where the latitude is 62.) dothe amount vnto 11. de-grees, and the distance in myles is 660 : nowe diuidyng 660. by 11, the quotient appeareth 60. agreably as it was in all the other woorkes.

Master. What if you dyd go farther northe, 19. degrees moare? I meane so farre Northe that the Pole were 81. degrees hyghe aboue the Horizonte, howe manye my-les thynke you woulde that place be from Southehampton.

Schollar. That can I quick-ly accompt by the Golden rule of proportion. The difference betwene those 2. places in degres is 30. then seyng I found before, that 11. de-grees gaue 660. myles, I sette the numbres thus in their forme of woorke, and then I multiplye

$$660$$
$$\underline{30}$$
$$19800$$

$$8$$
$$19800 \;(1800$$
$$xxx$$
$$x$$

$$11 \diagdown 660$$
$$30 \diagup$$

THE CASTLE OF KNOWLEDGE 80

multiply 660 by 30, whereof cometh 19800 : whiche I muſt diuide by 11, and the quotient wyll be 1800.

Maſter. Thynke you thys a true woorke?

Schollar. This woorke is true and without any doubte, ſo that the meaſure of myles in Englande were true, whiche wee take for our grounde.

Maſter. And if that meaſure bee not true, yet by that manner of woorkynge you maye attayne to a very true rate of myles betwene ſoutheHampton and Catneſſe.

Schollar. That is no greate matter, nother ſo harde to bee doone.

Maſter. And it is no greater matter, in bothe thoſe pla-ces to take the altitude of the Pole.

Schollar. That is true alſo.

Maſter. So that if this rate be not true, ther may be found a true rate by diligence.   Schollar. Yea ſurelye.

Maſter. And by that true rate you could fynde how ma-nye myles dothe anſwere to 30. degrees in the ſkye.

Schollar. Eaſilye.

Maſter. Well then : Take this for a true rate, tyll you can fynde an other more certaine. And nowe anſwere me : How manye myles are in compaſſe roude about the whole earth?

Schollar. Nay that is impoſſible for me to diſcuſſe yett, tyll I haue farther knowledge.

Maſter. Se how eaſye a thing ſeemeth impoſſible to you.

Howe manye degrees is there in the compaſſe of the whole ſkye?

Schollar. That can I certenlye ſay to be 360 : for as I lear-ned before, a degree is no ſtandynge meaſure, but a rate of proportion, and dothe betoken the 360. parte of anye cyrcle.

Maſter. You ſaye well. Now if the whole circumference of heauen be 360. degrees, I demaunde of you, howe manye myles doth anſwere to 360. degrees?

Schollar. That maye I doo as in the former woorke, ſettynge

The cōpas
of the hole
earthe.

$$1\,8\,0\,0$$
$$3\,6\,0$$
$$\overline{1\,0\,8\,0\,0\,0}$$
$$\frac{5\,4}{6\,4\,8\,0\,0\,0}$$

$$\textit{r}$$
$$6\,4\,8\,0\,0\,0 \;(\,2\,1\,6\,0\,0$$
$$\textit{333} \quad 0$$

ting the numbres according
to the rule of proportion.
$$\begin{matrix}30\\360\end{matrix}\angle\,{}^{1800}_{}$$
Then multiplying 1800, by
360, there ryſeth 648000, whyche I muſte
diuide by 30, and ſo the quotiente wyll bee
21600, whereby I knowe that 21600 myles,
doothe anſwere vnto 360. degrees in the
ſkye. And ſo it ſhoulde ſeeme that thoſe
are the iuſte numbre of myles aboute the
earthe.

Maſter. You neede to make no doubte thereof, excepte
you doubt whether there be any part of the earthe without
the circuite of heauen : or els that you doubte, whether the
earthe be in the middle of the worlde.

Schollar. The fyrſte doubte were to fooliſhe, and for the
ſeconde (all bee it I doubte nothinge of it) yet I adſure my
ſelfe by your promiſe, of the full proofe thereof in the next
treatiſe.

Maſter. And other doubte there canne be none, but this :
Whether the earthe and the ſkye bee bothe rounde. whyche
both I wyll ſo ſubſtantially proue vnto you, that no reaſo‑
nable man will doubt of it.

Schollar. Then am I certified in the poſſibilitie of
the moſte doubtefull concluſion of the three, whiche you
proponed : It maye pleaſe you to proceede to the other two.

The decla-
ration of
the ſeconde
concluſion,
for decliniġ
of ſhadows.

Maſter. You do conſider that this concluſion being true,
they that dwell 5400 myles from vs, doo dwell a quarter of
the earthe from vs.

Scholar. That muſte needes be ſo : for four times 5400.
doth make the whole circuite of 21600 miles.

Maſt. And ſo they ẙ dwel frō vs any māner of way, 10800
miles, thei dwel half the compas of the whole earthe frō vs.

Scholar. It foloweth ſo by the former reaſon.

Maſter. It is well knowen by the nauygations of the
Portingales and Spaniardes, that there is almoſt ſouth frō

vs.

vs, certain places inhabited about 6300. myles, as namelye
at the ſtreight of Magellanus. Alſo at the great forelonde *Magellanus*
of Affrike, commonly called the cape of Good hope, are *ſtreighte.*
there diuers regions repleniſhed with inhabitantes, and they *The cape*
be from vs. ſouthwarde aboue 5200. myles : then northward *of Good*
wee haue good knowledge of dyuers cuntries beyonde vs *hope.*
aboue 1200. myles, whiche bothe ioyned togither, do make
from the greate forelonde of Affrike aforeſaid in the ſouth,
vnto Wardehouſe in the northe part of Norwaye, aboute
6400. myles, whiche is more then a quarter of the compas
of the earthe : but from Wardhouſe to Magellanus ſtreight,
it is aboue 7500. myles, by which diſtaunce of myles, you
maye eaſilye gether how many degrees of the heauen eche of
thoſe places is from vs, and from the Equinoctiall.

Schollar. Therein I praye you, that I maye prooue my
newe cunninge. The cape of Good hope is from vs ſouthwarde 5200. myles, that is in degrees of the
ſkye 86 ⅔ accordinge to the former rate of
60 myles to eche degree. from whiche numbre of 86⅔, if I abate ſo many degrees as we
be northe of the Equinoctiall, which are 52
degrees, then doth there reſte 34⅔ degrees.
So that it appeareth hereby, that the ſayd forelonde is 34⅔
degrees ſouthe beyonde the Equinoctiall.

Maſter. Now for Magellanus ſtreight, prooue the lyke
woorke.

Schollar. It is 6300. myles ſouthwarde from vs : then by
the rule of proportion, agreablye to the former rate, it muſt yelde in degrees 105, oute of
whiche abatyng our diſtaunce northe from the
equinoctiall, (whiche is 52 degrees) and ſo remaineth 53. degrees. thereby I vnderſtand, that
they are ſo far beyond the Equinoctiall ſouthwarde. Now will I prooue for Wardehouſe, how farre it is
northe from the Eqinoctiall. It is from vs. towarde the

northe

northe 1200. myles, whiche muſt yelde in
degrees, after our former rate 20, from theſe
20. degrees I maye not abate 52 degrees for
our latitude, as I dyd before.

$$\begin{array}{l} 60 \\ 5200 \end{array} \diagdown$$

Maſter. It were againſte reaſon, ſeynge
that the latitude of Wardehouſe is greater

$$\begin{array}{l} 1200 \;(20 \\ 6\;0 \end{array}$$

then our latitude is, and lyeth on the ſame coaſte of the E-
quinoctiall : for in the former examples the two places were
on the contrarye coaſte of the Equinoctiall from vs.

Schollar. I ſee it well now, ſo that by reaſon I muſt needes
adde it to our eleuation, and ſo ther amounteth 72. degrees,
whiche is one degree more then you did affirme it to haue
in latitude, in your former declaration.

Maſter. The cauſe is this : that rate of 60 myles to eche
degree doth ſerue in goyng preciſely from ſouthe to north,
but nother is Wardhouſe iuſt northe from vs, but ſomwhat
towarde the eaſte. nother yet in the other two examples any
of bothe places was directly ſouthe from vs, for the Forelonde of Affrike beareth towarde the eaſte, and the Streight
of Magellanus bendeth towarde the weſte, yet for this tyme
it maye ſerue as well for our purpoſe, as if it were more preciſely doone.

An ordre in
teachinge.

Schollar. Yet I thinke in teaching there ſhoulde bee vſed
nothinge but certaine truthe.

Maſter. What ſo euer is taught to be retained for a truth,
oughte to be a very certaine truth in deede : and they do not
well that in ſuche manner doo teache fyrſte vntruthes for
truthes. but where inductiō is made by examples, it is often
tymes more or at the leaſte, no leſſe expedient to vſe examples not exactly true, then to take only precyſe true examples, for thereby it appeareth the proofe to bee of greater
force, if it will procede in an example whiche is not preciſely true. And in theſe examples we haue ſo large ſcope of triall, that we neede not ſticke for two or thre degrees, for I intende not to ſpeake particularly of any citye that is vnder
                                                              one

one certain degree, but of whole prouinces, whiche occupieth diuers degrees in their latitude : as you vnderſtand that
the whole iſle of Britayne doth occupy from 51 degrees, vnto 62, which containeth 11 degrees. But now to come to our
purpoſe : thus much you vnderſtãd, ỹ beyond ỹ equinoctial,
yea and beyond the tropike of Capricorne alſo, there be inhabitantes.

Schollar. Yea that ther be, aboue 29 degrees beſouthe the
tropike of Capricorne : for that tropike is but 23 degrees
and a half beyond the equinoctiall : and ther be inhabitants
53. degrees beyond the equinoctiall, as before is ſhewed.

Maſter. Well if there dwell men but 6 degrees beſouth the
tropike of Capricorn (for I ſayde before, I would not ſticke
with you for a fewe degrees, ſith I wold make my proofe the
more forceable) then I demaund of you, whiche way dooth
the ſonne ſtande from them at noonetide?

Schollar. It muſt needes be alwaies northe from them at
noone, as it is alwaies ſouthe from vs at noone, ſeynge they
are beyonde the ſouthe Tropike, towarde the Southe,
as we are beyonde the north Tropike towarde the northe.

Maſter. Then conſider two places that ſtande iuſte ſouth
and northe (bicauſe you like well a preciſenes in examples)
as Venice that famous citie ſtandeth north almoſt from the
cape of Good hope : Now conſider the matter thus : in theſe
two places there is one common meridiane line, ſith thei do
ſtand almoſt iuſte ſouthe and north the one from the other :
then when the ſonne is in the Meridiane line of Venice, is
he not alſo in the Meridiane lyne to them that dwell at the
ſayd Cape of Affrike?

Schollar. Yes trulye.

Maſter. Then thoſe twoo places haue their noone tydes
at one hower.

Schollar. So haue they.

Maſter. And at Venice theyr ſhaddowe goeth alwaies at
noone toward the north & neuer toward the ſouthe, bicauſe

H.i.   it is

it is far north from the northerly tropike, called the tropike
of Cancer, and fo is the forefaid cape of Affrike far fouthe,
beyonde the fouthe tropike, whiche is the tropike of Capricorne : wherefor (as you haue confeffed) their fhaddowe
at noone tyde, muft needes go all tymes of the yeare toward
the fouthe.

Schollar, So I fee that thofe two places haue a contrarye
propertye, touchinge their fhaddowes.

Mafter. That is parte of the thinge that I did intende to
fhewe vnto you : but yet they bothe do agree in this pointe,
that all times of the yeare their feuerall fhadowes do incline
towarde one coafte.

Schollar. That is true. for at Venice it goeth ftil north, and
at the cape of Good hope, it runneth alwayes fouthe.

Mafter. Thefe fort of people are named of the greke Cofmographers ἑτερόσκιοι, Heterofcij, bicaufe their fhadowes goeth ftyll toward one coafte.

Schollar. As though there were other people, whofe fhadowes did fometime go fouthward, and other tymes northward : I meane their fhaddowes at noone, for els all nations
haue in one daye, at diuers houres, much diuerfitye in theyr
fhaddowes.

Mafter. Ye vnderftand the time well. and you fhal perceiue
as wel, that ther be fuch places, which chaung their fhadows.
You confeffe that men dwel beyond the tropike of Capricorne fouthward : and other you know to dwel beyonde the
tropike of Cancer northward : & thinke you it not agreable
to reafon, that betwene thefe two peoples there do dwell dyuers nations in fo greate a plotte of grounde?

Schollar. I thinke yes. and I heare faye, by our owne cuntrye men, whiche trauaile to Guinea, that they wente beyonde the fonne, whiche alwaies I tooke to be a lye of libertye permitted to farre trauelers, but now I perceaue it maye
be true in one fenfe.

Mafter. Ther are 2. places of that name, and both are beyonde

ἑτερόσκιοι
Heterofcij
*Single fha-
dowed.*

THE CASTLE OF KNOWLEDGE　　87

yond the tropike of Cancer, toward the fouth, and the one
of them is almofte directlye vnder the Equinoctiall circle :
and bicaufe you haue named that cuntry whiche our nation
doothe well knowe, take it for your example. They of
Guinea beeynge nyghe vnder the Equinoctiall, haue the
Sonne fome tymes northe from them at noone, as when
he is in the tropike of Cancer : and other tymes they haue
the Sonne fouthe from them, when hee is in the Tropike of
Capricorne. and mufte not their fhaddowes chaunge in
like forte?

Schollar. It can not otherwaies be. And fo I fee, that when
it is midfommer with vs, then doth their fhadows go fouth
ward, to as many as dwell betwene bothe the Tropikes : and
in our myd winter, their fhaddowes goeth northward.

Mafter. Thofe people are named of the greekes ἀμφίσκια,
Amphifcij, bicaufe the noone fhadowes goeth both wayes,
fothe and northe.

ἀμφίσκια,
Amphifeij
Double fha
dowed.

Schollar. And farther I gather, that there is no quarter
in the Horizont, but their fhadowe runneth that waies fom
tyme in the yeare.

Mafter. You fay truthe. but the chief regarde is here gyuen to the fhadowe at nonetide, wherby you may conceaue,
that fometime they haue almofte no fhaddowe : for when the
Sonne at noone is righte ouer their headdes, then theyr
fhaddowe is ryghte vnder theyr feete, and not on anye
fyde.

Schollar. It mufte needes be fo. for feeynge the Sonne
is fometymes northe of them, and fometymes fouthe from
them, hee mufte needes twyfe in the yeare bee right ouer
their headdes, ones in going fouthward, and againe in com-
mynge northwarde.

Mafter. To helpe your memory and coniecture take this
figure for a prefidente and example, where I haue fet the line
A.C. for the horizont, and D.B.E. for diuers places of the
fon at Noone. Now if you call A. the north point of the hori-

H.ij　　　zonte

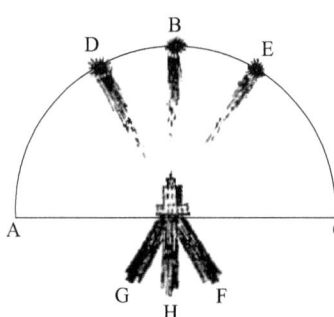

zonte, and C. the fouth pointe, then when the fon is in D. toward the north from their heds, their fhaddow goith to F. toward ẙ fouth. And when the fonne is in E. toward the fouthe, then is their fhaddowe in G. bēdīg toward ẙ north : likewaies the fonne be-ing right ouer their heddes in B, their fhaddow muft reft in H. ryghte vnder their feete. but I fee by your countenāce ẙ your mind woorketh in fome ftraung imagination : and I coniecture it to bee for that I haue drawen the fhaddowes beneth the Horizonte, as you take it.

Schollar. You haue truly coniectured my phantafye.

Mafter. Bicaufe this place ferueth not to declare condu-

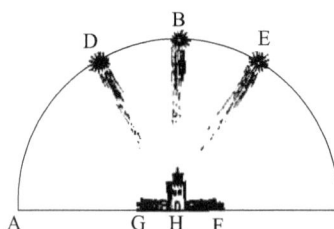

fions of bye matters, I wyll exhybite to you this other figure, where the fhaddowes doo run on the Horizonte, a-greablye to your phan-tafye, the letters of demonftration remay-ninge as they were beefore, and bothe thefe tende to one ende.

Schollar. But heere are but two fhadowes.

Mafter. Where wolde you haue the third fet?

Schollar. Right vnder the tower that giueth the fhaddow.

Mafter. But it may not reache from the foot of the tower, nother toward one coafte, nor other.

Schollar. No, that it maye not.

Mafter.

THE CASTLE OF KNOWLEDGE   89

Maſter. Then the foote of the tower doth couer it ſo that you can ſee no ſhaddow at all.

Schollar. That is moſt certaine.

Maſter. Yet remaineth ther an other ſort of people, which differ in one pointe from theſe other twoo ſortes, by reaſon that their ſhadowe in one daye runneth round about them, and goeth toward all coaſtes of the horizonte : wherefore the Greekes do call them περίσκιοι Periſcij.

Schollar. Is ther no engliſhe nor latin names for theſe ſortes of properties?

Maſter. The latin men borrowed of the grekes, both their knowledge and alſo many names of arte, bicauſe there is not the lyke grace of facilitie in compoſition, in the latyne tonge, as there is in the greeke tongue and therefore haue I geuen them no engliſhe names, bicauſe no one woorde can aptly expres theſe properties, excepte I woulde triflinglye make ſuche an immitation, to call theym, One ſhaddowes Two ſhaddowes, and Round ſhaddows : or els, which is not muche vnlyke, ye may call them Single ſhadowed, Double ſhadowed and Round ſhadowed.

Schollar. That immitation ſeemeth ſtraunge yet were it better to make new engliſh names, then to lacke words : therfore I will not refuſe to vſe them, till I can learn more apt names. but I praye you, where do thoſe men dwell, that haue their ſhaddowes runnyng ſo about them?

Maſter. Within the Polare circles. for all people whoſe zenith is within 23 degrees and a halfe of anye of bothe the Poles, haue their ſhaddowes running rounde aboute them. but as I ſhewed you before, the nearer they dwell vnder the Pole, the longer is theyr daye : and therefore the oftener doothe theyr ſhaddowes runne about them. for where the daye is but 24 houres longe, there the ſhaddowe runneth but ones aboute : and where it is halfe a yeare longe, there it runneth aboute 183 tymes : and in all other meane places raccodingly.

περίσκιοι
Periſcij
rounde ſha-
dowed.

H.iij.   Schollar.

Schollar. This is manifeſt ynoughe by your former declaration of the lengthe of the dayes, and the courſe of the
ſonne. And farther I perceaue that when they that be vnder
the Northe pole haue their ſhadowes thus runninge aboute
them, then they that dwell vnder the Southe pole haue no
ſhadowes at all, for it is continuall darkenes with them.

Lighte and Darknes vunder the Poles. Maſter. You ſaye well, concerninge the ſonne lyght, touching them that dwell directly vnder the Poles, but yet they
haue the lyghte of the Mone euery moneth more then 14.
dayes togither.

Schollar. Then do they not wante lyghte (thoughe they
lacke the ſonne) but only halfe a moneth togither, when the
Moone is in that halfe of the Zodiake, which is out of their
Horizonte.

Maſter. That is well conſidered of you. And yet euerye
moneth they lacke not lyghte, thoughe bothe the ſonne and
the Moone alſo bee oute of their ſighte : for as you ſee with
vs, that we haue lyghte before the ſonne riſing, and after the
ſonne ſetting, ſo haue they ſuche a lyghte by the beames of
the Sonne 50 dayes continuallye after they haue loſte the
ſighte of the ſonne, and ſo haue they the like lighte 50. daies
continuall, before the ſonne doth riſe to them.

Schollar. Then they wante not the ſonne lyght but only
82. daies, although they ſee not the ſonne in halfe a year, and
yet halfe that 82. daies they haue the mone in their ſighte, as
I perceaue by your former leſſons : for ſeing ſhe goeth about
the Zodiake euerye moneth, ſhe muſt needes bee halfe that
tyme in that parte of the Zodiake whiche is alwaies aboue
their Horizonte. This contemplation deliteth me muche,
to marke places abſente, as if I were preſent, and to ſee their
alterations by reaſon more certenly, then I can do by ſenſe,
if I were there preſently.

Maſter. Yet will I withdrawe you from this matter, tylle
an other more conuenient place : and now will I procede to
the thirde concluſion mentioned before : that is the generall
                                                              know-

knowledge of the times of the year, in all parts of the world. When the sonne is at the higheſt with vs, it is at the loweſt *The thirde* with diuers other nations, namelye to all them that dwell *concluſion* other vnder the Equinoctiall circle directly, other ſouthe *is declared.* from it : and therefore all thoſe nations haue mydde winter, when wee haue midde ſommer. But amongeſt them all there is one region, whiche is as farre beyonde the equinoctiall towarde the ſouthe, as we are towarde the northe.

Schollar. That region is about Magellanus ſtreight, as I gether by the ſeconde former concluſion.

Maſter. In deed the ſtreight of Magellanus is in that re‑ gion, for here I meane by a Region that whiche the Grekes do call a Climate, whiche is in forme lyke to thoſe Zones, *A Climate.* whiche I did deſcribe before, ſaue that there are more ſuche Climates or regions, then there are Zones : for the climates may well be accompted 48 betwene the twoo polare circles, *The nōber* whiche containeth but three of the Zones. but of thoſe cli‑ *of climates.* mats I will ſay no more at this preſent but that euery regiō where the longeſt day is half an hour longer or ſhorter then it is in anye other region, muſt bee accompted in a ſeuerall climate from it : ſo that vnder the equinoctiall the longeſt daye is but 12. houres, and with vs in the myddle of En‑ glande, it is about 18. houres : wherefore we muſt accompt that the myddle of Englande is in the 12. clymate from the Equinoctiall northwarde, and they that dwell 66. degrees and a halfe north, or ſouthe from the Equinoctiall, bicauſe their longeſt day is of 24. houres, that is twelue howers lon‑ ger, then it is in the myddle of the worlde vnder the Equi‑ noctiall (from which all thoſe accomptes of Climates do be gin) they muſt be iudged in the 24. Climate.

Schollar. Then are there 24. climates on eche ſyde of the Equinoctiall, betwene it and the polare circles, yet I remem bre that the common authors make mention but only of 7. on either ſide, whiche maketh but 14. in all.

Maſter. That ſhalbe anſwered anone, where I will ſet out

the ordre and reaſon of the diuerſity of both climates : but for
this time it ſhall ſuffiſe that you conſider this, that all places within one Climate, haue the tymes of the yeare alyke

The quali-
ties of con-
trarye cli-
mates.

exactely, and their dayes ſtyll of lyke quantitie the one to
the other, and they that dwell in the contrarye climate, as
many degres on the other ſide of the Equinoctial, thei haue
bothe the times of the yeare contrary, and alſo the quantity
of the daies diſagreable, for when it is ſommer in the one cli
mate, it is winter in the other : and when the daye in the one
dothe increaſe, the nighte in the other dothe increaſe after
the ſame quantitie iuſte.

Schollar. Then for example : In the cuntrye about Ma
gellanus ſtreighte, it is ſommer when wee haue winter : and
when our daye is at the longeſt then is their nyghte at the
longeſt.

Maſter. Truthe it is, and when wee haue ſpringe, then is
their harueſt : and ſo is it common to all them that dwell a
boue the earthe within thoſe twoo climates, yet is there this
difference, that in our climate and theirs alſo we maye ima-

Euery Cli-
mate hathe
4 quarters.

gine four quarters equally diſtincte : the firſte quarter being
that which we dwell in, and in the contrary climate, our meridian circle limiteth the firſt quarter, & alſo the third quarter in both places, ſo y̆ in this firſt quarter in both climates,
the times of the day and night ar alike : for when it is noone
to vs, it is noone to them : and when it is midnight to them
it is midnight alſo to vs.

Schollar. Then likewaies when the ſonne riſeth to them, it
riſeth to vs, and ſo ſetteth at one time in bothe Climates.

Maſter. Ye are far deceiued, for then of neceſſitie muſte it
folow, that their daye and ours at one time ſhould be of one
quantity, which is not true, as I ſaid before, but the reaſon of
that ſhalbe ſhewed anone. yet is it true, that their houres agre
with our houres if their meridian circle agre with ours. And
the ſame meridian circle vnder ground doth lighte in both
theſe climates, the 3 quarter alſo, wher it is noone when we in
the

THE CASTLE OF KNOWLEDGE   93

the fyrſt quarter haue mydnyght, and they haue mydnight
at our noone. Now may you eaſilye conceue by your owne
mynd, the places of the other two quarters.

Schollar. Ordre inforceth them, the one to be in our weſt,
and the other to be in our eaſte.

Maſter. That diſtinction is ſufficiente for you at this
time, and it is preciſely true, if you meane the eaſte, where
the Sone ryſeth at the begynninge of the Sprynge tyme,
or of the harueſt, wherfore for that time I wyll make myne
example : When the ſonne riſeth to vs in the ſpring tyme, it
is noone with them that dwell aboute Calecut, and when the
ſon is in our Meridian line, then doth he ſet to them : ſo that
whē the ſon doth ſet to vs, it is midnight to them about Ca   *Calecut.*
lecut, & thē is it noone to the famous cuntry of Peru : Again   *Peru.*
at that time the ſon riſeth to thē that be in the iſles of Moluc   *Molucca.*
ca. wherby you may gether that Peru & Calecut be in 2. contrarye coaſtes of the earthe, and therfore ſeeme to go wyth
their feet the one againſt the other, and their heddes the one
fromwarde the other, whiche ſorte of people therefore are
called of the Greeks and Latines alſo αντιποδε, Antipodes,   *Antipodes.*
as you myght ſay Counterfooted, or Counterpaſers. Now
to our purpoſe. all people that haue mydnight when other
haue noone, doo differ in ſonder by halfe the compas of the
heauens, one waye : yet may they not be called Antipodes,
except they differ in diſtaunce euerye waye a quarter of the
ſkye, and muſt haue one meridian circle. So that our Antipodes muſt be vnder our meridian circle, and muſt be halfe
the compas of that circle from vs.

Schollar. Then as wee are 52. degrees northe from the
Equinoctiall, ſo muſte they bee 52. degrees ſouthe from
the Equinoctiall, in that parte of the Meridian circle, whiche is vnder oure Horizonte, and then haue they mydde-
nyghte when wee haue noone : and hereby I perceaue that
they haue mydde nyghte when it is noone at Magellanus
ſtreighte.

Maſter.

Maſter. In deede it is daye then at Magellanus ſtreight, but not nighe noone, for Magellanus ſtreight is muche to farre toward our weſte : but for examples ſake that erroure maye be permitted, and eſpecially bicauſe there is no lond but ſea, where you ſhoulde meane that noone to bee : ſo can you giue it no propre name : but retaininge that name for example of the true place, you may conſider three ſortes of people, that is to ſaye, our ſelues, and thoſe that dwell by eaſt Magellanus ſtreight, vnder our Meridian cirde, which haue noone when we haue noone, and the thirde ſorte which are vnder the ſame Meridian, but haue midnighte when we haue noone, and are as farre ſouthe from the Equinoctiall, we we are northe, whome I named our Antipodes, and ſo ought they to be called in reſpect to vs, and we are Antipodes to theym alſo : But nowe comparinge theym with thoſe other by eaſte Magellanus ſtraight, they ar called eche to other πιϱίοικω Perioeci, as you may ſaye, lyke dwellers, bicauſe they dwell vnder one Meridiane cirde, and vnder one Parallele alſo, and be like in diſtaunce from the equinoctiall cirde.

Antipodes.

Schollar. There are manye places in euerye ſuche region or climate, but there are but two proprely vnder one Meridiane, and the one of them hath midnight when the other hath noone : ſo the tymes of the daye doth differ with them yet I perceaue that they haue the ſeaſons of the yeare agreable, bicauſe they dwell on one ſide of the equinoctiall. Then muſt it folowe that thoſe whiche vnto vs be Perioeci, are Antipodes to them that dwell by Magellanus ſtreighte vnder our Meridian.

Perioeci, like dwellers

Maſter. You ſaye well. and we vnto them by eaſte Magellanus ſtreighte, vnder our Meridiane, are called by the greekes and latines ἀντίχθονσ Antichthones, as you wold ſay Counterdwellers, or Counterclimates.
And thus haue you three ſortes of inhabitauntes by comparing the one with the other, wherof alwaies Perioeci (that

Antichthones, Counterdwellers.

is

is Likedwellers) haue like tymes of the yeare, but not of the daye. Antichthones or Counterdwellers, haue like times of the day, but not of the year. Antipodes or Counterpaſers, haue nother the parts of the year, nother of the day agreable togither, but cōtrary in both, how be it ther is a farther cōſideration for exactnes of this knowledg, which I will herafter declare to you in place more conuenient : but hereby maye you gather the diuerſityes of tymes of the yeare, and alſo of the dayes, accordinge to the diuerſitie of the inhabitauntes comparinge them all other to your owne cuntrye, or one of them to an other, as occaſion ſhall ſerue, and opportunitye of matter. And thus will I ende for this time, if I maye perceaue by your repetition of this thyrde treatiſe that you remembre all thinges therein declared.

Schollar. I were els to blame. but as I haue learned in it manye ſeuerall thinges, ſo for the ordre of the arte theſe I note as chiefe matters.

1   Firſte the diſtinction of the Plages of the worlde, accordingly as they be ſette forthe in the Horizont of the Sphere.

2   Then the Paralleles on earthe, agreable to the Paralleles in the ſkye, of like names, and diſtaunce proportionable.

3   Thirdly the diſtinction of the .v. Zones, by their qualities and limites, and of their inhabitantes.

4   The diuerſities of Spheres according to their diuerſe inclinations, but twoo are the generall diſtinctions, that is a Ryght Sphere, and a Bowinge Sphere.

5   Fyftlye, you gaue me a brefe ordre to take the heyghte of the Pole, or any other Starre or Planete.

6   Then folowed the diuers alterations of the Horizonte, as wel betwene Eaſte and weſte, as betweene Southe and Northe.

7   Seuenthlye, there was declared the cauſes of the diuerſities of the daies, fyrſte in diuerſe regions, and then in one region.

8   The difference betwene a Naturalle daye, and an Artificiall daye.

9   The quantitie of the longeſte daye in certen partes of the worlde, and namely vnder the Poles of the worlde.

10   How by this excellente Arte a man maye meaſure all the compaſſe of the earthe, and yet abyde ſtyll in one cuntrey.

11   A diſtinction of ſondrye inhabitantes, accordinge to the diuerſities of their ſhaddowes, whiche are three principallye.

12   Then laſtlye folowed an other diſtinction of inhabitantes, accordinge

The repetition of the thirde treatiſe.

to

to the agreeablenes and diuerſities of tymes of the yeare, and the
quarters of the daye, and theſe you named by three ſeuerall names
alſo, whiche are names of compariſon, bicauſe they take not thoſe
names, but in compariſon to other nations.

This I remembre to be the ſumme of this laſte treatiſe.

Maſter. You remembre it well, and vnderſtande it alſo
well, as it may appeare by your repetition. Therfore nowe
ſhall you depart for a time, and you ſhall reade ouer againe
your authors of the Sphere, whiche you did name before,
and now marke whether you can vnderſtande them, and at
your returne, I will inſtruct you more exactly in all the pre-
miſſes, and other diuers concluſions, whiche nowe I haue
omitted of purpoſe.

Schollar. I am moſte earneſtly bound vnto you for your
great gentlenes, whiche I pray god to requite, ſith I cannot,
and who wyll els I knowe not.

Maſter. Farewell then, and remembre your owne profit.

Schollar. The author of all profite, continew
and increaſe your profit, that you may
haue quiete time to trauaile for the
profite of manye.

THE FOVRTH TREATISE OF   97

THE CASTLE OF KNOWLEDGE

WHEREIN ARE THE PROOFES OF ALL

that is taught before, and other diuers notable concluſi

ons annexed therto. but nothing in a manner with

out demonſtration and good proofe.

SCHOLLAR.

F THE INEXPLICABLE BENEFITE of knowledge did not enforce me to for‑ gette all baſhfulnes, I myghte thinke it to muche ſhame, ſo often to trouble my Maſter frome his earneſt ſtudies, and to ſtaye him from his profitable trauell with mine importune crauynge of knowledge, namely ſithe I canne not recompence anye parte of hys paynes : yet hys gentlenes is ſuche, that hee ſeeketh more the profite of other, then his owne pleaſure or peculiare commoditie : and therfore will I boldly entre into his houſe. Are you at home ſyr?

Maſter. I am alwaies at houe for my friendes, if I bee not with them from home : yet ſome times I can not be at home for my ſelfe.

Schollar. The leſſe for me and ſuche as I am, that often trouble you more for our owne commoditye, then for your gayne.

Maſter. I ſeeke to gaine no more then competentelye maye ſerue my neceſſarye vſes, with conueniente regarde to my charges : but if I offende anye wayes in couetinge monnye, I adſure you it is to beare the charges in ſet‑ ting forth ſuch monumentes of knowledg, as were meruai‑ lous profitable for all men, very pleaſant to many men, & yet eſtemed only of wiſe men. but ſith I cānot do the good that I wold, and other want will which haue goodes in exceſſe, I muſt do as many other doth, wiſh good to all men, & helpe

I.i.   them

them as I canne. And for your parte I looke none other re-
compenſe but this, that you alwayes be thankefull to your
Maſter. and as hee helpeth you freelye, ſo doo you healpe
other againe, and hyde not the knowledge priuately, whiche may profite many publikely. but now to your matter :
haue you peruſed the authors of the Sphere which ar commonly readde?

    Schollar. To reade them all, it were to muche for my
lyfe tyme, and the profite not ſo greate, as I heare manye
menne ſaye : for as the noumbre are infinite, ſo the later wryters doo moſte commonlye but repete that, that
twoo or three of the auncientes haue written before. wherfore as I learned that the beſte wryters of them for my
ſtudye, were Proclus, Ioannes de Sacro boſco, and
Orontius the Frenche man, ſo I haue readde them, and out
of them haue I collecſted a table of theyr moſte notable
matters, whyche as yet I vnderſtande not, or els doo
deſyre to heare the demonſtrations for their proofe.

    Maſter. You haue doone well in bothe pointes. for as
the numbre of writers are infinite, ſo haue I founde great
tedious payne in readinge a greate multitude of them.
Notwithſtandyng as you ſhall hereafter ſeeke further knowledge, ſo muſte you reade more wryters in that matter :
wherefore amongeſt a greate noumbre woorthye the readinge, I wyll name a fewe vnto you, whyche I wiſhe you
to ſtudye : and the reſydue I leaue to your owne diſcre-
tion. Cleomedes the greeke authour, is very woorthye
to bee often readde : but beſte in hys owne tongue, for the
latine booke is muche corrupted. Alſo Eudide his booke
entituled Phænomena, and Stoffler his commentaries vppon Proclus Sphere : whyche booke I wiſhe were well re-
cogniſed (as it hathe greate neede) then myghte it ſerue
in ſteede of a greate numbre of other bookes. Dyuers
Englyſhe menne haue written right well in that argument :
as Groſtehed, Michell Scotte, Batecombe, Baconthorpe,
and

THE CASTLE OF KNOWLEDGE 99

and other dyuers, but fewe of their bookes are printed as yet, therefore I will ſtaye at thoſe three for this tyme. As for Plinye, Hyginius, Aratus, and a greate manye other, are to bee readde onlye of maſters in ſuche arte, that can iudge the chaffe from the corne. and Prolemye that wor-thye writer and myracle in nature, is to harde for younge ſchollars, except they be fyrſte inſtructed not onlye in the principles of the Sphere, but alſo well traded in Euclides his Geometrye, and alſo well exerciſed in the Theorykes of the Planetes. But nowe let me ſee the table that you haue collected.

1   The ordre and mouinges of the nine Spheres.
2   The ſpaces of their reuolutions by their propre motions.
3   The forme of heauen is rounde, and his mouynges circulare.
4   The earthe is rounde in forme, and the water alſo.
5   The earthe is in the myddle and Centre of the worlde, and is but as a pointe in compariſon to the Firmamente, and doth not moue anye waies.
6   The compaſſe of the earthe, and the diameter of it, what they make in common myles.
7   Of the circles in heauen what is theyr iuſte quantityes, their numbre, their ordre, their diſtaunce, and their offices.
8   Whye the Zodiake hath that name, and whether anye ſuche formes bee in the ſkye.
9   The diuers ſignifications of a figure, and the declyninge of them. There are two Horizontes, one ſenſible, and the other onlye iudged by reaſon, and what the quantities of them bothe are.
10   The Greekes and the Latines doo not agree in the deſcription of the circles Arctike and Antarctike, and what are theyr reaſons.
11   Whether there bee anye dwellers in the Vntemperate Zones.
12   What bee the circles Verticall and circles of Heighte, the circles of ho-wers, and of the twelue houſes.
13   Of the ryſinge and ſettynge of the Signes and other Starres, bothe in the Ryghte ſphere, and alſo in the Bowing ſphere, after the Aſtro-nomers.
14   Of the Latitude of the Sonne and the twelue Signes from the eaſte and weſte.
15   Of the riſinge and ſetting of the ſtarres, after the mynd of the poetes.
16   Of the diuerſitie of Naturall daies, as well as of Artificiall daies in di-uers partes of the earthe.
17   The diuerſities of howers, wherof ſome ar equall, and other vnequall

I.ij.

accor-

accordinge to the courſe of the ſonne.

18   The heighte of the ſonne aboue the Horizonte at all howers, and in all regions.

19   The diuerſyties of ſhadowes, wherof ſome be called Ryght ſhadows, and other be called Turned Shaddowes.

20   The diſtinction of the circles Paralleles neceſſary in Coſmographye, with the proportion of their degrees, to the degrees of the Equinoctiall.

21   The diſtinction of Climates and the numbre of them, and howe large in breadth eche of them is.

22   Of the Longitude and Latitude of regions and other places, and how bothe theſe ought to be taken.

23   The deſcription of the Mylke waye in the ſkye, whiche is commonly called Watlynge ſtreete, and what is the cauſe of that coulour in it.

24   The numbre and names of the chief ſignes and figures that be in the ſkye, and whye they be ſo called.

25   Of the circles and mouinges of the Planetes, and namely of the eclipſes of the Sonne and the Moone.

Theſe be the titles of ſuch matters as I haue noted in them moſte meete for this tyme, ſyth manye other thynges are ſufficiently taughte in the former treatiſes, and ſome other thynges, namely in Orontius booke, appertaine to Coſmographye, whiche I perceaue by your ſayinges, you mynde to reſerue for a peculiar treatiſe of that matter, and therfore I haue omitted them here.

Maſter. So myghte you haue doone ſome other thynges alſo, whiche you haue noted here : howe be it I will vſe my libertye therin, to expreſſe in conuenient largenes thoſe thinges, that be meet for this place, and the reſt will I touch with as conueniente briefnes : referringe the other to theyr more conueniente places.

Schollar. Syr I know right well, that your iudgement is as well to be folowed in the ordre of teaching, and choiſe of matter, as it is to be eſteemed in the teaching and explication of all doubtefull caſes.

Maſter. In ordre of teaching is more credit to be gyuen to a maſter, then in affirming of anye doctrine : for the ordre
is by

THE CASTLE OF KNOWLEDGE   IOI

is by longe experience beft knowen of fuch men : but for affirming of any doubtefull doctrine, no man ought to faye
any more then he can fhewe good reafon, for thapprouyng
of the fame. And now to your matter. although you folow
the ordre of Ioannes de Sacro bofco in many of your propofitions, yet will I beginne with your thirde propofition,
and referre the twoo firfte to a more meete place, fythe the
proofe of them can not well bee vnderftande, withoute a
great numbre of other cōdufions, which muft fyrft be proued. And for to begin with the declaration of the roundnes of the fkye, and his circulare motion, I thynke it good
to folowe that ordre whiche mouyd men fyrfte to obferue
this kinde of arte.

At the fyrfte beginninge of the worlde, when this arte *The firfte*
was vnknowen, menne marked the ryfinge of the Sonne *occafion to*
and the Moone, and other notable ftarres, as the Broode *thinke the*
henne, whiche is called of many men the Seuen ftarres, and *be rounde.*
other like : and perceauinge them to rife alwaies aboute the
eafte, and fo to afcende by lyttle and lyttle to the Southe,
from whence they dydde defcende againe foftely to the weft,
where they dydde continuallye fette : and the nexte daye a-
gain they perceaued them to begin their accuftomed courfe
and fo continued like as before : wherin although they fawe
fome diuerfitye, yet they perceaued that diuerfitye to bee
vniforme, and after a yeare to retourne to the olde ftate
agayne. by this occafion they beganne to ymagine that
thys manner of mouynge coulde not bee but in a rounde
and circulerre forme, and alfo in a rounde and circulerre bodye.

Then to vnderftande this matter the more exactlye, they *The fecond*
obferued the mouinges of fuche ftarres as neuer go vnder *occafion.*
ground, which be about ẙ north pole : & ther thei perceaued
by diligēt marking of thē, efpecially in ẙ long winter nights,
& that at fundry times, ẙ thei turned round about one point
in the fkye : and thofe ftarres that were nighe to that pointe

                        I.iij.   did

dyd make but a lyttle compas in their mouinge, and the farther that any ſtarres were from that pointe, the greater was

The thirde occaſion. the circle of their reuolution. Then thirdelye they marked
certaine notable ſtarres, whiche did riſe and ſet, but yet were
not farre from thoſe other ſtarres, whiche do neuer riſe nor
ſett, and they might wel perceaue that they did continue but
a lyttle while vnder the Horizont out of ſight, wher as contrarye wayes, thoſe ſtarres that were farther from that point
or Pole, did remaine longer time vnder the Horizont, out
of their ſighte, whereby they were inforced to thinke, that
theſe varieties and formes of mouynge coulde bee in none
other manner of body then in a rounde forme, and that the
ſame mouynge was circulare and rounde, as it did manifeſtlye appeare in the northe parte of the ſkye, where the ſtarres
continually moue rounde aboute one pointe, and do neuer
ſet vnder the Horizont. And that point about whiche they
noted this motion to bee, they called (as reaſon inforced

A Pole. them) the Pole of the worlde.

Schollar. What doth that word ſignifie?

Maſter. It hath his name of turning : as you wolde ſaye,
a Turne point. and it doth betoken the ende and extreame
pointe of any Axetree, howe be it by ſpeciall prerogatiue
the name is appropried to the endes of the Axetre of the
worlde.

Schollar. This picture dooth ſome
what repreſent the motiõ of the ſtarres
aboute the north Pole.

Maſter. You ſay truth. how be it aptly it can not be perceaued in flat forme
but in a roũd body, as a globe is : but in
that point (me thinketh) ther is no better inſtrument then the ſkye it ſelfe, wher
everye man maye learne that lyſteth to marke, and there bee
certaine notable ſtarres in that place and namelye Charles
wayne, whiche is called alſo the greate Beare, whoſe motion

is ſo

is ſo euidente, that euery childe may marke it: And twiſe in
the yeare, that is in the middle of February and in the middle of Auguſt, they ſerue for a iuſte horologe : ſo that the
finger in a clocke doth not more aptely pointe the howers,
then doth that figure of Charles waine. *Charles waine.*

Schollar. There can bee no more apte declaration of the
roundnes of the heauen, and of his circular motiō, then the
ſight of thoſe ſtars which moue ſo roundly, and kepe their
quarters in heauen ſo preciſely. and yet I haue hearde of certaine great clerks, that in no caſe thoughte it reaſonable to
affirme ſuche a forme of roundnes, or ſuche a round motion in heauen : buſt moſte of all I meruaile of that famous
man Lactantius Firmianus, which doth affirme (as I haue
hearde) that the heauen is not rounde, but flat and playne. *Lactantius Firmianus his erroure.*

Maſter. Many ſcrupulous diuines by myſſe vnderſtan-
dynge of ſcripture, haue abhorred the ſtudye of Aſtronomye, and alſo of philoſophye. and often tymes doo more
ſharply then diſcretely raile at theſe bothe, and yet vnderſtande they not any thinge in eyther of them bothe. ſuche
men are to haſtye to bee good iudges, that will ſo quickely
pronounce ſentence, before they haue anye good euidence,
and will determine the caſe, before they vnderſtand the mat
ter. for how can anye man vnderſtand well or iudge rightly
ẙ thing that he knoweth not? yet ſuch drowſy dreamers haue
oftentymes deceaued many wiſe men, with their appearante
reaſons, but yet none but ſuch, as either were giuen to hate
the name of philoſophy, or els at leaſt had no time, or none
habilitie to gette vnderſtandinge in it. By ſome ſuche men I
may think that Lactantius was ſeduced : and the more eaſily,
for that he had conceaued a deadly hatred againſt all philoſophers and againſt philoſophy itſelfe : but I wil let him and
his folowers paſſe, and retourne to the matter. *Lactantius opinion of the forme of heauen. lib. 24. c. 3*

Schollar. Yet if it pleaſe you, I wolde gladly hear his reaſons, that he maketh for approuing his opinion, ſeyng hee
is named ſo greate an oratour and ſo famous in learnynge,

104   THE FOVRTHE TREATISE OF

that many men will beleue him without any reafon.

Mafter. Who fo euer wyll beleue him in this point, muft
do it without reafon : for he alleageth no reafon for his pur-
pofe, but taketh it as a certaine truthe, thereby to improue
the opinion of the Antipodes, as I will more largely declare anone in proouing the roundnes of the earthe. But feynge he coulde bring no reafon for his opinion, you fhall
heare fome reafon againft his phantafye, and then iudge as
you can.

That the
fkye is not
flatte.

Firfte I reafon thus : If the heauen be flatte and plaine as
a borde, then howe fo euer it ftande, one parte of it mufte
needes be nearer to the earth then any other parte of it. and
that parte by all lykelyhod muft be right ouer our heddes,
is not that fo?

Schollar. I can not imagin els any forme of fituation : and

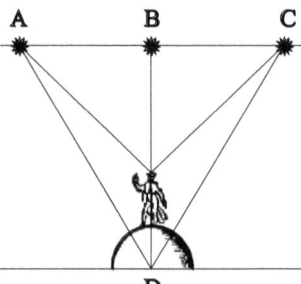

that doth appeare partly in
this figure, where A.B.C.
ftandeth for the fkye, and
lyeth flatte ouer the earthe,
whiche is heere reprefented
by D : and now I fee that B,
whiche is righte ouer D, is
muche nearer to it then A.
or C, or anye other poynt
in that flatte plaine forme,
whiche is fette to reprefent the flatte fkye.

Mafter. Nowe then what will Lactantius fay, or any man
for him? doth this heauen moue or not?

Schollar. He can not deny that which we maye fee with
our eies, that bothe the Sonne, the Moone, and all Starres
doo moue euery hour continuallye.

Mafter. Yet peraduenture he might faye, as fome other
like contemners of philofophy haue faide, that the ftarres
and Planetes do moue in the fkye, as fifhes do fwimme in
the water : and that they go forwarde thoughe the heauen
ftand

ſtande ſtyll.

Schollar. I remembre I haue hearde of that ſayinge, and that a famous writer of late doth maintaine that opinion.

Maſter. What will they ſaye then, dooth keepe the ſtarres in ſuche a iuſte ordre and equalitye of diſtaunce? whiche ne-uer altered any one witte ſyth the beginning of the worlde, is it poſſible that the ſtarres ſhuld moue in the ſkye as fiſhes doo ſwimme in the water, or as birdes flye in the ayer, as ſom terme it, but that the ſtarres muſte ſtragle in their courſe, as the fyſhes do, and as the byrdes alſo do?

Schollar. I haue ſeene both fyſhes in the water, and foules in the ayer, to keepe a meruailous certene courſe in their fly-ing and ſwimming, and namely fiſhes that go in ſculles, as herringes commonlye doo, and other fyſhes diuers times, and wilde geeſe alſo and ſtorkes in their flyinge, whereof I haue often muſed.

Maſter. You maye often ſee ſuche notable ſightes : yet if you marke them, you ſhall ſee muche alteration in their fly-inge, as well as in the ſwimming of the fiſhes : whereby you may think their ordre not to be conſtant, but ſomtimes one flyeth a lyttle faſter, and an other a lyttle ſlacker : and ſome time they ſwarue on the one ſide, and ſomtime on the other. but were it not a fonde ymagination, to thinke that ſtarres doo flye and folowe one guide as byrdes doo, and in 5000. yeare ſpace to keepe their places ſo preciſely, that they varye not one minute of a degree?

Schollar. In deed it were meruailous, and ſo are all Gods woorkes.

Maſter. Yet is there one inuincible reaſon againſte that opinion, gathered of the figure of the Milkye way in hea-uen, whiche many men in England do call Watlyng ſtreete, comparing it to one of the greate highe waies in Englande that is called Watlyng ſtreete. This Mylkie way if it ſerued for none other purpoſe, yet doth it ſeeme woorthy the no-ting, for the exact confutation of the ſaide opinion, and for

The Mylky way called of the grekes Galaxia.

                                                that

that caufe it myghte feeme to bee made by God, which hath
wroughte manye meanes to leade men vnto truthe. This
way is in the fkye itfelfe, as all men hath confeffed, and their
eyes doo teftifye, and the ftarres that bee in it are always
feene to keepe their places in it : fo that it mufte needes
folowe, that the fame waye doothe mooue with the ftarres, and then confequentlye the fkye mufte needes moue
alfo.

Schollar. Yet it may be faid, that the ftarres which bee in
it doo moue alwaies fo certainly in it, that it maye feeme to
moue, as though it ftande ftill.

Mafter. Did you euer marke the fame Mylke way?

Schollar. Yea verily, and that often.

Mafter. And did you perceaue in it any boughts, corners,
partitions, or fuche other like markes, wherby you myghte
knowe one part of it from an other?

Schollar. That haue I done alfo, in fo muche that in fom
places it feemeth to be diuided into two waies.

Maft. That is true. And think you if the ftarres did moue
in it, and it ftande ftill, that thefe ftarres which now be by the
partition of thofe branches, mufte not within foure or fiue
howers be paffed farre from that place?

Schollar. It fhuld fo folowe, yet that is not fo : for I haue
marked the contrary oftentymes, that they keepe thofe places ftyll.

Mafter. Then do not the ftarres moue from their places,
but as thofe places moue with them.

Schollar. It appeareth now to plaine to bee made doubtfull any more.

Mafter. Yet will I prooue it better. Dydde you euer
marke anye notable place of that Mylke waye at the be-
ginnynge of the nyghte in the eafte, or in any other coafte
of hauen?   Schollar. Yea forfouthe.

Mafter. And haue you marked whether that place hathe
gone anye farther weftward that nyghte?

                                                    Schollar

THE CASTLE OF KNOWLEDGE   107

Schollar. I haue marked it well, and haue perceaued that it hathe moued a greate waye from his firfte place : and who fo euer lyfteth to trye it, let him at fixe of the clocke in the deepe winter marke any notable places in it, and at tenne of the clocke the fame nyght, hee fhall perceaue it to haue gon weftward more then a quarter of the fkye.

Mafter. Your woordes are true, meanynge a quarter of the fkye aboue your Horizonte : and by this you fee, it can not bee auoyded, but that the fkye dooth mooue as well as the ftarres.

Schollar. It is mofte manifeftly proued, fo that Lactantius himfelfe can not denye it, onleffe he will deny that hys owne fenfes may iudge infenfible thinges.

Mafter. Then if the heauen be flat, as he doth imagyne it to be, and it doth moue weftwarde, as all men dooth fee, other he mufte fay that the fkie is infinite in length, and that wee neuer fee any parte of it againe after it is ones paft our fighte : and therby affirme, that there be infinit many fonnes and as many moones, and an infinite numbre alfo of all other Planetes, and of all feuerall kinde of ftarres, or els hee muft declare whiche waye that the Sonne, the Moone, and the other ftarres doo com into the eafte againe.

Schollar. He can not faye that they come backwarde the fame waye that they went forwarde, for then wee fhoulde fee them in their retourninge : and to faye truthe, there can bee none other forme of mouinge, but in rounde forme, that may bringe them into the eafte againe : But peraduenture he may fay, that though the fkie be flat and plain in forme, yet it hath a rounde motion.

Mafter. Some other man may fay fo : for he thinketh the contrarie as his woordes importe, for in reprouing Aftronomers, hee faith : *Ex motu fyderum opinati funt coelum volui.* By the mouing of the Starres they imagined that the heauen doth turne rounde. by which wordes hee feemeth to meane that the ftarres moue, but not the fkie.

Schollar

Schollar. That is fully improued before.

Mafter. If it were not, I myghte reafon with him thus :
Seyng he affirmeth as reafon inforceth him, that the ftarres
do moue, and will not confeffe that the fkye turneth round,
then (as I declared before) one parte of the fkye whiche is
ouer oure headdes, is nearer to the earthe then the bothe
endes be.

Schollar. That appeareth plaine, excepte hee wolde faye
againft all reafon, that the earthe were as large as the fkye.

an argumēt
againſt the
flatneſse of
the ſkye.
The maior
or maxime.
Mafter. Yet thoughe hee woulde faye fo, my reafon fhall
proceede in full ftrengthe, fyth fome partes of the fkye by
his meaninge mufte needes bee farther from vs then fome
other. Therfore I frame my reafon thus : All thinges that
men can fee, feeme greatteft when they bee nygheft vnto
menne, and the farther they bee from their fight, the leffer
they fhewe.

Schollar. I thynke no man fo childifhe to denye that. for
euery hower our fighte doth approue that it is fo : if we fee
a man a farre of, he feemeth no bygger then a lyttle child :
and a greate fhippe farre in the fea, dooth fhewe no bigger
then a crow fometimes.

The minor.
Mafter. Then takinge that for a maxime in argumente,
I annexe this minor, that the ftarres mouynge in that imagined flat fkye, are moft nygheft to vs, when they bee ouer
our headdes : and they are fardeft from vs, when they be in
The conclu
ſion.
the eafte or in the wefte : wherefore I inferre the condufion,
that the ftarres mufte feeme greateft, when they be ouer our
heddes : and they mufte feeme muche leffer, when they be in
the eafte or wefte.

Schollar. This condufion is plainlye falfe. for our eyes
doo teftifye the contrary, fyth alwaies the fonne, the moone
and the ftarres doo feeme greateft at the ryfinge in the eaft,
and at their fettinge in the wefte. And they fhewe fmalleft,
when they be nygheft ouer our headdes.

Mafter. If the condufion be falfe, and the argument good
as La-

THE CASTLE OF KNOWLEDGE   109

as Lactantius can not comptroll it, then I maye obiecte to him his owne rule : Necesse est falsa esse, quae rebus falsis congruunt. It can not be chosen but those muste be false sentences that doo agree with false matters. and so muste they needes bee false premisses, that do inferre a false conclusion.

Scholar. In good faithe I thinke nother Lactantius, nother any man els is able to auoide this reason, except he will auoide that fonde opinion of imagining a flatte skye, and the standing of the same vnmouable : yet if anye man wolde saye, that the heauen were square, or of any other forme of diuers angles, as here you se many varieties in these figures.

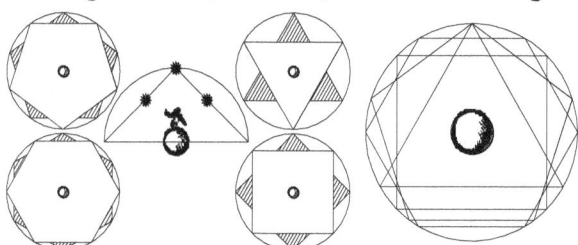

An other reson by auoi ding of emptines whiche nature cannot bere.

How might I aptly reproue their opiniō, if thei will affirme farther, that the skye with suche a forme doth moue round? for by so saying they mighte auoide the danger of this last inconuenience.

Master. While they mighte seeme to auoide one danger, they fall into an other : as for a proofe. I tourne those figures round, wherby in deed it appeareth, that euery part of them keepe styll theyr owne distaunces vnchangeably frome the centre, but yet is one parte more nerer the centre then an other parte is, and euerye parte in their turning seemeth to describe a circle about the centre, eche circle in hignes accor ding to the distaunce of that parte wherby it is described, and so the greatest circles are made by the extreame angles, of euery figure.

Scholar. All that is easily perceaued, at the first sighte in tourning the figures aboute.

K.i.   Master.

Maſter. Then if the heauen bee cornered, it maye haue no leſſe roome to moue in, then the compaſſe of the vtter-moſte circle doth require.

Scholar. That appeareth certaine, for els it woulde ſtaye by thoſe corners, or els break the corners in the tourning, wherof nether is to be fantaſied but of fools, whoſe thoughts are pardonable in all thoſe that refuſe not their cōmon fe-lowſhippe, but not in other, although for their woorthines they might be Wardens of that company.

Maſter. Then if for their motion they require ſo large a circle, as may compas their corners, there appeareth voyde roome againſt euery ſide, in which roome what ſhal be ſet to auoide emptines, which nature can not beare?

Scholar. Let them anſwere that lyketh that phantaſy, for I can imagine nothing, except I ſhuld name Ayre, but that by his nature can not aſcend ſo highe.

Maſter. You geſſe well, that it muſte be ſome ſubtile and liquide thinge, that might change his place as faſt as the hea uens do turne : for in turning, the corners will come anone where the emptines is now, and ſo ſucceſſiuely eche chaunge place with other. but Ayer you ſay cā not come thither, ſith it may not aſcend ſo highe : the lyke may you ſaye of fier and water, and muche more of thearth. Againe if they could aſ-cend, how ſhuld they pearſe through the ſubſtance of the hea uens? beſide that being elementes, and therefore corruptible and ſubiecte to daily alterations, they are vnmeet to be mat-ched with the vnchangeable ſubſtance of the heauens.

Scholar. This is reaſon inough againſt that imaginatiō, ſith nature can not ſuffre it to bee emptye, and nothinge els but part of the ſkye can ſupplye it.

The thirde reaſon for apt mouing. Maſter. Yet conſidre farther : ſyth the motion of heauen of all other muſte bee iudged the moſte ſwifteſt, whiche in 24. howers dooth runne ſo large a race, that is manye folde greater then the compaſſe of all the earthe, ſo that euery ho-wer it runneth many thouſand miles, dooth not this ſwyfte

mo

THE CASTLE OF KNOWLEDGE   III

motion require that forme, which is of all other moſt apte
for mouing? & doth it not repugne to ſuch formes as be full
of corners, & therfore vnapt to moue ſwiftly or vniformly?

Sc. It appeareth plain madnes to dream ones the contrary.

Maſt. Then all men know that as cornered bodies be moſt
vnapt for to run, ſo is a round globe moſt apt for all other.

Sc. Euery cõmon turner can ſkil in ẙ reaſon, & know ẙ a litle
altering of the one ſide, maketh the boul to run biaſſe waies.

Maſter. If the reaſon be ſo plaine that common artificers
can ſkyll of it, it were to great a folly for learned menne to
doubte of it.

Scholar. They that doubt of it, neuer waied their opinion
with any reaſon, as I maye thinke, for theſe reaſons ſuffice to
perſuade any man.

Maſter. Yet ones againe way this for the forme of heauen,
ſith it incloſeth all thinges, and is the greateſt of all other,
were it not meete that it ſhuld haue the greateſt forme which
is moſt large and apte to compas and incloſe all other? *The fourth reaſon for capacitie.*

Schollar. It is bothe meete and neceſſary alſo.

Maſter. Then is it well knowen of yonge ſchollars in geometry, that as of all flatte formes of like circumference, the
circle is the greateſt, ſo of all ſounde formes of lyke circuite
the Globe is moſte largeſt, and therefore moſte apteſt for
the forme of the ſkye, whiche incloſeth all thynges that man
canne ſee.

Sch. I myght be aſhamed to demaunde anye more profe
for the roundnes of heauen or his circulare motion, yet are
the reaſons ſo pleaſante, that I delite muche in the hearinge
of them, and therefore canne bee content to imploye as
muche time in hearing them, as you thinke good to beſtow
in framynge them.

Maſter. I coulde occupye you ſo a greate tyme : but I
thinke it not beſt to ſtaye thereon to longe, ſyth wee haue
many other matters to prooue, and at other tymes we maye
talke hereof againe. Theſe reaſons whiche you haue hearde

do proue not only that the motion of heauen is round, but alfo that the rounde forme doth beft agree to the fkye, for largenes of capacitye, for aptenes in mouing, for auoyding of emptines, and for the iufte appearance of the ftarres in vniforme bignes, whiche I thinke fufficiente for this time.

Schollar. There be twoo thinges by the waye which I defire muche to heare more largely declared : the one is for the appearance of ftarres, whiche feeme mofte greateft at theyr rifinge and fettynge : the other is, for the auoydinge of emptines, whiche as I haue often hearde, fo woulde I gladly ones vnderftande.

Mafter. The firfte of them appertaineth to perfpectiue, and the fecond vnto naturall phylofophye, fo that bothe doo requyre an other place and tyme : yet bicaufe I haue alleaged it for this prefent matter, although the reafons why it is fo, may not well here be repeted, yet that it is fo, fhall be

All thinges shew great through vapoures or myste. brefely declared. In a myftie morning as you walk, all things

that you fee, feeme greater through the myfte, then in deede they be. a pennye in the water feemeth broader then it is, and the deeper that it lyeth, the greater it appeareth : fo the Sonne and the Mone and all other ftars being nigh to the earth, do fhew through the vapours that afcend frō the ground, and therfor appear greater then they be : & if the vapours be many, the ftarres fhew the bigger : the caufe is, the interruptiō and reflectiō of the fight beames by the vapours & the water. & like is the caufe in feing throughe glaffe, which occafioned weke fights to feke aid of fpectakles.

Sch. Many vfe that aide, that know not the reafon thereof.

Nature abhorreth emptines. Mafter. So manye drawe water at a plompe, that knowe not the caufe, why the water dothe afcend, whiche is onlye

na-

natures worke to auoide emptines. And many men vſe bel-
lowes to blow the fier, whiche know not the reaſon of their
firſte inuention, and therfore can not mende them if they be
hard to draw. many men alſo draw waters by fountaines higher then the ſpringe, yet few of them do knowe what is the
reaſon of their woorke, and therefore fewe canne amende it,
if the faulte be any thinge doubtefull. A greate numbre of
other lyke thinges could I ſhewe, where natures abhorful-
nes to permitte any emptines, doth cauſe ſtraunge effectes,
in thinges that are vſed of many men, and well knowen of
fewe men. But as it appertaineth not to this place to diſ-
courſe largely in thoſe matters, ſo an other tyme ſhall ſerue
for them. And nowe lette vs proceede in oure purpoſed attempte, to ſee what proofes I can bringe for the roundenes
of the earth : wherein I will beginne with a diſtribution diſ-
iunctiue, containynge many opinions touching the forme *Diuers opi*
of the earth : and eche of them will I ſubſtantially improue, *nions of the*
ſaue that onlye whiche affirmeth it to bee rounde, and that *forme of the*
will I ſo fullye approoue, that I doubt not but you ſhall *earthe.*
thynke your ſelfe fullye ſatiſfied. Som menne conſideringe
that as for the ſkie no forme was ſo meete as a round form,
bycauſe of his ſwifte mouinge, ſo for the earthe whiche
ſtandeth ſo ſteddilye, they iudged no forme ſo meete as
a Cube forme, which they eſteemed moſte
ſtable of all other : and therefore manye *Why fortun*
aunciente Philoſophers by the forme of *is pictured*
a Cube dydde ſecretely ſignifie conſtancy *ſtanding on*
and ſtablenes : and contrarye waies by the *a globe.*
forme of a globe they expreſſed changable
alteration, and continuall mouing.

Scholar. That I may perceaue by the placing of Fortune on
a rouling globe, in token of hir inconſtancy & voluble chan
ginge. And therefore haue I often phantaſied, that dice, *Why dice be*
whiche is the image of Fortunes inconſtancye, and ſerueth *made in*
only for fortunes playes, myghte beſte haue beene made *cubik form.*
K.iij.   in

Robert Recorde

in forme of a Globe, for they are vnconſtant as fortune hir ſelfe.

Diuers for-
tune.

Maſter. Ther ſeemeth in Fortune two diuers natures, the one is lyghte and alwaye flickerynge, the other is heauy, and therefore more ſtable, ſo that ofte tymes we ſee them that haue a lyghte and pleaſaunte fortune, as lightlye leeſe, that they lyghtly gayned : but where heauye fortune ſetteth hir foote, ſeldom can ſhe be remoued, hir ſteppes are ſo ſtayed : but to expres more exactly the nature of the cube reſembled in the dice, bothe in forme and in effecte, you ſhall marke well the meaning of that olde prouerbe : Iacta eſt alea, The dice is caſte. or the lotte is drawen. or fortune is paſt. by whi- che ſaying is declared, that the thinge that is ones done, can neuer againe be vndone, although it may be altered, and ſo cõſtancy in that appeareth moſt certein. for as your chance on the dice beyng ones caſte, you muſte be content to ſtand to it : ſo fortune when it is paſte, can not bee altered. And that is the cauſe why all men vſe to ſaye, when they expreſſe their ſtay in lyuing : Suche is my fortune. Yet many learned men put difference betwene chaungable chaunce, and ſtable fortune, callyng the firſte Fortuna, and the other Fatum : ſo that deſtiny is ſtable, though fortune chaung right often. But thus I forget our purpoſed intent, with ſo many digreſ- ſions of other bye matters.

Schollar. I founde no faulte nor thought no tyme loſte, ſyth the matter is pleaſaunte and ſomewhat to our purpoſe.

Maſter. Well, this was their imagination, that thoughte the earthe to be of a cubyke forme, for that they iudged it the moſt ſtedfaſt form.

The ſecond
opinion.

Then an other ſorte deuiſed a three cornered forme like

A rygge forme.

he rygge of an houſe where tone ſyde lyeth flatte, and the other two leane a ſlope. And thys forme they iudged better for twoo cauſes. Firſte they thought that it
was

was more fteddy then a cube form, bicaufe it hath a broader foote, and a leffer toppe : and fecondly for that they thought it a more apte forme to walke on, and more agreable to the nature of the earth, wher fome times there ryfeth highe hils, and fometime againe men may fee greate vales defcendyng.

Schollar. This imagination is groffe inoughe.

Mafter. And fo groffe is the iudgement of them that fo-lowe not, or fearche not for true reafon, but content them felues with a lyght conceaued fantafye.

Schollar. And in this they be deceaued, that they accompt this form more apt to walk on : for the flat of the cube is plai ner, & therfore more apte to walk on, then is a flope ground.

Mafter. If the fyxte parte of the earthe were onlye inha-bited, then woulde it appeare fo in deede : but if you go any farther, then haue you vnapte plaineffe to walke on in theyr imagination, whiche go fo downe righte, that they do feare fallynge. Againe they thinke this Rigge forme meeteft for the ftanding of the fea, and for running of riuers : for in the fyrfte forme, if the fea fhould refte on the ouermoft plaine, then wolde it ouer runne all that plaine, and fo flowe ouer all the earthe : where as in this fecond forme it mighte refte about the foote of the earthe, and yet the flope rifyng wyll not permit it to ouer runne all the earthe. And fo for riuers if there by no flopenes (as in a cube there is none) then can not the ryuers runne well.

A thyrde feƈte thinkinge to amende thefe bothe, imagined the earthe to be plaine and flatte : for fo they fantafied that it wold reft mofte fteddilye, and fo was it very eafy to walke on.

A playne Flatte.   *The thyrde opinion.*

Schollar. We are more beholdynge to thofe men, for deuifing our eafy wal-kinge, then we are bounde to them for their wife doƈtrine.

Mafter. The fourthe feƈte, fearyng leaft by this opinion they fhoulde leefe the fea and all other waters, imagined the

The fourthe opinion.

K.iiij.   forme

forme of the earthe more apte to holde water, and deuifed it hollow lyke a bolle.

Schollar. Thofe men were verye ftudious for ftaying of water, more then they were for framyng of their wittes.

Mafter. Yet his vaine follye didde feeme to them greate wifedome.

Schollar. Saue that I do credite your report, I wolde neuer haue thoughte, and muche leffe haue beleued, that euer anye fuche madde imaginations hadde beene phantefied of anye men.

Mafter. Who lyfteth to fee the monftruoufe opinions of fuche dreaminge doters, maye reade them often touched in Ariftotle his naturall bookes, and aboundantly in Plutarche his boke De philofophorum placitis. and in Galene and Eufebius in bokes of the fame matter peculiarly writen. But thefe 4 opinions which I haue here reherfed, are briefly noted in the firfte boke of Cleomedes fphere, though not in like ordre : and faue that in the feconde opinion I iudge his printe corrupt, and that for πυραμοαδης, I do reade and tranflate περιφοαυλης: as it may well be gathered by his owne confutation, which will not agree fo well for confuting al ftiple formes or fpire formes, but as mens iudgment ought to be free, fo if any mā lift to folow ỹ print, I wil not withftād him.

Schollar. Although fome of thefe opinions are fo groffe that they neede no confutation, yet I praye you repeate the confutations that Cleomedes doth vfe.

Mafter. I am well content, and better pleafed to alleadge them in his owne name, then to afcribe them to my felfe, for diuers caufes. Firfte he beginneth with the thirde opinion, *he reprofe* and reproueth it thus. If the earthe were flatte and plaine, *of the third* then fhould all nations haue one horizonte : for in a plaine *opinion.* flatte forme, there can be no iufte caufe of alteration of the Horizont.

Scholar.

THE CASTLE OF KNOWLEDGE   118

Scholar. That foloweth mofte certenly.

Mafter. Then muft the Sonne and Moone and all other ftarres rife to all people, when they rife to anye one, and fo mufte they fette (eche one in his courfe) to all men at one inftante.   Schollar. That will followe alfo.

Mafter. If the Sonne rife to all men at ones, and fette like-wayes at one time, then mufte the daye beginne to all people at ones, & all nations muft haue night at one time precifely.

Schollar. That is falfe as all men confeffe : for at Hierufa-lem (whiche is well knowen) it is day thre houres foner then with vs, and fo is it nyghte fooner by thre howers alfo. But in Calecut (as learned men affirme, and trauelers thither, do confirme) it is daye 6. howers foner then with vs, and it is night 6. howers foner to them againe then to vs.

Mafter. Your fayinges are true if they be well taken : but and if this conclufion bee falfe, as it is in deede, then mufte that opinion be falfe, whereof this conclufion is inferred.

Schollar. So doth it well folowe, and is fully prooued.

Mafter. One ftronge reafon for the varietie of howers is gathered by the eclipfes duly obferued, and namely of the Moone,, for as it happeneth at one inftance of time, fo is it not one hower to all nations. As for example : This year of *Examples* 1556, the eclipfe of the Moone fhall be with vs the 17 day of *of eclipfes.* Nouembre at 3. of the clocke in the morninge, and to them at Calecut it fhall be at 9. of the clocke in the morning : yea we fhall fee the Moone in the fouthweft, and they fhall not fee her at the fame inftant, for fhe will be to them vnder the horizonte in the northweft. like waies in the yeare of 1562. there fhall be a great eclipfe of the Moone with vs, whiche fhall endure aboue three houres and an halfe, and yet fhall they at Calecut fee no part of it, by reafon that the Moone fhall be farre vnder their horizont before that eclipfe begin. And in lyke manner this lafte yeare 1555. was there a greate eclipfe of the Moone the fifte daye of Iune, at three of the clocke in the morning, yet in Calecut there was none eclipfe
                                                             feene

feene then, for the Moone was fet vnder their horizont two howers almoft before the eclipfe began. But in the yeare of 1551. when we had the eclipfe of the Moone at 9. of the clock at night, the 20. day of February, they at Calecut fawe that eclipfe at thre of the clocke in the morning the nexte daye, as the Portingales that were there can teftifye. Wherby it is manifeft, that their Horizont doth not agree with ours, and thereof doth it folowe that the earth is not flatte. But nowe to returne to Cleomedes againe, (vnto whofe wordes I haue added but the examples of the eclipfes) his feconde reafon againft the flatneffe of the earth, is this.

An other re profe of the flatnes of the earthe. If the earth were flatte and plaine in forme, then the Pole muft needes appeare at one height to all parts of the world, and the artike circle (which inclofeth the ftarres that neuer fet) fhuld be but one to all nations. But bothe thefe thinges appeare plainly falfe : for as vnto vs about London the Pole is not fully 52. degrees highe, fo if you go northward, you fhall fynde the Pole to rife higher and higher, till it bee fully 90. degrees highe. and in going fouthward, the eleuation of the Pole waxeth leffer and leffer, till you come to the middle of the earthe vnder the equinoctiall, where the pole is of no height, but is equall with the Horizont. Alfo in all thefe pla ces, you fhall haue feuerall arctike circles.

Scholar. That muft needes folow the diuerfitye in the ele-uatiõ of the Pole, as it hath been fufficiently declared before.

Mafter. As the firfte improbation doth reproue the flat-nes of the earth betwene eafte and wefte, bicaufe it regardeth chiefly the rifing and fettyng of the Sonne and other ftarres, and their courfe betwene eafte and weft, fo this fecond con-futation improueth the opinion of plaineffe betwene fouth and north. So doth it folow, that the earthe is flatte nother one way nother other, but bothe waies hath fome certain ri-fing, which anon I will proue to be a iufte roundenes.

The thirde confutation. A thirde reafon is alleged by Cleomedes, touching the e-qualitie of daies to all nations, which fhoulde of neceffitye
folow

THE CASTLE OF KNOWLEDGE 119

follow if the earthe were flatte, and all people had one horizonte, but bicaufe it is fo little difagreable from the fyrfte
reafon of one Horizonte, and one tyme of rifinge and fettinge of the fonne, I haue ioyned them both in one, as before it dothe appeare. Thefe thre reafons are plaine inough.
The fourth reafon whiche Cleomedes doth make, is not fo
eafye, yet is it as certaine as any of the other : and therefore I
will fhewe you what it is, feyng you defire to heare his owne
arguments, although I determined before to allege fuch reafons only, as myght appeare eafy to vnderftand.

Scholar. If it be not ouer muche obfcure, it may pleafe you
to declare it in the mofte playneft forme ye can.

Ma. I will only alter his ordre in the propofitions, adding
that wich is not eafye to be gathered, to make it the eafier to
your vnderftanding. This is it.

If the earth were plaine, it fhoulde folowe, that the whole *The fourth*
diameter of the world from one fide of the fky to the other, *confutation*
fhoulde be but 100000. furlonges, that maketh 12500 miles, *of the plain*
which faying appeareth fo abfurd, that no man will graunt *nes of the*
it. but if any man wold do it, this argument folowing fhall *Earthe.*
côfute him. Firft therfore I reafon thus. If the earth be plain,
then al places in the earth ar as far a fonder, as their Zeniths,
or Verticall pointes be in heauen. This maxime muft I adde
vnto Cleomedes, to make his reafon the more plaine.

Scholar. But this maxime do I not vnderftande, wherfore
I befeeke you both to proue it, and declare it.

Mafter. I am content.

You knowe by the former treatifes, that the Zenith is the
pointe right ouer the headde of any people, whofe Zenith
it is : whereof it mufte folowe that euerye diuers place in
earthe, mufte needes haue a feuerall Zenith in the fkye.

Scholar. That is plaine.

Mafter. Then imagining the earth to be flatte, the lynes
that dooth afcende from any twoo places, vnto theyr Zenithes in the fkye, mufte needes be paralleles, as here in this
picture

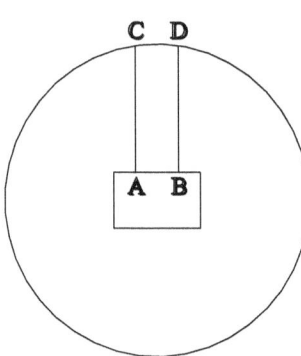

picture doth partly appear. for if the cirdle be set for the skye, and the flatte square within it for the earthe, then take two places in the earth, as A and B. the zenith to A is C, & must needes be right ouer it, and therfore the line that is drawen from A to C, must be a iust plumb line, & perpendiculare to the flatte earth. And likewaies the ze- nith to B is D, which muste needes be righte ouer it, and therfore the line that goeth frō D to B, must of neceffitye be a perpendiculare and plumbe line to the flatte earthe alfo. Then if bothe thofe lines be per pendicular to one flatte plaine, or to one line standinge for that plaine flatte, all the angles that they bothe doo make with the thyrde lyne AB, muste bee righte angles, accor- dinge to the definition of a perpendiculer line. Nowe if all their angles be right, then are they all equall accordynge to the fourthe grauntable requeft in the feconde booke of the Pathway, that all righte angles be equall eche to other. And if all their angles be equal, then muft their matche angles be equall of force : wherby it doth folow accordinge to the 18. Theoreme of the feconde booke of the Pathway, that thofe two perpendicular lines be paralleles, feyng that on 2 righte lines, as A C and B D, there is drawen a thyrde ryghte line A B, croffewayes, and maketh twoo matche corners of the one lyne, equall wyth the lyke twoo matche corners of the other lyne.

Scholar. Hereby I haue not onlye gotten the vnderftan- ding of your proofe, but alfo I perceaue a farther vfe in the Theoremes of the Pathway, then I knewe before.

Mafter. I will profecute my proofe. Syth thofe twoo
lines

lynes bee paralleles, and equallye diſtaunte, then is there
as muche ſpace betweene A and B, as there is betweene
C and D.

Scholar. Thus is your maxime ſufficiently proued, and
fully declared : for AB. betokeneth the diſtaunce of the two
places in earth, and CD, ſtandeth for the diſtaunce of their
zeniths in the ſkye.

Maſter. Nowe therefore will I retourne to Cleomedes
argument. They that dwell at Lyſimachia (in Grece) & thei
that dwell at Syene (in the ſouthe parte of Egypte) haue betweene them in diſtaunce 20000 furlonges (that is 2500 miles) wherefore it muſt folowe that their zenithes in the ſkye
be no farther a ſonder, ſeyng they be limited by two perpen
diculers equallye diſtaunte : but it is well knowen by good
proofe of inſtrumentes, that Syene is vnder the Tropike
of Cancer directly, and Lyſimachia is vnder the hedde of
the North dragon, which 2 places in the ſkye are iuſtly proued to be a ſonder the 15 part of the whole compas of heauen, that is the firſt part of the diameter of the ſkye. Wherfore if 20000 furlonges be the firſt parte of the diameter, the
whole diameter muſt be but 100000 furlonges : & the whole
compas of the ſkie muſte be but 300000 furlonges, and of
theſe furlonges it is prooued, that the earthe contayneth
in compas 250000. ſo is the heauen lyttle bygger then
the earthe in compas. whiche abſurditie maye eaſily be confuted by the Sonne, whiche in compariſon to the ſkye, is
a verye lytle parte of it, and yet is bygger than the earthe
mannye folde : whereby anye manne maye ſee what ab-
ſurditye foloweth that opinion, to thynke that the earthe
is flatte.

Scholar. I doo metely well vnderſtand this reaſon, but I
ſhuld better haue conceaued it, if I had knowen the two places whiche hee alleageth for examples ſake.   *A like rea-*
                                                               ſon.
M. Then will I for your pleaſure make ẙ like argument by
example of 2 places which ar better knowen to engliſh men.

                    L.i.   You

you knowe the caſtle of Arundell.

Scholar. The name is auncient and famous.

Maſter. And Newe caſtle vppon Tine is well knowen to
you alſo.   Scholar. So is it.

Maſter. To go the next waye betwene theſe two places it
Arundel ca is 270 englyſh myles. And the Zenith of Arundell caſtle
ſtle. (whiche is the iuſte point of the latitude of it) is 50 degrees
and 30 minutes, as ones I remembre I tooke note of it in ri-
ding that waies. The Zenith alſo of Newcaſtle is from the
equinoctiall 55. degrees, ſo is the difference betwene their ze
niths 4 degrees and 30 minutes. Now (as I haue declared be
fore) If the earthe be flatte and the perpendicularre lines bee
paralleles and equidiſtant, that go vp from theſe two places
to their zeniths, then is 4 degrees and 30 minutes, iuſt equal
in quantity to 270 myles.

Sc. That is true, as it is proued before in the third treatiſe.

Maſter. You are farre deceaued : it is declared there,
that 270 myles in earthe, muſte anſwere in proportion to
foure degrees and an halfe, and not that they are equall
togyther.

Scholar. I perceaue mine owne negligence in markinge
the propretye of ſpeache. I ſhoulde haue ſayd, that as foure
degrees and an halfe is the eight ſcore part of the whole com
pas of heauen, ſo 270 myles is the eighte ſcore parte of the
circuite of the earthe.

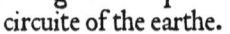
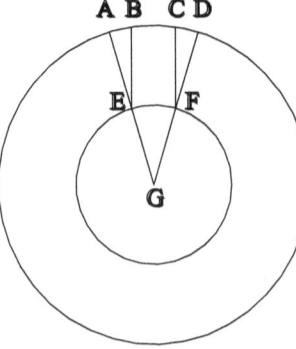
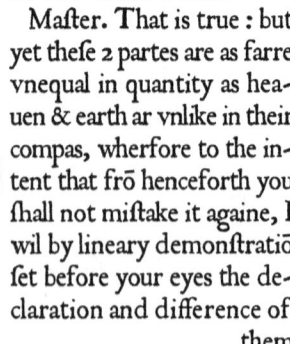

Maſter. That is true : but
yet theſe 2 partes are as farre
vnequal in quantity as heauen & earth ar vnlike in their
compas, wherfore to the intent that frō henceforth you
ſhall not miſtake it againe, I
wil by lineary demonſtratiō
ſet before your eyes the declaration and difference of
them

them bothe more plainly then curioufly.

Here in this figure you fee two circles drawen vppon one centre, their common centre being G, from which there are drawen to the vttermoft circle two right lines G A, & G D, thefe lines do croffe the leffer circle in 2 pointes E and F, frõ whiche two pointes I haue drawen two paralleles, vnto the circumference of the greater circle, whiche two paralleles be BE, and CF. Nowe may I fay, that bicaufe thefe two circles be made vpon one common centre, and twoo lynes drawen from that centre to the circumference of the both circles, bicaufe AGD is one common angle in them bothe, ther-fore are there arche lynes indofed betweene thofe two ryght lynes lyke in proportion.

Scholar. I perceaue it well : fo that if the arche lyne AD in the greater circle, be the fyxte parte of it, then is EF the arche lyne of the leffer circle, the fyxte parte of his owne circle, in lyke manner. but yet that arche of the leffer circle is not fo greate as the lyke arche in the bygger circle.

Mafter. Then what faye you of the arche BC, in com-parifon to the arche EF, whiche bothe arches are betweene twoo lines paralleles?

Schollar. They mufte needes bee equall, feynge there is iufte as muche diftaunce betweene EF, as there is be-tweene BC.

Mafter. So maye you nowe perceaue what difference it is to faye, that two arches of two feuerall circles, are like in pro portion : and to faye that they are equall in quantity.

Schollar. Nowe I perceaue it plainly, that although 4 de-grees and an half (as your former reafon did import) be like in proportion to the whole circumference of heauen, as 270 miles are in comparifon to the compaffe of the earthe : yet it foloweth not that they fhould be equall togither.

Mafter. But fuppofynge the earthe to bee flatte, then it foloweth as I haue declared beefore, that they are equalle in quantitye, feeynge bothe beetoken the distant of one couple of paralleles. And thē it foloweth, that feinge 4 degrees & a half is the four score part of the compas of heauen, if I multiply 270 myles (whiche is equall to it) by 80, therof will amounte the numbre of myles that make the compaſſe of heauen, whiche are 21600 myles. Nowe to know the diameter of it, I take the two receaued numbres for the proportion betweene the circumference of a circle and the diameter of it, whiche are 22 and 7, (as in the Pathway is declared more largely) and by the rule of proportiō I work

270
 80
21600

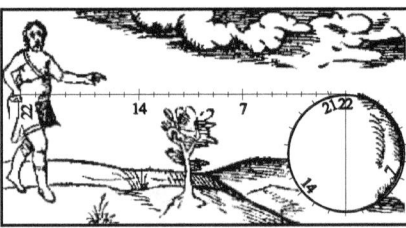

in ſaying : if 22. giue 7, what ſhal 21600 yelde? and there amounteth 6872 $\frac{8}{11}$, whiche muſt be ȳ whole diameter of the ſkie, if the earth were flatte.

Scholar. That is to greate an inconuenience for any man to af firm. for therby I ſe it wold folow that if we go any waye from our owne cuntry, 3436 miles, we ſhal com hard to the ſky, which is to childiſhe a fantaſye, ſith not only reaſon, but dayly trauell decla- reth the contrarye. Againe I re- membre that in the thirde treatiſe you declared that the

22 ⟋ 7
21600 ⟍ 6872 $\frac{8}{11}$

21600   1
    7   1321
151200   39666
            151200 (6872
            22222
            222

earthe was ſo muche in compaſſe, whiche muſte needes bee many fold leſſe then the heauens, whiche ar ſo farre diſtaunt from the earthe on euery ſide.

Maſter. Thus are all Cleomedes reaſons againſt the flat- nes of the earth fully alleaged, & ſomewhat largely declared : Now wil I proceede to ȳ confutatiōs which he vſeth againſt ȳ other opiniōs, folowĩg his own ordre. wherfore next doth
folowe

folow the confutation of them which fay that the earth is ho
lowe like a bolle. Againft whofe phantafticall imagination *The confu-*
he reafoneth thus : If the earthe were hollowe as a bolle, then *tation of*
fhould the Sonne, the Moone and all Starres in thier rifing *the fourthe*
appeare foner to them that dwell in the wefte, then to them *opinion.*
that dwell in the eafte : whiche thinge is contrary to daily ex-

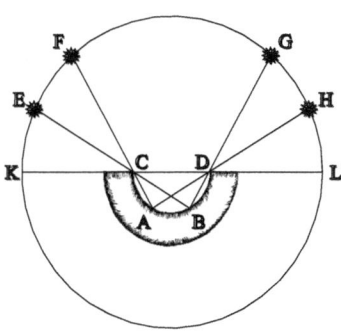

perience. For declaratiō
of which faying by line-
ari demōftration I think
good to drawe a figure,
wherin you may aptly fe
the force of his reafon.
The vttermoft circle of
ẙ figure doth reprefent
the fkye, and the inner
moft half circle ftādeth
for ẙ imagined holow-
nes of the earthe, & the
halfe roundelet AB, reprefenteth the maffy fubftance of the
earth, the right line KL, expreffeth the diameter of ẙ world,
and therfore the right Horizont of the earthe, K beinge the
eaft and L the weft. Now for explication of Cleomedes reafon : If the earthe were holow, as here the forme of it is drawen, then when the Sonne is rifen, in the eafte aboute E, it
wold appeare to them that dwell in the weft by B, & not vnto
them ẙ dwell in ẙ eaft by A. for the brow of the holow groūd
by C, doth hide the Son yet frō them, fo ẙ he muft afcend as
high as F, before they ẙ dwel in the eaft by A may fee hym.
Again when ẙ Son goeth doune, by this opiniō he fhuld fet
to them that dwel in the weft by B, as fone as he came to G,
by occafion of the browe of the ground by D. and yet they
that dwell in the eafte by A, fhould fee him a great while lon
ger : for that browe of grounde by D, wyll not yet hynder
their fighte, vntill he be defcended as low as H. So fhoulde
they that dwell in the weft fee the Sonne fooneft in the mor

L.iij.   nynge

126   THE FOVRTHE TREATISE OF

ning, and they that dwell in the eaſt ſhoulde ſee him lateſt at eueninge.

Schollar. This thinge is ſo falſe, that euery chylde knoweth the contrarye.

Maſter. Yet of that opinion dooth there folowe farther *An other reproofe of the ſame opinion.* inconueniency, as Cleomedes doth ſhew : for by this fanta-ſye, they that dwell in the ſouthe ſhould ſee the northe Pole more higher aboue ground, and ſo ſhould haue a larger arctike circle, then they that dwell in the northe, as by the ſame figure it may be declared.

Scholar. I perceaue it well : for if I make K to be the ſouth, and L the north, then it appeareth in this form of the earth, that they which dwel in the ſouth by A, may ſee as low as H : and they that dwell in the northe by B, canne ſee no farther northe then G. whiche is ſo farre againſt reaſon and daylye experience, that it muſt needs appeare to be a vaine fantaſy, that bringeth forthe ſo mad and monſtruous concluſions.

Yet an other confutatiõ of the ſame opiniõ. Maſter. Yet doth there folow more ſonde concluſions of it : for by this opinion all nations that dwell within that holownes, ſhould ſee leſſe then halfe the ſkie, leſſe then halfe the Zokiak, and leſſe then halfe the Equinoctiall, wherof it wold follow (beſide other abſurdities) that they ſhuld haue their nighte commonly longer then their daye, bicauſe that parte of heauen which they ſe is leſſe (eſpecially to them that dwell in the botome of that holownes) then that part which is vnder their horizonte : Yea they that dwell in the botome of that holownes, canne neuer haue their daye ſo longe as their nighte, bicauſe they do ſee ſo litle a portion of the ſkye. As a man that is in a deepe trenche or in a pitte, can ſee but a litle of the heauens. And thus hath Cleomedes ſufficientlye confuted thoſe two opinions : whiche kinde of confutation Ptolomye doth vſe alſo againſt bothe thoſe opinions.

Ptolomye. Scholar. Then muſt they needes be good : for as I heare all learned men ſay, Ptolemye is the father of that arte, and proueth all his woordes by ſtrong and inuincible reaſons.

Maſter.

Mafter. No man can worthely praife Ptolemye, his trauell being fo great, his diligence fo exacte in obferuations,
and conference with all nations, and all ages, and his reafonable examination of all opinions, with demonftrable con-
firmation of his owne affertion, yet mufte you and all men
take heed, that both in him and in al mennes workes, you be
not abufed by their autoritye, but euermore attend to their *Autority of writers.*
reafons, and examine them well, euer regarding more what
is faide, and how it is proued, then who faieth it : for autori-
tie often times deceaueth many menne, as here by and by in
Cleomedes it fhall appeare, whofe argumentes in confuting
the other two opinions ar nothing fubftantiall : which chan
ced other bicaufe he fawe the fondenes of thefe opinions fo
great, that he fought no great reafons to confute them, other
els haftinge in his writinge caufed him to vfe the leffe diligence in framynge his reafons. but nowe will I repeat them. *Cleomedes*
     If the earth were of cubike forme, then fhould all nations *argumente*
haue fyxe howers daye only, and 18 howers nyght, feing ther *againft the*
be rounde about the cube four fides, fo that on eche of them *firft opiniõ.*
the Sonne fhoulde fhine 6 howers only : this is a very weake
argument.
     Schollar. Yet vnto me it feemeth a ftrong reafon : for fe-
ing that the Son doth go round about the fkie and aboute
the earth alfo iuft in 24 howers, it muft needs folow that he
fpendeth only 6 howers in euerye quarter : and a cube hathe
but four fydes in his compaffe (althoughe it haue 6 fides in
all) wherfore in mine opinion it is well conduded, that euery
one of tofe four fides, doo fee the Sonne 6 howers iuftlye.
     Mafter. Often haue I readde in Galene, and more often
haue I feen it by experience, that better it is for men to want
all arte of reafoninge cleane, then to haue fuche confidence
in a meane knowledg therof, that may occafion them to deceaue them felfe, and to feduce other. You are fully perfwa-
ded that this agrument is good : whereby it appeareth that
you efpied not the want of that meane propofition, whiche
L.iiij.   fhould

ſhould make the argument good, which muſte be this : that
euery quarter of the ſky, agreeth to one quarter of thearth.

Schollar. That not only I thinke to be true, but your ſelfe
affirmed it alſo before this time, as a true ſentence.

Maſter. And ſo will I do ſtill, affirming it of the true form
of the earthe, but not of this imagined cube forme.

Scholar. Why, is there anye difference in the quarters of
any formes? is not a quarter of a cube the fourth part of it,
as well as a quarter of a Globe is ẏ fourth part of the globe?

Ma. Yes, but yet doth not the quarters of the cube ſo agree
with the quarters of a globe, as the quarters of two globes
agree togither.

Scholar. That I vnderſtand not.

Maſt. Then will I declare it manifeſtly by lineary demonſtration. Marke theſe fi

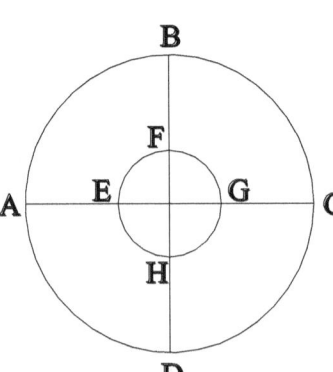

gures. Here you ſe firſt
for the true opinion, 2.
circles drawen one with
in the other vpon one
centre, and the ſame are
diuided into four quar
ters ech of them, ſo that
the four quarters of the
leſſer circle, E F G H,
do anſwere agreably to
the four quarters of the
greater circle A B C D,
but in the ſecōd figure,
where the cube is made in lue of the earthe, the quarters do
not agree, as you may perceaue by the draught of the right
lines, agreable to eche ſide of the cube : for euery ſide of the
cube hath almoſt halfe the circle aboue his horizontall line.
Wherfore if you will haue a cube drawen in a globe, in ſuch
forte that the quarer of the one in cōpaſſe ſhall agree to the
like quarter of the other, that cube muſte be ſo great, that his

corners

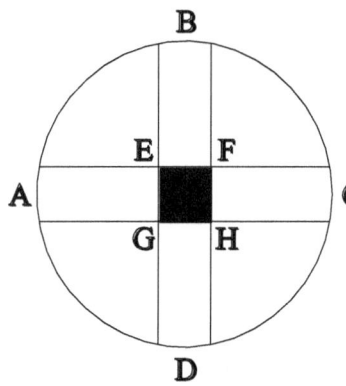

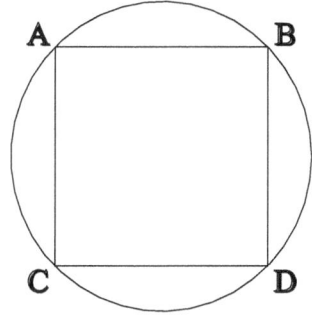

corners may touch the globe on eche fide, fo mufte it bee as greate a cube as maye bee made within that globe. And I am fure you will not fay that the earthe is fo great in comparifon to the fkye.

Schol. Now I fe mine owne erroure, and the fault of Cleomedes argument.

Mafter. And if anye man wold excufe Cleomedes, he muft fay, that Cleomedes did make ẙ reafon againft fuche : as affirmed two errours at ones, that is the cubike form of the earth, & the greatnes of it alfo to bee fuche, as mighte touche the fkye with euery corner : but if this had been his meaninge he might eafily haue expreffed it fo : but what fo euer he ment he framed the confutation of the fecond opinion in the like forte. for this is his argument.

If the earthe be of a three cornered forme, then fhuld the Sonne fhew 8 houres iuftly on eche fide of it, and fo wold it be to al people 8 houres day, & 16 houres night : which thing is to appearant falfe : fo can not that opinion be true. for declaratiō of this argument I haue drawen firft a circle for the fky, and then a fmall triangle forme D E F, vnto whofe thre sides

Cleomedes confutation of the secōd opinion.

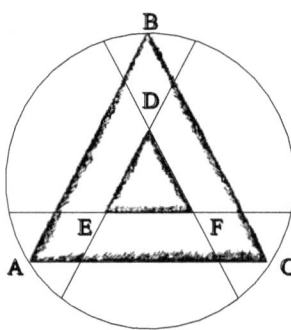

ſides I haue drawen 3 ſtreight lynes, repreſenting three ſeuerall horizontes. but it appeareth at the firſte ſight, that eche of thoſe horizontes doo contayne aboue them almoſt halfe the ſkye. So that in this quantitye of the earth, Cleomedes reaſõ taketh no place, nother generally in any other but one, where the three corners of the earth may touch the ſkye, for whiche forme I haue drawen the greate triangle A B C.

Scholar. Yet although Cleomedes argumentes bee not ſufficient to confute their opinion, that would ſay the earth were of any of theſe bothe formes, their opinion is falſe neuertheleſſe. thinke you not ſo?

Maſter. Yes verely : for a weake confutation of an vntruth doth not make that vntruth to become true. And bicauſe you ſhall not thinke that theſe opinions haue anye ſure grounde, I wyll repeate Ptolemye hys confutation of them both, by one vnfallible reaſon.

Ptolemy his confutation of the firſte and seconde   You ſee in bothe theſe imagined formes of the earthe, that there can be no more horizontes, then there be ſides in the fygure.

Scholar. That is certaine : for all that dwell on one plain ſide, muſt needes haue one horizont : wherfore if the forme of the earth wer four ſquare in his compas, then could ther bee but fower Horizontes, that waye : I vnderſtande it betweene eaſte and weſte, and in all varieties there canne be but ſyxe, ſyth a cube hath but ſyx ſydes : lykewaies in the thre cornered forme, there canne be but three diuers horizonts betwene eaſte and weſt.

Maſter. You ſaye well. And ſeeynge all that dwell
on one

on one plaine fyde haue all one horizonte, they mufte haue
day all at one inftant both for the fonne rifinge and alfo for
the fetting, fo can ther be no more variety in the beginning
and ending of daies, then there are fides in the figure of the
earthe, whiche by the firfte opinion muft be but 4, and but
3 by the feconde opinion, whereas the contrary is well knowen by dailye experience, as well as by reafon, that euerye
15 degrees in diftaunce weftwarde maketh the daye an hour
later : and contrarye waies euery 15 degrees of diftaunce eft-
ward, caufeth the daye to be rather by one howers fpace.

Sch. That is proued alfo before, in confutation of the third
opinion, and namelye by examples of eclipfes. But what if
any wolde affirme that the earth were made of many flattes,
as of 24 (for an example) betwene eaft and weft, then fhuld
there be no more horizontes, then there bee howers in one
naturall daie, and yet fo the difference of howers could not
confute them.

Mafter. You muft thinke that learned men canne as well
marke the difference in euerye minute of an hower, as the
common people can obferue diuerfities in howers : yea the
learned obferuations are more exactly taken thē the 60. part
of a minut of an hower, wherfore feyng it is fo well proued
by fondry obferuations, and efpeciallye by eclipfes, bothe
of the fonne and the moone, that euerye mile diftaunce betwene eafte and weft, dooth make a feuerall horizonte, there
can bee no other forme of the earthe aptlye affigned, but a
rounde circular forme. And by the lyke reafon, by the ordrely afcending of the Pole, in goinge northward, and by
the vniforme defcending of it in going fouthwarde, it muft
needes appeare that there can bee none other forme of the
earthe betweene fouthe and northe, but a rounde forme alfo.

Scholar. Nowe canne I ende your argumente of the di-
ftribution difiunctiue, whiche maye be framed thus.

The earth muft haue fome forme, either cubike, thre cornered, flatte, or holow, or fome fuche lyke, other els a round
forme

*The colle-
ction of the
arguments*

by disribu tiō disiun- ctiue. forme, but his forme can not be cubike, nor threcornered, nother flatte, nother holow, nor anye suche lyke, as before is fully prooued, wherefore it muste needes be rounde.

Master. It foloweth well. for it is not possible that in any other imagined forme of the earthe, the horizontes should alter toward euery coaste so vniformely, and the dayes differ so proportionably, the Pole to be eleuate so ratably, or to be depressed so ordrely, and all other appearances to answer *A roller forme.* so agreably. Yet some men (as Ptolemy doth reporte) had inuented an other forme lyke a roller, or a rounde pyller, whose endes shoulde lye north and south, by whiche forme althoughe they thought none of the varieties of appearances myghte bee hindered, yet in that forme the eleuation of any one of the Poles could haue but two varieties, for euer more it muste appeare nother ouer their heddes, as to them that dwell on the flatte eandes of that roller, or els to all o- ther that dwell about the compas of the roller, it muste still appeare in their horizonte, so shoulde ther bee no starres about either Pole alwaies appearant aboue ground, nother all wayes hydde vnder grounde, but all starres should ryse and set to all them that dwell about the roller. And againe they that dwell on the flatte endes of the roller, shoulde haue but one Horizont, so large in distaunce of ground, as the whole thicknes of the earthe is : all whiche imaginations are bothe well knowen to be vaine, & also easye to be confuted by the former reasons, which serue so largely, that you can ima- gine no forme other then round, but those reasons will con fute it. wherefore your argument doth proceede well.

That the water is round by di uers profes. Yet farther for the roundenes of the water also, and name- ly of the sea, you maye frame argumentes by the lyke forme of appearances : for where so euer you bee on the sea, you shall see halfe the skye iustlye, and the farther west that you go, the later dooth the Sonne rise : and contrarye waies the farther easte that you saile, the sooner in the morning will the Sonne appeare to you. whereof I will declare vnto you

a no-

THE CASTLE OF KNOWLEDGE   133

a notable example, and a iuſte proofe.

Imagine a ſhip ſwift of ſaile to be at the cape of Cornwall _An exãple of the roũdnes of the ſea by a ſhip- pes courſe._ ready to make ſayle towarde the weſte directly, and to haue a greate gale of winde, it is poſſible that ſhe maye run 240 myles in 24 howers : for I haue beene at the triall of a greater courſe, therefore I ſpeake (as men ſay) within my boundes : after which rate ſhe ſhall runne in 16 howers 160 myles. Now let hir hoiſe ſaile at the ſonne riſing, and let the time of the year be ſomwhat before midſommer, or little after, when the Artificiall day from ſonne riſing to ſonne ſettinge, is 16 ho- wers longe : by this meanes at the end of 16 howers, ſhe ſhall be weſt of the cape of Cornwall where ſhe began her courſe 160 myles : and then ſhall the ſonne be at ſetting to their ſight that dwell at the ſaide cape, but the ſhippe ſhall haue the Sonne aboue foure degrees hyghe at that inſtaunte, by reaſon that ſhe dydde runne with the Sonne, and that the roundenes of the ſea doth chaunge the horizont ſo many degrees in 160 myles.

Scholar. Althoughe this example bee pleaſaunt, yet it paſſeth myne vnderſtandinge, ſith that I beleued hitherto, accordinge to your former doctrine, that 160 myles would not haue altered any waies three degrees, ſeyng 60 myles do anſwere to one degree.

Maſter. That ſayinge is true all wayes for the eleuation of the Pole, for going betwene ſouth and northe in all pla- ces, but for going betwene eaſte and weſte, it ſerueth onlye for the myddle of the worlde, that is vnder the Equino- ctiall circle : and in all other places, the farther you bee from the Equinoctiall, the fewer myles anſwere to eche de- gree, by reaſon that the paralleles growe leſſer ſtyll to- warde the Poles : yet the leaſte of theym is dyuided into thre hundreth and ſixtie degrees as well as the greateſt, whereof hereafter I will inſtructe you more exactelye. in the meane ceaſon, you ſhall vnderſtande, that for the lati-

M.i.   tude

tude of the cape of Cornewalle, euerye degree requyreth

How many myles ann-fwere to a degree at the fouthe coafte of Englande.

onlye 37 myles : whiche beynge multiplied by 4, maketh but 148 : and therefore I fayd aboue 4 degrees did anfwere to 160 myles, as the truthe is.

Scholar. Nowe I perceaue fomwhat better the reafon ther of by the proportion of the parallele circles in the Sphere. and furely this proofe is pleafante, and eafye inoughe to bee tried.

A lyke exã ple of a fhip pes courfe.

Mafter. A lyke example may this be. Suppofe at the fame tyme of the year when the day is at the longeft, that there is a fwifte fhippe at the wefte pointe of the ifle of Iflande, wher the longeft day is 20 howers from Sonne rifing to fonne fet-ting, in thofe 20 howers, that fhippe might fayle weftwarde 200 myles. Then confidering that at that latitude whiche is aboue 63 degrees, there anfwereth but 27 miles to a degree. when the fhip is at the ende of his courfe, the fonne will fette to them that bee in Iflande, and then fhall the fhippe haue the fonne 7 degrees and almoft a halfe, aboue the horizont, (which maketh halfe an hower in time) fo that by the round nes of the fea, they haue chaunged their horizont fo much in twentye howers faylinge. Nowe turne his courfe and let the fhippe haue like wind homeward againe the nexte daye, and let him make faile at the fonne ryfinge, then fhall it bee after fonne fet halfe an hower, before fhe fhall ariue at the for mer porte : by reafon that the fonne ryffe halfe an hower later to the fhippe, where fhee was in the wefte, then it dyd to them at Iflande : and therefore mufte it fet halfe an hower rather at Iflande, fo hathe the fhippe lofte halfe an hower, by comming eaftwarde againft the fonne.

Scholar. I vnderftand that. As 15 degrees doth anfwer to an hower, fo 7 degrees and a halfe maketh halfe an hower : wherefore if the fhyppe fayle iufte twentye howers, and that artificiall daye is iuft 20 howers longe, then fhall they come to their port in Iſland halfe an hour after fon fetting, bicaufe

it was

it was halfe an hour after Sonne riſing in Iſland, before they began to make ſaile.

Maſter. This varietie coulde not happen, except the water alſo were rounde as well as the earthe. And for farther proofe of the roundnes of the ſea, daily experience doothe teache vs, if we wold diligently obſerue it, howe that when a ſhippe doth draw towarde londe out of the maine ſea, the lowe grounde doth not appeare at the firſte vnto the ſhippe but the toppes of high hilles and cliffes : like waies they that be on the londe and looke to the ſhippe, they ſee the toppe of the ſhip firſte, and after that the maſtes, ſayles, and ſhrou des before they can ſee the hulle, and body of the ſhip. Now I demaund of them that thinke the water to be flatte, what is it that letteth the ſyghte, ſo that it canne not as well ſee the lowere grounde from the ſhippe, or the hulle of the ſhippe from the londe? *An other proofe that the water is rounde.*

Scholar. They can name nothing but water : for there is nothinge els betwene them, hable to ſtay the ſight. But then peraduenture they will ſaye, it is the waues of the ſea, whiche riſe verye highe often times.

Maſter. That were to childiſh an anſwer, ſith the lyke doth appeare, and that moſt exactlye, in a greate calme, when the ſea ſeemeth as plaine and as ſmothe as a borde : ſo that they muſte ſhewe ſom ſuch thing, as is higher between them then any of both theyr ſyghts, when the ſea is as quiete as can be.

Scholar. Then is there nothinge but water. But then it ſeemeth to me, that if the water did riſe rounde, the farther the ſhippe were from the lande the higher ſhe ſhould be, and therfore the better myghte be ſeene.

Maſter. Your imagination hath ſmall ground of reaſon : for although the earthe and the water both ioyntlye and ſe- uerally bee rounde of nature, and therefore haue in deed no place hygher then other in their circumference, yet all vul- gar men ſhall thinke by apparance that that place is higheſt wher thei ſtand, & that frō them on eche ſyde ther is a round

M.ij.   deſcent

defcente, vntill by imagination they come to the right contrarye pointe where their Antipodes be, whome they fhall
think to be right vnder thē, wher as thofe Antipodes haue
the contrarye imagination, that they dwell on the higheft
parte of the grounde, and that their fea is hygheft, and fo
bothe defcendeth compaffedlye vnto the contrarye poynte
to them again. and thus euerye other forte of people think
that they dwell on the higheft parte of the londe, and alfo
of the fea, (if they dwell on the fea) and they fhall thynke
that bothe the fea as well as the londe doothe defcende from
them eche waies. As in this circularre forme of the earthe

and fea, the menne that dwell
by A, thinke them felues to
dwell hygheft of all other, fo
that on eche fyde of them the
londe & fea feemeth to defcend,
& therefore they iudge the fhip
that is by B, to bee lower then
they, where as that fhippe, con
trarye waies, feemeth to them
that be in it, to bee on the hygheft parte of the worlde : and
therefore they thinke that the londe by A, is lower then they
are. Againe they that dwell by C, and the fhippe that is by
D, are of like imaginations, eche in his fantafie thinking him
felfe hygheft, and the other lower. And fo of them that
dwell by A and by C, eche meruayleth how the other canne
go, and his headde downewarde : yet in deed none is lower
then other, fith eche of them is equallye diftaunte from the
centre of the earthe, whiche is the loweft place of all other.
and therfore no waye is accompted lower except it be nearer
to that centre. wherby alfo it may appeare contrary to your
fayinge, that although the fea bee rounde, yet fhall not the
fhip feem to afcend ftill, but rather feem to defcend, thoughe
in deed it doth none of both, but moueth circularly about ẙ
centre

centre of the world, fo that it can not aptly be called a right motion, but a compaffed motion that a fhippe maketh, faue that it is tollerably to be borne in vulgare fpeache, bycaufe euery fmall arche of a great circle feemeth to be a right lyne to the fyght of the eye. And in this figure is fomwhat repre-fented the declaration how the compaffed form of the water doth let the fight to fee the fhip, and likewaies how that thei on the londe may fe the toppe of the fhip when they can not fee the hulle, and they in the hulle of the fhip can not fe thofe places on the londe, whiche other in the top of the fhip may fee, by reafon that their fight is aboue the height of the wa-ter. And this may ftande for a conuenient proofe.

Scholar. So dooth it appeare manifeftly, now that my for mer mifconceaued fantafye is reproued. And fo I remember when I haue loked after a fhyp that departed from the porte where I ftoode, firft I loft the fighte of the hulle as thoughe it had fonke into the fea, and yet I faw the toppe ftill. but at lengthe I lofte the fighte of it alfo, as thoughe all had fonke into ỹ water. Which by your declaratiō I perceaue doth folow of the roundnes of ỹ water : for other reafon I can find none.

Mafter. Although you could fynd other reafons neuer fo many, yet this reafon doth enforce that effect. this is ỹ reafon that Ptolemy, Cleomedes, and after them Ioannes de Sacro bofco, and other alfo do alleage, but the fame Iohn hathe an other reafon more phyficall thē geometricall, borowed out of naturall phylofophy, which is this : Seing that the water is a body of vniforme fubftance, the partes of it muft be of lyke condition as the whole bodye is : but the partes of water dooth all wayes couette a rounde forme, (as wee fee in euerye droppe that falleth from any thinge, or ftandeth on any thinge) wherefore of iufte congruence the whole body of the fea and water muft needs couet the fame forme.

A phyficall reafon for the roundnes of the wa-ter.

Schollar. In deede all droppes that fall from the ayer in a mylde rayne, when menne maye marke it, doo fall in a rounde forme, and fo the droppes that fall from the

M.iij. eaues

eaues of the houſe, or from anything els, yea and the drops
of dewe that ſtande vppon anye leaues of herbes, or other
lyke thinge.

Maſter. For a farther experience, fyll anye veſſell brym
full of water, and you ſhall perceaue by tryall, that the water is higher ouer the myddle of that veſſels mouth, then it
is by the brimmes. And againe pour out water on a borde
or on a ſtone, and you ſhall ſoone ſee that it will ſhewe in a
round forme, and will be deeper in the middle, then it is by
the ſides.

Eraſmus
Rheinhold.
Yet farther reaſons there be alleged, whiche were to tedi-
ouſe to repeate : but twoo of them I can not omytte, whiche
are declared by Eraſmus Rheinholt a manne not onlye of
greate learning, but alſo of as greate honeſty in ſeekinge to
profite all men by his trauaile, although ſometime hee wan-
ted leaſure to examine ſome of his writinges, as it may appeare by one of thoſe two reaſons, whiche is this.

An other
reaſon.
By the longe courſe of euerye greate ryuer (ſayth hee) it
maye appeare that the water doth couet a rounde forme, els
could it not ſo much riſe in roundnes, as it doth in running
ſo longe a courſe. for example he bringeth the courſe of the
greate ryuer Danubius, which ſpringeth in the Alpes, bee-
ſyde Vlma in Swicerlande, and entrith into the ſea Euxine,
aboue Conſtantinople, whiche is from Vlma 312 germanye
myles, that is 20 degrees, whiche is the eightenth part of the
whole circuite of the earthe : whereby it muſte needes folow
that the myddle of that ryuer is higher then the fountaines
or the mouthe, by 13 germanye myles (that is 52 englyſhe
myles) in plumbe heighte. for declaration whereof hee maketh this demonſtration linearye, ſuppoſynge A E B C, to
be as one of the greateſt circles about the earthe, whoſe centre is D. this circle muſt be imagined ſo to paſſe agreably to
the courſe of Danubius, that A maye repreſente the fountaines of it, and B the mouthe of it, ſo E ſhall ſtand for the
myddle parte of the riuers courſe. and A E B, for the whole
                                                              courſe

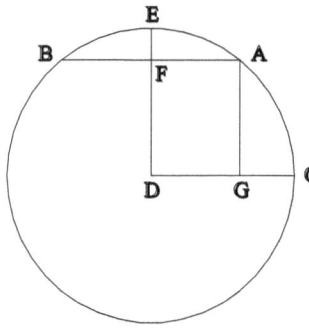

courſe. Now is it ſayd before, that betwene A and B
are 20 degrees, then if you
draw a right line from the
one to the other, as heere
you ſe A F B, it will be lower vnder the myddle of
the arche, by the length of
the line E F, whiche is almoſte the 60 parte of the
ſemidiameter of ẙ earthe,
and maketh iuſtlye 52 en-
glyſh myles, ſumwhat leſſe then 57 : whiche is the 60. part of
the ſemidiameter of the earthe.

Scholar. This reaſon ſeemeth pleaſaunte, but I perceaue
not the reaſon of the iuſte quantitye of the lyne E F.

Maſter That dependeth of the arte of Sines and Cordes
and is very certaine without any ſenſible errour, of whiche
in an other place ye muſte learne the vſe. And in deed as you
ſaye, this reaſon is pleaſaunt and the author muche to bee
prayſed and loued, and as muche is it to be lamented, that
the ſhortnes of his life would not permitte him to haue re-
cogniſed his workes againe : wherfore that he can not do by
preuention of deathe, I truſte ſome of his friendes will do :
for althoughe they be but litle faultes, yet pittye it is that in
ſo good woorkes there ſhoulde remaine any litle ſpottes, as
in this argument there are two, which yet hinder not the argumente. And althoughe it might bee truely ſayde that the
heighte of the myddle of Danubius is not 52 myle, and is
but 36 mile, yet is the forme of his arguemente good, for
that height is ſufficient to proue that the middle appeareth
muche higher then the fountaines of it : the cauſe of this o-
uerſyght was, that hee did eſteeme the courſe of Danubius
to runne by one of the greateſt circles of the earth, which is
not ſo : for it hathe in latitude from the equinoctiall 46 de-

M.iiij.   grees

grees, so muſt the parallele of his courſe bee litle more then
two third parts of the greateſt circle : but as this is ſomwhat
to ſtraunge for you yet beyng vnexpert in the arte of Cordes and Sines, and in the knowledge of Coſmographye,
ſo I wyll lette it paſſe with this lyghte admonyſhmente,
wyſſhynge that hee hadde alſo more aptelye expreſſed hys
meanynge, and the vſe of his termes, for auoidinge of ſlan-
dero_uſe tongues, for it myghte nowe bee anſwered hym,
that Danubius is no hygher in one place, then in an other,
ſeeynge all diſtaunce of heighte is to bee accompted from
the centre : and the middle of the riuer by E, is no farther from the centre D, then is the fontayne A, or the
mouthe B.

    Schola. Marye that obiection is certaine, and therefore is
his errour manifeſt, and his argument of no force.

    Maſter. You triumphe to muche before the victory. his

Erasmus Rheinholt excuſed.

argument is better then you do conſidre it : his intent was to
proue that the water doth not run by a right line and doun-
warde ſtill, as the vulgare ſorte doothe imagine, but that
it runneth circularlye. wherefore it foloweth well againſt the
vulgare opinion, to ſay that the water of Danubius is hygher in the middle of this his courſe, by ſo manye miles in
height plumb vpright, then it ſhuld be by their imaginatiõ
So is there none other fault in this point, but the want of di
ſtinction of the true opinion of highnes and lownes, from
the wronge takinge of the ſame names, wherby thoſe which
do not know his great learning, and myght happen to hear
his argument, wold iudge that other he were wonderfullye
deceaued, other elſe that he did to much abuſe hys tearmes :
but if deathe hadde not preuented him, hee woulde haue
declared his meaninge, I doubte not, as I haue declared it.

Erasmus Rheinholt his ſeconde argument.

    Nowe to hys ſeconde argument. he proueth that there
can be no ſuch holownes in the ſea, as there is betweene two
hylles : for ſeeynge the ſea is a heauye bodye, and preſſeth
towarde the centre of the worlde, euerye parte of it
                                    will

THE CASTLE OF KNOWLEDGE 241

wyll doo the lyke if it be not ftayed. And the water beynge a lyquide and fluxible bodye, can not be ftayed by his owne partes : wherefore it foloweth that there can remaine no va-lyes nor dales, nor hollowe partes in it, but it fhall quickly be fylled with water. and therfore wee fee, that nothinge can be more plainer then is the toppe of water, fyth euery part fo exactly ioyneth with other, in fyllinge vp all vnequalitie : whereof it foloweth, that if the toppe of the water be iufte equall and lyke diftaunte from the loweft part of the world, (which hath been often declared to be the centre of ẙ earth) then mufte the face of the water needes be round, according to the definition of a circle.

Scholar. That foloweth well in deede : for as eche parte of the circumference in a circle is equally diftaunt from the cen tre, fo if all partes of the face of the water be equally diftant from the centre, it muft needes be circular, as the circumfe-rence of a circle is. But if it be fo round, and ought to haue his place aboue the earthe, how doth it happen that it doth not couer the whole face of the earthe? and fo fhoulde there be no earth feene.

Why the water doth not couer all thearth.

Mafter. Haue you forgotten what you readde in Ioannes de Sacro Bofco, for to anfwere that queftion?

Scholar. In deede he fayth that the other three elementes doo compas the earthe round about, faue that for the pre-feruation of man and beaftes, the drineffe of the earth doth withftande the moyfture of the water.

Mafter. That reafon fauoreth more of the determinati-ons theological, then of the demonftrations mathematical, wherfore I will adde therto a proof by good demonftratiõ that it can not compaffe the earthe rounde : for whiche pur-pofe firfte I faye, that the water beinge indofed within the boundes of the earthe, can not be fo greate as the earthe is. Againe confidering that one portion of water being mixed with 4 tymes fo muche earth, wold make it all fofte and flab-by, it may not be thought that the water of the fea and of

That the water
I.
can not cõ-pass thearth
II.

all

III.

IIII.

V.

all ryuers and springes ioyned togither, is so muche as the firste parte of the earthe. Farthermore if you consider the firme stablenes of the earthe, and the vnstable swaruynge of the water, you wolde thinke that if the water were able to matche the twentith parte of the earthe, it woulde make the earthe more vnstable then the nature of the earthe, and the preseruation of earthly creatures could beare. Yea it would be a weak ground to bear so wondrefull a waight as it doth, if the quantity of water were notable, in comparison to the quantity of the earth. Yet now for farther triall, suppose (as I thinke it true) that on the flatte face and circumference of the earthe, there is as muche water as londe, so mighte it appeare that the water were as muche as the londe, as manye men doo affirme.

Scholar. And moste part of learned men (as I haue heard say) do vouche that as a moste certaine truthe.

Master. It is true, as I iudge also, yf they meane lyke cosmographers that halfe the face of the earthe (as I sayde) is couered with water, but then imagine what depthe maye that sea be of.

Scholar. No manne can tell.

Master. Yet by triall of mariners it hath been founde in fewe places, a hundreth fathomes deepe, whiche is litle more then the tenthe parte of a myle.

Scholar. That not withstandinge, it maye bee deeper in some places.

Master. For a suppofition, imagine it were in all places a myle deepe, takinge one place with an other.

Sch. I thinke that to to muche a great deale, consideringe that all knowen partes are not in the deepest, accomptinge one place with an other, as good mariners can testify, aboue 40 fadome, and so groweth shallower still to the shore.

Master. The more that that suppofition excedeth truth, the stronger shall the proofe be of the smalnes of the water in comparison to the earthe.

                                                   Schol.

Scholar. Then for trials fake, I fuppofe it were fo.

Mafter. How deepe thinke you now the earth to be?

Scholar. I remember you faide before, that 57 myle was but the 60 parte of the femidiameter of the earth : then muft the whole earth be in thicknes 6840 myles.

Mafter. That is agreable to that rate : but as I fayde before, the diameter is 6872 $\frac{8}{11}$. And nowe if you abate one fifte parte of that depthe, the reft will make the fide of a cubike forme, almofte as great as the globe of the earthe : as it appeareth in the workes of Geometrye.

Scholar. The fyfte parte of 6872 is 1374. which beyng deducted from 6872 there refteth 5498.

Mafter. That numbre is fomewhat to lyttle, but 5541 is very nigh the fide of a cube, equal to the globe of the whole earthe, therefore multiplye it cubikly, as you haue learned in Arithmetike, and then fhall you fee, howe manye miles fquare are in the whole globe of the earth.

Schol. If 5541 be multiplied by it felfe, it maketh in fquare numbre 30702681, which being multiplied again by 5541, doth yeld 170123555421. which is the cubike numbre to 5541. and fo confequetly muft it be that cube whiche is equall to the earthe, in his whole globe.

$$
\begin{array}{r}
5541 \\
5541 \\
\hline
5541 \\
22164 \\
27705 \\
27705 \\
\hline
30702681 \\
5541 \\
\hline
30702681 \\
122810724 \\
153513405 \\
153513405 \\
\hline
170123555421
\end{array}
$$

Mafter. So is it very nighe. But now for the quantitye of all the fea, this way muft you worke. Firfte to know all the plat face of the earth, you muft mul

$$
\begin{array}{r}
21600 \\
6872\frac{8}{11} \\
\hline
15709 \\
43200 \\
1512 \\
1728 \\
1206 \\
\hline
148450909
\end{array}
$$

tiply his circumference by his diameter, as it is declared in the Pathwaye, and fo will there amounte 148450909 : whiche is the full platte forme of all the face of the earth : wherof prefuppofing (as the truth doth inforce vs) that halfe the fame is fea and water : then dooth it followe, that the whole platte face of the fea and water is

7422-

74225454 myles and a halfe in all togither, which is not the
2000 parte of the earthe.

Scholar. But muſte not this numbre be multiplied by the
depthe of the ſea?

Maſter. Seynge that depthe is not in one place with an o-
ther aboue one myle, and 1 dooth nother multiplye nor diuide, it will remaine as it is.

Scholar. Then dare I thinke farther, that the depthe of
the ſea beynge not a quarter ſo muche generallye, the earth
muſt nedes bee 10000 tymes ſo greate as the ſea, and all o-
ther waters.

Maſter. Your woordes erre not muche from the truthe :
and therfore by this reaſon it doth appear, that the water being ſo little in compariſon to the earth, can not aptlye compas the earthe. And by this it appeareth alſo how childiſh-
lye they doo erre, that thinke the water to bee tenne tymes
ſo greate as the earthe : for if it were but twiſe ſo greate as
the earthe, it muſte of neceſſitye couer all the face of the
earthe : yea I will ſaye conſtantlye, if all the water were
as muche as the hundreth parte of the earthe, it would
ouer runne all the earthe, and couer it cleane : whiche I
maye eaſilye prooue, but not brieflye : and ſeeynge the
ſame thinge is all readye declared in the Pathwaye, I will
omytte it heere, ſyth it is a more appropried proofe for
Geometrye, then for Aſtronomye : and nowe will I returne
to the proſecutinge of our former matters, accomptynge
this ſufficiente for the declaration of the roundnes of the
earthe and alſo of the water ſeuerallye. and now wyll I adde
one reaſon to approue that bothe they do make one perfect
rounde globe.

*That the
earthe and
water to-
gither doo
make a per
fett globe.*
Euerye groſſe and ſounde bodye doth gyue a ſhadow like
vnto his owne forme : the earth is a groſſe and ſound body,
therefore muſte it gyue a ſhadow lyke hys owne forme : but
in all eclipſes of the Mone, which are cauſed by the ſhadow
                                                              of

THE CASTLE OF KNOWLEDGE   145

of the earth, his fhadowe is alwaies conftantly round, whether the fhadow doo runne eafte, wefte, fouthe, or any other
waies mixtly : wherfore it foloweth, that ỹ forme of the earth
is round, whiche giueth that rounde fhaddow.

Scholar. How fhall a man vnderftand that the fhadowe
of the earthe is rounde?

Mafter. In the edipfe of the moone, other all the mone is
darkened, or els but one part of hir : If all the mone be darkened, then doth the darkenes begin on the eafte fyde of the
moone in circularre forme, and encreafeth ftill in the fame
forme, tyll all the whole moone be edipfed, and then decreafeth the darkenes againe, fo that the wefte fyde of the mone
is darkened, but the darkenes vadeth by lyttle and litle, and
yet ftyll in circularre forme. And if the moone be darkened
only in one parte, whether it be the fouth part, or the north
parte, yet ftill is the fhadowe round in forme : where as if the
earthe were fquare or cubike, other three cornered, or of o-
ther fuche forme, the fhaddow wolde fo appear in the mone
as by the thirde and fourthe figure, you maye partlye
perceaue.

Examples of the firfte forme where all   *Eample of the thirde and fourthe*
the moone is eclipsed at the full eclipse.   *formes.*

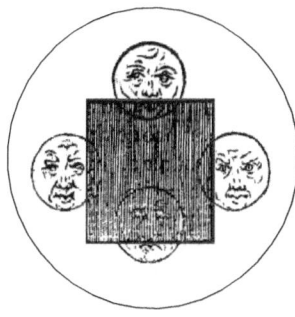

N.i.   But

Examples of the thyrde and fourth   Examples of the other two fortes, of one
              formes.   parte eclipsed.

The fouthe parte.

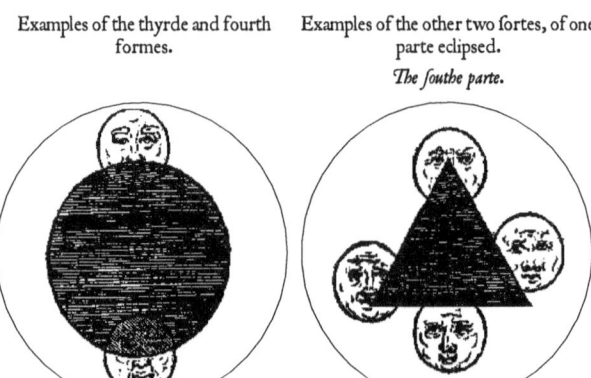

The northe parte.

That the earth is but a pricke in respecte of the skye. But I will omitte this matter tyll anone, bicaufe it is not eafye to vnderftande without farther explication of other matters incident therto. And bicaufe I haue begon to fpeak of the fhaddowe of the earthe : I will alleage one argument more, taken by the fame fhaddowe to approoue the fmalnes of the earthe in comparifon to the fkie. wherfore thus I frame mine argument.

The Sonne is but a very fmall portion in comparifon to the whole fkie, and yet the Sonne is manyfolde bigger then the earthe : wherfore the earthe mufte needes bee but a verye fmall thinge in comparifon to the heauens.

Scholar. Your argument is good, and the maior is manifeft to euery mans fight : but how do you proue the minor?

Mafter. Euery darke body giueth fhadow accordinge to the quantitie that it beareth to that fhyning body, which giueth the light, fo that if the fhining body be equall to y̆ dark body, thē doth the fhadow run in form of a piller, or of a rol ler, like byg at both the ends : but if the bright body be grea ter then the dark body, then doth the fhadow growe leffer & leffer

THE CASTLE OF KNOWLEDGE   247

leſſer in ſpyre forme, or taper faſhion, and at lengthe doth ende in a ſharpe pointe. Contrarye wayes, if the lyghte bodye be leſſer then the darke bodye is, then doth the ſhaddow grow greater and greater, ſtill as it goeth from the dark body, and is ſmalleſt at the beginning, contrary to the taper forme, whiche is greateſt at the beginninge : and this forme maye be called maundforme, or bell forme, bicauſe it is like a maunde baſket, or a bell.

Examples of theſe thre diuers ſhaddows.

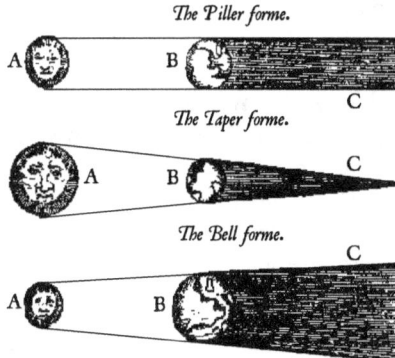

The Piller forme.

The Taper forme.

The Bell forme.

A repreſenteth ẙ ſon or other lyght body. B the earth, or any dark body, and C the ſhadow.

Scholar.   This may ſtand as a ſure maxime, ſith both reaſon & ſenſe doo teſtify it to be tru.

Maſter. Then do I infer farther : that if the ſonne were leſſer then the earthe, the ſhadowe of the earthe would grow greater and greater, and would be infinite in lengthe : wherby it wold darkē the moſt parte of the ſtarres, euery night : & very often it wold ſhadowe ẙ mone, and that for a lōg ſpace togither. as you may gather by this figure, wher A repreſēteth ẙ ſon in leſſer form then the earth, which is ſignified by ẙ circle marked with G, & ẙ ſhadowe that cōmeth by this form, is marked with

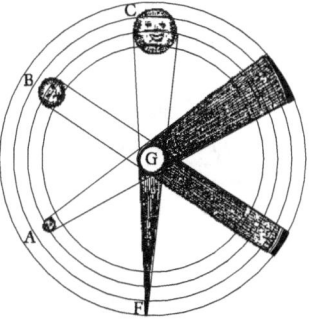

D, which occupieth a great part of the ſkye, and therefore

N.ij.

woulde darken all the ſtarres in ſo muche ſpace of the ſkye, which is nyghe hande a quarter of that hemiſphere that is aboue our horizon. And as the ſhaddow tourneth about accordyng to the motion of the Sonne, ſo in four and twen-tye howers all the ſtarres that be nyghe vnto the zodiake, ſhould ſuffre eclipſe : whiche thinge is contrary to dayly ex-perience, for wee ſee there (about the zodiake and againſte the ſonne) the ſtarres very bright.

Scholar. This reaſon doth ſuppoſe, that the ſtarres do re-ceaue their light of the ſonne, which thinge was not yet pro-ued by you, althoughe I thinke it to be true, yet in a good argument, no doubtfull ſentence may be alleged.

Maſter. Then ſeing this place doth not conueniently per mit ſo longe a digreſſion to prooue that, I will vſe the mone for an example, which appeareth ſo manifeſtlye to borrowe her lyghte of the Sonne, that according as ſhe receaueth the lyghte from him, ſo dooth ſhee appeare greater or leſſer in lyght, according to hir diſtance from him. and when ſo euer ſhe commeth into the ſhaddowe of the earth, ſhe leeſeth her lyght, other fully or in part, accordingly as ſhe paſſeth and toucheth the ſhaddowe of the earthe. wherefore as longe as the moone ſhoulde be within that ſhaddow, ſhe muſt needs be in the eclipſe : and the ſhaddowe beinge ſo great, ſhe ſhuld be eclipſed not only euery moneth at the full, but ſhe ſhould continue almoſte foure dayes to gither in that eclipſe, ſeing that ſhaddowe dooth occupye as muche of the ſkye, as ſhee doth moue by hir propre courſe in foure dayes.

Schol. That abſurditie is to manifeſt to graunt vnto : and yet the greatnes of the ſhaddow inferreth no leſſe, ſyth it oc-cupyeth ſo muche of the ſkie.

Maſter. The like inconuenience will follow, if the ſon and the earth were both of one greatnes, as are B & G in the for-mer figure, for ſo wolde the ſhaddow run of one bignes like a roller, as is repreſented by E, and wold darkē diuers ſtars, and namelye all that bee in the myddle of the Zodiake, and

the

THE CASTLE OF KNOWLEDGE   249

the moone fhould both oftener be edipfed (then in deed fhe is) by the greatnes of the fhaddowe, and wold tarry longer in the edipfe, by that fame reafon, then good reafon wold allowe. But feing we perceaue no ftarres directlye againft the fonne to be edipfed, nother yet the mone, in fuche forme as that pyllerlyke fhaddow would caufe, we muft needes thinke that the fhaddowe is muche abated, beefore it come to the fphere of the moone, and is cleane confumed before it come at anye of the ftarres, whiche kinde of abatement could not be, but where the light is much greater then is the body that maketh the fhaddow, as is C in comparifon to G.

Scholar. So muft it followe, that feyng the Sonne is the lyghte body, and the earthe giueth the fhaddowe, of necef-fitye the Sonne mufte be greater then the earthe.

Mafter. Yea in deede, and that manye folde.

Scholar. Then of more force mufte the earthe bee a verye fmall body in refpecte to the whole fkye, which is infinitely greater then the fonne, as euery childe may perceaue.

Mafter. Yet haue I farther matter of profe, that the earth is not only a very fmall bodye in regarde to the fkie, but is without anye vewe of greatnes in that comparifon.

If the earthe had anye notable quantitye in refpecte of the fkye, then mufte the diameter of the earthe haue as greate a quantitie, in comparifon to the diameter of the fkie. for as in twoo circles the proportion of the diameters is equall to the proportion of the circumferences, fo is the proportiō of the fhorter to the longer, greater then is the proportion of their two platte formes : but in two globes the proportiō of the fhorter diameter to the longer, is muche greater then is the rate of their platte formes : and yet muche more great ter then the proportion of the leffer globe to the bygger. *The fecond reafon for the quanti-tie of the earthe.*

Scholar. That is fufficiently proued in Geometry, wher-fore you may proceede with your conclufion.

Mafter. If the diameter of the earth haue notable quantity in cōparifon to the diameter of the fkie, then the ftars which
N.iij.   are

ar ouer our headdes, be nygher vnto vs by a notable quantitie, then when they be in the eafte, or in the weft.

Scholar. In deede they are nearer by the femidiameter of
the earthe : whiche of it felfe mufte needes bee accompted a
notable quantitie.

Mafter. But if it fhall be fo accompted in regarde to the
halfe diameter of the fkie, then muft the ftars ouer our heds
feeme bigger by a notable quantitye, then when they are in
the eafte or wefte.

Scholar. That reafon is not only approued by Geometrye, but alfo by cōmon fight and daily experience, that the
nigher any thing is to the fighte, the greater it feemeth : and
the farther from the fighte, the leffer it fheweth.

Mafter. There is no fuche diuerfity perceaued in the quantitie of the ftarres, but that they appeare ftyll conftantly of
one bignes : wherfore it muft follow, that their diftance is all
one in all partes of the fkye, and then doth not the femidiameter of the earth make anye notable diuerfitie in diftance :
wherefore it muft be thought that the quantitye of it is not
fenfible in comparifon to the femidiameter of heauen, nother the circumference of it in comparifon to the circumference of the fkye, and muche more may not the whole quan
titye of it bee accompted fenfible in refpecte to the whole
quantitie of the worlde.

Schol. That foloweth well : for as I learned in Geometry,
if the diameters of any two Globes, be in fuche proportion
that the greater do contain the leffer a thoufand times, then
be their circumferences in the fame rate : but the platte forme
of the greater, is 1000000 folde greater then the leffer : and
the whole fubftance of the bigger globe, doth containe the
fmaller globe, 1000000000 tymes.

Mafter. Vndoubtedly it maye bee perceaued by fight as
well in dialles, as other greater inftrumentes made for obferuations, that the femidiameter of the fonne his fphere is
more then a thoufand times longer then the femidiameter
                                                      of the

THE CASTLE OF KNOWLEDGE   **251**

of the earthe, els wolde not the fhadowes agree fo exactly as
they do : for they moue as duely and ordrely about the centre of all fuche inftrumentes, as if their centre were the very
centre of the world, which thinge could not be, if thofe two
centres dyd differ notably, in refpecte to the fphere of the
Sonne. And if it were not, that an introduction dooth not
admitte the exacte proofes of the arte, I could herby declare
the proportion of thefe two femidiameters fo exactly, that
you fhould confeffe that proofe to bee righte certaine and
good. But now wil I procede to the declaration of this third
reafon by linearye demonftration, although it be fomwhat
obfcure, without other helpe.

In this figure, which reprefenteth the three notable circles
in a diall, that bee made by the *The thirde*
courfe of the Sonne, in the thre *reafon.*
notable places of the zodiake,
that is in the two tropikes and
in the equinoctiall, the vttermofte arke **BLC**, reprefenteth
the tropike of Capricorne, and
is heere made no byger, then
the qnarter of a circle, bycaufe
the Sonne doth fhine but fyxe
howers vnto vs, when hee is in

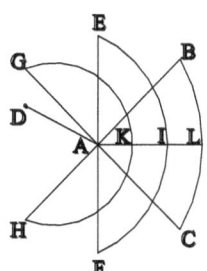

ẏ figne. the equinoctiall is fet as halfe a circle, bicaufe the fon
being in it, doth fhine to vs 12 howers, and is here limited by
E I F. The tropike of Cancer containeth thre quanrters of a
circle, bicaufe that when the Sonne is in it, then is there 18 ho
wers from Sonne rifing to fonne fetting : and that circle here
is fignified by G K H. The centre of this diall is A, and the
ftile that giueth the fhaddow is DA, whofe toppe being D,
doth defcribe thofe cantylles of circles, in fuche precifenes,
as if that diall ftood in the centre of the earth. and like waies
the diftinction of the howers is fuche exactlye in that diall,
as if the centre of the diall, wer the very centre of the world.

N.iiij.   Schol.

152   THE FOVRTHE TREATISE OF

Scholar. I do conceaue good reafon of profe hereby, but yet I thinke I fhall perceaue muche more, when I fhall vnderftande the iufte vfe of thofe dials, as well as of other feuerall inftruments of lyke vfe.

Mafter. You fay truthe : and therefore wyll I paffe from this thirde reafon, and come to the fourthe proofe, whiche is thys.

The fourthe reafon for the smalnes of thearthe. If the earthe were of anye bygnes in comparifon to the worlde, then fhoulde his femidiameter beare fome vewe of byggeneffe to the femidiameter of the fkie. and fo confequently the horizont that we haue on the ouer parte of the earthe, fhould not diuide the fkie into two equall partes, for that part which fhuld be vnder the horizont, would alwaies be the greater, and the leffer parte aboue the horizonte, as

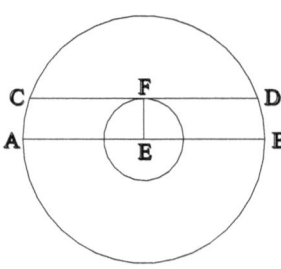

in this figure it doth appear. where A C D B is the circle of the fkie, and the leffer circle is the earthe, the centre E, being cōmon centre to them bothe. and E F is the femidiameter of the earthe, as E A is ỹ femidiameter of the fkye. Nowe if E F bee notable in quantitie in comparifon to E A, then will the line C F D (beyng the horizonte on the toppe of the earth) differ notably from the line A E B, beynge the diameter of the worlde, and the horizonte to the centre of the earthe. And fo fhall not that horizont C F D diuide the worlde into two equall halues, but the ouer part aboue the horizonte fhall be leffer then the other parte that is beneth the fame horizonte, whiche thing is contrary to daily experience, and to all obferuations : for we may fee in the longe winter nights thofe ftarres that be in the horizont in the eafte at the beginning of the nyght, to be in the fame hoirzont in the wefte, at the ende of twelue howers : and con

trarye

THE CASTLE OF KNOWLEDGE   153

trarye waies thofe ftarres that did fet in the weft, when thofe
other did rife in the eafte, fhall rife againe when the other do
do fet. And fo of the fonne and the moone when they be in
contrarye pointes of the Zodiake.

Scholar. That is at the full of the moone.

Mafter. In deede then are they right oppofite the one a-
gainft the other : but if the moone be at the full, long before
the fonne fetting, then will fhe rife fomewhat after the fame :
and contrary waies if fhe be at the ful after the fonne fetting,
then will fhe rife fomwhat fooner, by reafon that fhe moueth
eaftwarde euery hower 33 degrees. And although vnto them
that be meanly acquainted with the motions of the planets,
the declination of the moone and hir latitude, may occafion
fome doubtefulnes to rife, yet vnto the learned, thofe many
folde varieties in the motion of hyr and thother planets, do
confirme the principles of aftronomy more adfuredly : but
this will I omitte tyll an other more conuenient tyme.

Scholar. This is well proued nowe, that the earth in com
parifon to the whole world is but as a pricke or a mote, and
lykewaies in comparifon to the other fpheres.

Mafter. You mufte except the fpheres of the thre planets
whiche bee beneth the
fon. for vnto them the
diameter of the earthe
beareth a notable quan
tity : for the femidiame-
ter of Venus Sphere,
is but 167 tymes fo
long as the femidiame-
ter of the earth : and the
femidiameter of Mer-
cury his fphere is fhor-
ter muche, for it is litle
more then 64 times the
femidiameter of the
earthe

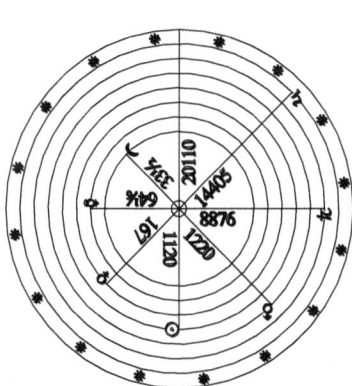

earthe, but the moone hath hir femidiameter only 33 tymes
and a halfe longer then the earthes femidiameter : all which
proportions with the refidue, I haue fet forth in this figure,
wherby you may perceaue, that vnto ỹ femidiameter of ech
fphere, is annexed the numbre that importeth howe often it
containeth the femidiameter of the earth. that is to fay : the
fonne his femidiameter containeth it 1120 times, Mars 1220
times, Iupiter 8876 tymes, Saturne 14405 tymes and the
eight fphere or ftarry fkie, 20110 tymes.

Sch. I remembre that Faber on the Sphere doth accompt
thofe diftances by miles, which is a pleafant matter to read.

Ma. In that place Faber foloweth the accompt of Alphra
ganus the Arabitian, which fpeaketh of myles much longer
then the Italian myles be : for 6 of the Italian miles do make
but 5 of Alphraganus miles : of which diuerfity at an other
tyme I will inftructe you, namely in the treatife of Cofmographye : where I wyll fet forth diuers varieties and appearante repugnances of fondry writers, for the meafuringe of
the earthe : and proue it to be a difagrement more in wordes
then in meaning : and to come by reafon of their diuers miles, or other inconftant meafures. And bicaufe you like that
table fo well, lo heere is an other drawen accordinge to the
rate of 60 myles to eche degree. But heere by the compas is
vnderftande the inner concauitie of eche fphere.

The eyght Spheres.	The myles that theyr femi-diameter containeth.	The myles of euery fphere in compass.
☽ The Moone.	115278	724604 $\frac{4}{7}$
☿ Mercurye.	220500 $\frac{2}{33}$	1386000 $\frac{4}{235}$
♀ Venus.	573872 $\frac{8}{11}$	3607200
☉ The Sonne.	3848367 $\frac{3}{11}$	34189737 $\frac{1}{7}$
♂ Mars.	4192363 $\frac{7}{11}$	26352000
♃ Iupiter.	30501163 $\frac{7}{11}$	191721600
♄ Saturne.	49500818 $\frac{2}{11}$	311148000
The eight fphere.	69105272 $\frac{8}{11}$	434376000

And

And his conuexitie or vtter compas is equall to the conca-
uitye of the nexte fphere aboue it.

Scholar. If the whole circuite of the fkye bee 434376000
myles, and the fame compaffe is 360 degrees, then mufte it
needes follow, that euery degre of that fky
contayneth iuft 1206600 miles, as by diuifion it may be fufficiently well proued. But
howe is this fuppofition of diftaunces approued to be true?

Mafter. That profe dependeth of more
knowledge, then this introduction teacheth, and therefore
muft be referred to a higher treatife. But in the meane ceafon admitting this fuppofition, you maye eafilye tell, howe
manye myles the fonne and the moone are in breadthe, fee-
inge eche of them is accompted about 31 minutes by theyr
diameter, eche in the myddle of his owne fphere.

Scholar. Nowe I vnderftande the forme of woorkinge
for tryall of this matter. Fyrfte I mufte fearche how manye
myles make a degree in eche of thofe fpheres, and then take
a parte proportionable of that nũbre
agreable to 31 minutes & a halfe. Therfore to begyn with the fonne. As his
whole fphere in the middle is in compas 25270868 myles, fo tryinge it by
diuifion, I fynde that euerye degree in
that fphere doth containe 70197 miles
nygh hande. Then fay I by the golden
rule, if 60 minutes (whiche make one
degre) do require 70197, what doo 31
and a halfe make? After iufte multipli
cation and diuifion, as that rule dooth
importe, I fynde the whole diameter
of the fonne to containe in myles,
36853 : where as the earth (as before is
noted) dooth containe in his diameter
but

but 6872 myles. So that therby it appeareth, that the fonne is more then 5 tymes fo broade as the earthe is ouerthwart.

Mafter. That is well limited. for els if the flat of the greateft circle of the whole earthe myght appeare vnto vs, as the flatte forme of the fonne doth, the flatte forme of the fonne ought to be accompted about 29 times fo great as the earth is, in lyke forme. And the whole globe of the fonne mufte needes be about 155 tymes fo greate as the earth in his whole Globe.

Scholar. I perceaue that dooth followe by twoo rules of Geometrye, wherof the firfte is this.

In what proportion fo euer the fides of any twoo fquares be, thofe fquares are in the fquare of that proportion : fo that if the fides be as 2 to 1, the fquares are as 4 to 1 : and if the fydes be as 3 to 1, the fquares are as 9 to 1. &c. The feconde rule is this : In what rate fo euer the fydes of any cubes be, the cubes do beare the lyke rate cubikly multiplied. as if the fydes be as two to one, the cubes are as 8 to 1 : and if the fydes be as thre to one, the cubes are as 27 to 1. &c.

Mafter. This is well applied of you, that you can frame your common rules in Geometry to fuche fpeciall matters. And nowe may you proue the lyke in the moone.

Sc. You fay, that the circumference of the fphere of ẙ mone is 724604 myles, and $\frac{4}{7}$: then diuidyng it by 360, ther wil amount the quantitie of one degree : whiche yeldeth in this rate 2012 myles and $\frac{72}{90}$: but accomptinge the breadth of the moone 31 minutes and a halfe, the myles that anfwere vnto it, are but 1057 : wherby it foloweth, that the diameter of the earthe being 6872, is 6 times and a halfe greater then the diameter of the moone. And therfore the flatte of the earthe in his greateft circle, is aboue 42 tymes fo greate, as the like flatte forme in the moone : and the whole globe of the earth is 273 tymes fo greate, as the whole globe of the moone.

Mafter. In this accompt you take the innermoft circumference of the fphere of the moone, and in the like accompt

                                                                        manye

manye other take the vttermofte circumference, but it appeareth more reafonable to take the myddle diftaunce bee-
tweene them bothe, which is 1055302. (as 1386000
here by example dooth appeare) and in 724604
that place of diftaunce to take the rate of ——————
hir diameter.   2110604

Scholar. So it feemeth moft indifferent $\overline{\quad 2\ \ 222\ \ 2\quad}$
reafon. And then the meafure of one degree wyll be 2931 $\frac{71}{180}$
and of that there will aunfwere to the diameter of the mone
(being accompted 31 minutes and a halfe) 1539 myles. Nowe
if I diuide the diameter of the earthe (whiche is 6872) by it,
there wyll be in the quotient 4 and a halfe almoft : fo wyll it
appeare that the diameter of the earth is 4 times and a halfe
almofte fo longe as the diameter of the moone : and the flat
of the earth 20 times fo large as the flat of the moone. And
the whole earthe nynetye tymes fo greate as the globe of
the Moone.

Mafter. Yet according to the common accompt, the earth
is but 39 tymes fo muche as the moone : but hereof and of
many other thynges that feeme aboue the reache of mannes
witte, I will an other time inftructe you farther. for it is no
meete mater for an introduction. And thys is broughte
for exaumples fake onlye, that you myghte vnderftande
the ordre of fuche forte of woorkynge, and therby learne
to trye your authors fayinges. But nowe it is tyme
to proceede to other matters, and to declare the true place
of the earthe, and to prooue that it ftandeth in the
myddle of the worlde, whiche thinge althoughe it may fu- *That the*
ficientlye bee gathered by that that is written beefore, yet *earthe is in*
I wyll declare certayne inuincible reafons for confutation *the middle*
of them that myffeplace it. And to begyn with all, there *of the*
can be but three dyuerfities of places in generall, without *worlde.*
the centre of the worlde : for other it mufte bee befide the
Axe tree of the worlde, and yet equallye diftaunte from
bothe the Poles, or els it mufte bee on the Axe tree of
                                        O.i.   the

the worlde, and yet nearer to one Pole then to an other : or thyrdlye it mufte bee befide the Axtree of the worlde, and alfo nearer to the one Pole then to the other. befide thefe three varieties there is lefte but one more (whyche is the true placynge of it) and that is to be on the Axe tree of the worlde, equallye diftaunte from bothe the Poles : wherefore if the firfte three opinions bee reproued as falfe, this fourthe muft needes remaine as onlye true. And nowe for the confutynge of the three fyrfte opinions I will vfe Ptolemyes argumentes, augmentyng them with a larger explication.

The confu-   If the earthe were out of the centre of the worlde, and yet
tatiō of the   ftode in the middle betweene bothe the Poles, then fhoulde
firft opiniō.   not the Horizonte cutte the fkye into twoo equall halues. And thereof woulde followe, that in the righte fphere the daye and the nyghte fhoulde not be of one lengthe. As for

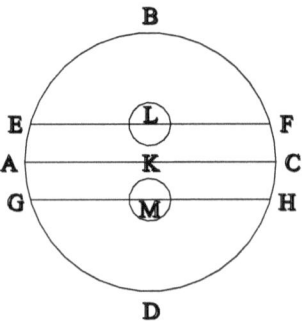

example : If you would ima-gine the earthe to ftand as L dooth in this figure, then woulde the Horizont be the righte line E L F, and fo the parte that is vnder the Hori-zont is greater then the other parte of the fkye aboue the Horizonte : wherefore in the ryghte Sphere the nyghte mufte needes alwaies be lon-ger then the daye. but if you would imagine the earth to ftand where M, is fet vnderneth K, which is the verye centre of the worlde, then woulde that Horizonte G M H, whiche anfwereth to that centre, be vn-der y̆ true horizont of the centre of the world, that is y̆ right line A K C. And fo fhoulde the nighte alwaies in the righte fphere be fhorter then the daye, bicaufe the greater parte of the fkye is aboue the Horizonte, and the leffer parte vn-

der

der it. And by the like reaſons in al other bowing ſphers ther
ſhoulde bee no equalitye betweene the daye and the nyght :
and if there were any it ſhould not be in that time when the
ſonne were in the iuſte middle betweene the twoo Tropikes,
(that is vnder the Equinoctiall line) bicauſe that the Equinoctiall line is not equally parted by the Horizont, but the
greater parte is aboue the Horizont, after the one ſuppoſition, and after the other ſuppoſition it is vnder the Horizont of the earthe.

Scholar. This I doo vnderſtande well, accomptinge the
circle A B C D, to repreſent the Equinoctiall lyne.

Maſter. And farther you may perceaue (as all men, in all
ages, and in all nations do confeſſe) that the increaſe of the
dayes from the ſhorteſt to the meane, and from the meane
daye to the longeſt are not onlye agreeable betweene them
ſelues, but are lyke alſo exactlye to the decreaſe of the daies
from the longeſt to the meane, and from the meane to the
ſhorteſte. whiche thynge coulde not bee, excepte that
the myddle circle betweene the twoo Tropikes (whiche is
ryghtlye called the Equinoctiall circle) were equallye dyuided by the horizonte into twoo iuſte halues. And farther : ſeeyng there can be no poſition of ſuche obliquity (except it be righte vnder the Pole) but ſome one circle of the
Sonnes courſe muſt be diuided equallye into two partes by
the Horizonte, ſo that when the Sonne were in that circle,
the daye woulde be equall with the nyght : which thing as all
nations confeſſe, happeneth at one tyme to all menne, and
that is when the Sonne is in the beginning of Aries or Libra, preciſely vnder the Equinoctiall lyne : wherefore not
onlye that circle dooth ryghtly agree with hys name, but
alſo it foloweth that the ſame Equinoctiall line is equallye
parted into twoo iuſte partes by the Horizonte. And therefore the earthe muſte needes bee iudged to bee in the centre of the worlde.

Farthermore, if the earthe were ſuppoſed to bee to-

160　　　THE FOVRTHE TREATISE OF

*An other cōfutation of
that firſte
opinion.*
ward the eaſte or toward the weſte, from the myddle of the
world, (as in this figure it is ſet toward the eaſte, which is limited by A) thē as the ſpace
toward the one ſide is ſhorter thē the ſpace to the other
ſide frō the earth, ſo the ſtars
woulde ſeeme bigger in that
nearer part, and leſſer in that
farther parte.

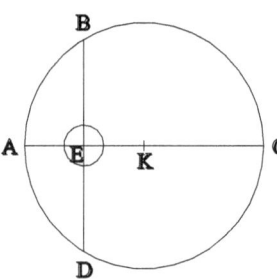

Sc. Which thing is before
reproued, and by daily experience may be confuted.

Maſter. Therfore can not
it be a true opinion, that inferreth ſo falſe a conduſion. And
yet there woulde follow of it more abſurditie : that from the
morning vntill noone ſhould bee ſhorter tyme, or els longer then from noone vntill nyght.

Scholar. That muſt needs folow alſo, ſeeyng that noone
is that time of the daye, when the ſonne is in the cirde which
goeth right ouer our headdes from ſouth to north, whiche
here in this figure is repreſented by the right line B E D, as
I gather by your former doctrine.

*An abbridged argument of all
the premiſ
ſes.*
Maſter. You geſſe well. and by the contrarye of all theſe
you may conclude thus : that ſeyng the tyme before noone
is equalle to the tyme after noone, and the ſtarres appeare
nother bygger nor leſſer in the weſte, then they doo in
the eaſte : And that when the ſonne is in the Equinoctiall
lyne, the dayes are equall to the nightes, it foloweth cer-
tainlye, that the earthe canne bee no wayes out of the Axe
tree of the worlde.

*Againſt the
ſecond opinion.*
And now for the ſeconde opinion I reaſon thus.

If the earthe were on the Axe tree of the worlde nygher
to the one Pole then to the other, then woulde the Horizonte onlye in the righte Sphere dyuide the ſkye into
twoo

THE CASTLE OF KNOWLEDGE   261

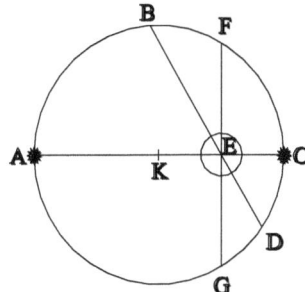

twoo equall partes, and in
no forme of bowing fphere,
as by this figure you may ga
ther, wher E ftandeth for the
earth, and A E C for ẙ right
horizont. B E D and F E G
for two oblique horizontes,
in 2 feuerall bowing fphers :
and K limiteth the centre of
the worlde.

Scholar. Here I fee mani-
feftly that only the right horizont dooth diuide the greater
circle (whiche is fette for the fkie) into 2 equall partes, and
none other : wherby it would folowe, that wee whiche dwell
52 degrees northwarde from the Equinoctiall lyne, fhoulde
fee muche leffe then halfe the fkye : but that is falfe, as it hath
beene often tymes proued, wherfore I perceaue that opinion can not be true.

Mafter. Yet an other argumente againfte that opinion, *An other*
may this be. Yf the earthe were nygher to the one Pole then *argumente*
to the other, when the Sonne is in the iufte eafte, the fhad- *againft the*
dowes of anye thinges in earthe, woulde not runne full *fecond opi-*
wefte : but all fhaddowes in earthe runne full wefte, when *nion.*
the Sonne is iufte eafte : (and contrarye wayes) therefore
canne not the earthe bee nygher to one Pole, then to the
other.

Scholar. This argumente is good, and the minor is
well knowen to euerye fenfible man : fo is there no doubte
but of the maior.

Mafter. For the proofe of it, I fette this figure.
Wher the great circle A B C D betokeneth the Horizont,
and the leffer circle E F G H, ftandeth for the earthe. The
centre of the worlde is E : the eaft is D : and the wefte is B :
the fouthe is A : and the northe is C. In the earthe the
lyne F G, ftandeth as a Parallele, wyth the ryghte
                                    O.iij.   line

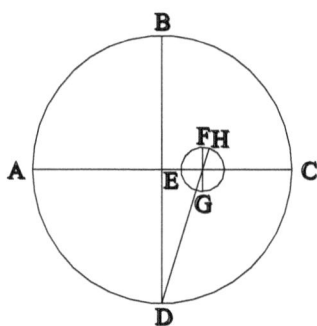

line B E D, and the righte
line D H rūneth croffe thē
bothe, and maketh an an-
gle on the centre of the
earth, equal to the angle by
C D : whofe largenes is agreable to the imagined diftāce
of the centre of the earthe
frō the centre of the world.
wherfore the greater that ȳ
diftance is, the larger is the
angle of that declination, and the leffer diftaunce, caufeth a
leffer angle : but yet if the diftaunce be any thing, then will
that angle of declination be notable inoughe.

Scholar. The refte is eafye to confidre : I meane that all
fhaddowes runne in a right line from the lyght bodye, that
caufeth that fhadow : fo that the fonne being in D, which is
the iufte eafte, wolde caft the fhaddowes in the earthe, not to
F (which is the weft in the earth) but to H, which is almofte
northweft : and therefore is your maior duely proued, and
the feconde opinion fully confuted : but how may the thirde
opinion be anfwered?

Againft the thirde opi-nion.

Mafter. The thirde opinion is, that the earthe ftandeth
out of the axe tree of the worlde, and alfo nearer to the one
pole then to the other : fo doth it containe both the other o-
pinions : wherfore feyng they both are reproued, this third
mufte needes feeme falfer then ony of them bothe, bycaufe
it indudeth all the vntruthe of them bothe. And therfore to
conclude with Ptolemye, the increafe and decreafe of dayes
coulde neuer be fo ratable and iuftly proportioned as they
be, if the earthe ftoode any where els, then in the very centre
of the worlde. And farther more the eclipfes of the moone

A confir-mation.

fhuld not happē, (as now they do) at the precife hour of ful
oppofition, if the earthe were not in the very centre of the
worlde : for confidering that all the thre bodies of the Son,

An other reafon.

the

THE CASTLE OF KNOWLEDGE   263

the moone, and the earthe mufte needes be in one right line
(as in the doctrine of thofe eclipfes it is taught) there is no
place in the worlde, where the earth may ftand in that right
line common to all fuche eclipfes, but only the centre of the
worlde : as for examples fake I haue noted 4 feuerall eclipfes

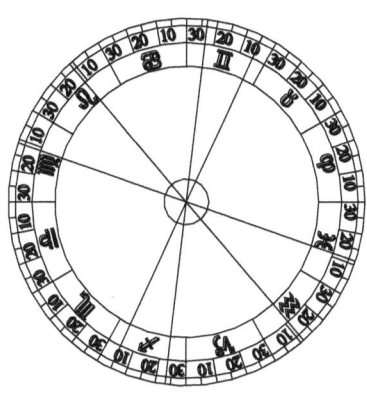

of the moone : the firft
was in ẙ year of Chri-
ftes incarnation 1551,
the 20 day of Febru-
arye, when the Sonne
was aboute the 12 de-
gree of Pifces, and the
moone aboute the 12
degre of Virgo. The
feconde eclipfe was in
the yeare of 1553, the
fonne being in the ele-
uenth degree of Leo,
and the moone in the
eleuenth degree of Aquarius : The thirde eclipfe happened
on the fifte daye of Iune, 1555, the fonne being in the 23 degre
of Gemini, and the mone in the 23 of Sagittary. The fourth
eclipfe, fhalbe this yeare 1556, the 17 daye of Nouembre, at
whiche time the fonne fhalbe in the fifte degre of Sagittary,
and the moone in the fifte degree of Gemini. Nowe if you
lyfte to take more examples, for farther tryall you maye fo
doo. yet two feuerall eclipfes ferue as well for this proofe as
10000. And then drawing lines for eche eclipfe frõ the place
of the fonne to the place of the moone, all thofe lines mufte
needes paffe by the earthe, and there is none other pointe,
whereby they all (or any two of them) can paffe, but onlye
the centre of the Zodiak, (which is the centre of the world)
therefore mufte that centre of neceffitie bee accompted the
place of the earthe. And this may fuffice for this time touchinge the earthe and his accidentes, principallye appertai-
O.iiij.   nynge

ninge to Aftronomye : for althoughe manye other thinges
are to bee confidered in it, they appertaine rather to philofophers or Cofmographers, then to Aftronomers, and
namely in the doctrine of the principles. As touching the
diftinction of the zones, I haue fayde fomwhat before, & fom
what more wil I fay anon. But as for the quietnes of the earth
I neede not to fpende anye tyme in proouing of it, fyth that
opinion is fo firmelye fixed in mofte mennes headdes, that
they accopt it mere madnes to bring the queftion in doubt.
And therfore it is as muche follye to trauaile to proue that
which no man denieth, as it were with great ftudy to difwade
that thinge, which no man doth couette, nother any manne
alloweth : or to blame that which no manne praifeth, nother
anye manne lyketh.

Schol. Yet fometime it chaunceth, that the opinion moft
generally receaued, is not mofte true.

Mafter. And fo doo fome men iudge of this matter, for
not only Eraclides Ponticus, a great Philofopher, and two
great clerkes of Pythagoras fchole, Philolaus and Ecphan
tus, were of the contrary opinion, but alfo Nicias Syracu-
fius, and Ariftarchus Samius, feeme with ftrong arguments
to approue it : but the reafons are to difficulte for this firfte
Introduction, & therfore I wil omit them till an other time.
And fo will I do the reafons that Ptolemy, Theon & others
doo alleage, to prooue the earthe to bee without motion :
and the rather, bycaufe thofe reafons doo not proceede fo
demonftrablye, but they may be anfwered fully, of him that
holdeth the contrarye. I meane, concerning circularre motion : marye direct motion out of the centre of the world,
feemeth more eafy to be confuted, and that by the fame reafons, whiche were before alleaged for prouing the earthe to
be in the middle and centre of the worlde.

Scholar. I perceaue it well : for as if the earthe were alwayes oute of the centre of the worlde, thofe former ab-
furdities woulde at all tymes appeare : fo if at anye tyme
the

Whether
the earthe
moue or
not.

the earthe fhoulde mooue oute of his place, thofe inconueniences would then appeare.

Mafter. That is trulye to be gathered : howe bee it, Copernicus a man of greate learninge, of muche experience,
and of wondrefull diligence in obferuation, hathe renewed
the opinion of Ariftarchus Samius, and affirmeth that the
earthe not only moueth circularlye about his owne centre,
but alfo may be, yea and is, continually out of the precife cētre of the world 38 hundreth thoufand miles : but bicaufe the
vnderftanding of that controuerfy dependeth of profoun-
der knowledg then in this Introduction may be vttered con
ueniently, I will let it paffe tyll fome other time.

Scholar. Nay fyr in good faith, I defire not to heare fuch
vaine phantafies, fo farre againfte common reafon, and repugnante to the confente of all the learned multitude of
Wryters, and therefore lette it paffe for euer, and a daye
longer.

Mafter. You are to yonge to be a good iudge in fo great
a matter : it paffeth farre your learninge, and theirs alfo that
are muche better learned then you, to improue his fuppofition by good argumentes, and therefore you were beft to
condemne no thinge that you do not well vnderftand : but
an other time, as I fayd, I will fo declare his fupofition, that
you fhall not only wonder to hear it, but alfo peraduenture
be as earneft then to credite it, as you are now to condemne
it. in the meane ceafon let vs proceede forwarde in our former ordre, wherin by ordre of your table I fhould fpeake of
the circles in heauen, both of their numbre, how many they
be, and alfo of their quantities, how great they are, which is
to be vnderftand in cōparifon to the Equinoctiall, or fome
other greate circle. Then of their ordre, and their diftance
a fonder : and likewaies what is their offices, wherunto they
ferue. of all whiche thinges, although I haue all ready fayde
inoughe for fo briefe an Introduction, yet bicaufe in theyr
numbre there may be fome difagreement, and in their quan
tityes

*Of the cir-
cles in the
fkye.*

tities. Diſtance and ordre there maye bee ſome varietie, at
the leaſte in diuers places, therefore I will ſpeake a little of
Equinoćtial. them againe. Firſt for the equinoćtiall, there is but one thoroughe all the world, and he is equally diſtaunt from eche
Pole, and therefore is called the Girdle of the ſkye : hys
office was declared beefore to bee the lymite of the myddle of the world, in whiche the Son maketh the dayes equall
to the nyghtes. Alſo hee declareth the true eaſte and weſt,
and is not onlye the common meaſure wherby all other circles are iudged in quantitye, but alſo it is the true meaſure
of motions celeſtiall, and the very rule to iudge all aſcenti-
the tropiks. ons by, as hereafter more largely ſhall appeare. Nexte vnto
this circle are there 2 Tropike circles, one on eche ſide of it,

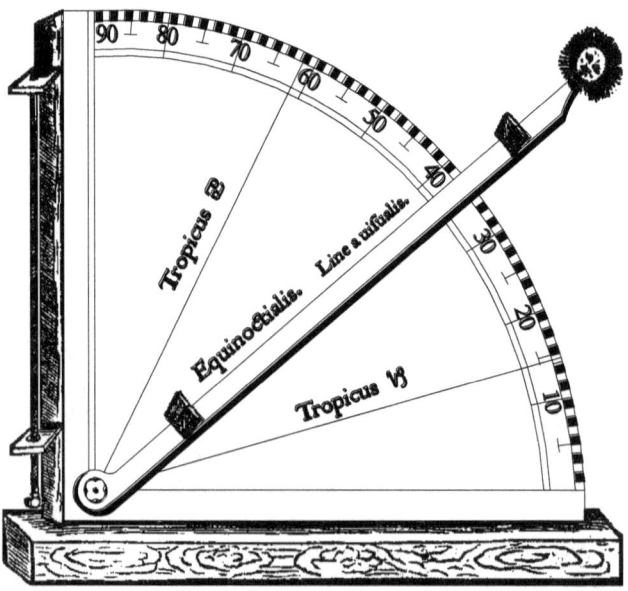

THE CASTLE OF KNOWLEDGE   158

whofe diftaunce a fonder may well be marked by a quadrant
fet fo in place conuenient, that it may ftand iuft plumbe with
the flatte of the horizont, and be tourned full fouthe. Then
obferue many daies aboute the middle of Iune the hygheft
point that the fonne wyll afcend vnto, and fhine duely thoroughe thofe two fightes in the ruler, mouinge it hygher or
lower, as occafion ferueth, tyll it ftande exactely pointinge
the heyghte of the Sonne at no one beynge at the higheft.
The lyke obferuation fhall you make diuers dayes before, at
and after the myddle of Decembre, tyll that you be affured
of the iufte heighte at noone of the fonne, beynge at the loweft then toward the fouthe. The pointes of thefe two obferuations well marked in the edge of the quadrante, are the
true places of the two Tropikes : and the diftaunce of thofe
two markes a fonder by numbre of degrees, is the very true
diftaunce of the twoo Tropikes. In the iufte myddle between thefe twoo tropikes is the place of the Equinoctiall
circle. Example. With vs. where the pole is 52 degrees highe,
the winter tropike wyll be 14 degrees and a halfe aboue the
Horizont. the fommer tropike 61 and a halfe. and the Equi-   *The greteft*
noctiall iufte 38 degrees in height. And the numbre of   *declination*
degrees that are betweene this Equinoctiall and any one of   *of the fonne*
the tropiks is named the Greateft declination of the fonne,
whiche in our time is about 23 degrees and 28 minutes. The
other pointes of declination of the degrees in the ecliptike
line from the equinoctial circle, bicaufe they be many in nūbre and diuerfe in vfe, I thinke it good to expreffe in a table
which hereafter fhall ferue you for fundry vfes.

Scholar. The like table is in Orontius.

Mafter. Not euen the lyke, as by conferring you maye
perceaue : but for the vfe of it, take what degree you lift
of anye Signe, and by this table you maye knowe his dedination from the Equinoctiall circle. The Signes are written partelye on the headde of the table, and partelye
on the foote of the fame. The degrees in the fyrfte
                                                    columne

168   THE FOVRTHE TREATISE OF
THE TABLE OF DECLINATION
PARTICVLARLY FOR EVERY DEGREE
of the Ecliptike lyne, and so for the Sonne.

deg.	Aries, Libra,		Difference min.	Taurus, Scorpius.		Difference	Gemini, Sagittarius.		Difference	deg.
	degr.	min.	min.	deg.	min.		degr.	min.		deg.
1	0	24	24	11	50	22	20	23	12	29
2	0	48		12	11		20	35		28
3	1	12		12	32	20	20	47	11	27
4	1	36		12	52		20	58		26
5	1	59		13	12		21	9		25
6	2	23		13	32		21	20	10	24
7	2	47		13	52		21	30		23
8	3	11		14	12	19	21	40	9	22
9	3	34		14	31		21	49		21
10	3	58		14	50		21	58		20
11	4	21		15	9	18	22	7	8	19
12	4	45		15	17		22	15		18
13	5	8		15	45		22	23	7	17
14	5	32	23	16	3		22	30		16
15	5	55		16	21		22	37		15
16	6	18		16	39	17	22	44	6	14
17	6	41		16	56		22	50	5	13
18	7	4		17	13		22	55		12
19	7	17		17	29		23	1	4	11
20	7	50	22	17	46	16	23	5		10
21	8	12		18	2		23	10		9
22	8	35		18	17		23	13		8
23	8	57		18	33	15	23	17	3	7
24	9	19		18	48		23	20	2	6
25	9	41		19	2		23	22		5
26	10	3		19	17		23	24		4
27	10	25		19	31		23	26	1	3
28	10	47	21	19	44		23	27		2
29	11	8		19	58		23	28	0	1
30	11	29		20	10		23	28		0
degr.	degr.	min.		degr.	min.		degr.	min.		deg.
	Virgo. Pisces.		Difference	Leo. Aquarius.		Difference	Cancer. Capricorn.		Difference	

columne

THE CASTLE OF KNOWLEDGE   169

columpne doo ſerue for the ſignes that bee on the heade of
the table, and the degrees in the laſte columpne doo ſerue
for the ſignes in the foote of the table, and the common angle againſt the ſigne : and the degree that you ſeeke for, doth
containe the degrees and mynutes of the declination due
to it.

Scholar. I perceaue it well : if I would knowe howe muche
the tenth degree of Leo doth decline from the equinoctiall,
I muſt looke in the columpn ouer Leo right againſt the nū
bre of tenne in the laſte columpne, where I fynd 17. 46.

Maſter. That is 17 degrees, and 46 minutes, which is the
declination of the 10. degree of Leo from the equinoctiall
circle.

Scholar. I muſt alwaies vnderſtande that 60 minutes do
make a degree : ſo theſe 46 minutes are $\frac{3}{4}$ of a degree and $\frac{1}{60}$
more. But what is the vſe of this table?

Maſter. That ſhall you knowe in the next treatiſe. in the
meane ceaſon to procede with the parallele circles : there foloweth next, the Arctike and Antarctike circles, whiche are
in numbre two, and there office is to encloſe thoſe ſtarres,
whiche euer appeare aboue our horizont, or neuer appeare
aboue the ſame, as before is declared : but bycauſe euerye
ſeueralle Climate hathe thoſe cyrcles diſagreeynge frome
other Climates, therefore theyr diſtaunce frome the o-
ther cyrcles Paralleles canne not bee certaine, (but for one
region certaine) nother yet theyr quantities, nother theyr
ordre : for where the eleuation of the pole is leſſe then 66
degrees and a halfe, there are thoſe circles leſſer then the tro
pikes, and are in ordre betwene them and the Poles, beinge
alwaies diſtaunt from the Pole iuſt ſo many degrees as the
Pole is in height aboue the Horizont in that region.

Scholar. It canne not bee other waies. And therefore it
foloweth, that where the pole is more then 66 degrees and a
halfe in heighte, there the Tropike is aboue the Horizonte,
as at Wardehouſe you declared it to be : and therefore

*The Artik
and Antar
tik circles.*

P.i.   in

in that climate the Arctik circle is greater then the Tropike of Cancer.

Of the fiue zones a-gainst the Greekes.

Master. Hereby appeareth the ouersighte of moste parte of the Greekes in limiting the Zones : for they appoint the Arctike and Antarctike circles for boundes of the Temperate Zones on the one side, and the Tropikes on the other side : wherof neither bounde can be well admitted, after their own explication of the qualities of the Zones. for if the temperate Zones shall be called those Zones that be inhabited, as they do so name them, then bycause there was knowen inhabitauntes innumerable besouthe the tropike of Cancer, it muste needes followe, that the tropike canne be no bounde of the temperate Zone : but yet otherwaies accomptinge the distinction of the Zones, not by that they are inhabited or vninhabited, but by the varietie of the motion of the sonne in respect to them, and by other accidents of shaddowes, there maye be good reason to make the tropikes boundes of the temperat zones : mary there is not the like reason for the Arctike and Antarctike circles. for confutation therfore of that opinion, I make this argument.

An argument in Ferio.

No vncertaine and variable boundes can limite anye certaine place : the temperate Zones are places certaine, and the Arctike circle with the Antarctike are chaungable, and vncertain limites, Therfore can not they be the boundes of the temperate Zones.

Scholar. This is a good argument, made in Ferio, the sowerth moode of the fyrste figure. And the maior is moste true, sith nothing can more disagree, then certain and vncertain, stable & vnstable, being contraries togither. The minor hathe 2 parts in it, which both seeme as true : for as long as the Sonne keepeth one yearely course, so longe the regions muste remaine as they were, and that is for euer, other styll temperate, other styll vntemperate. And so is that part of the minor true. The other part for the inconstancy & changablenes of the circles arctik & antarctik, must needs be true

by

THE CASTLE OF KNOWLEDGE   178

by their definitions, approued of the fame Greekes : for euery region hath a feuerall Actike circle. Wherfore I meruaile
muche that the Greekes beynge fo wife men, and fo greatly
learned, fhuld be fo muche ouerfeen and fo foroly deceaued :
but peraduenture ther are but few of that opinion, and fuch
as were leafte learned.

Mafter. Parmenides, Ariftotle, Cleomedes and Produs
may not be accompted vnlearned, and yet they with manye
other haue written that as truth. But hereby may you perceaue what folly it is, whē men receaue any doctrine as true,
and do not well weigh it, but credite the autority of the firft
teacher. So it appeareth in this matter, that bicaufe Parmenides, whiche was a great Philofopher, had fyrft taught that
diftinction of the zones, all the refte did folowe his opinion
as a plaufible doctrine, without examination of it, till Pofidonius began to efpye that errour & to confute it : as Strabo dothe declare in his fecond boke of Geographye, which
place in the latine tranflation is fo euell expreffed, that no
fentence in it importeth anye fence : wherefore as well for the
commoditie of you as of other, I will fumwhat amend that
place, wifhinge them that haue leafure and learning to help
to amend many other faultes of that good booke and other
lyke. The Latine tranflation is this.

Ad Septentriones, neqz penes omnes exiftentem, neqz eifdem vbi- *A place of*
cunqz. Quifnā temperatas quae immutabiles funt diuideret? Cum igi- *Strabo a-*
tur non penes vniuerfos fit feptentrionales effe, nihil effet ad argumē- *mended.*
tum. fi enim penes habitatores temperatae omnes, ad quos dicitur, fo
los temperata? Quod autem non vbiqz eodem modo, fed mutari, bene
comprehenfum eft. ipfeautem in zonas partiens, quinqz ad coeleftia
quidem vtiles effe afferit. Ex his duas circumftantes fubter polofvfqz
ad eas quae feptentrionales habent tropicos, diuer farum vmbrarū effe
ab aliis duabus, quae deinceps funt vfqz ad habitantes fub Polis. Quae
vero inter Tropicos eft, vtrinqz vmbras habere.

Scholar. Other the matter is very obfcure, or els there
wanteth lyghte in the declaration of it.

Ma. Ther is little fence in all thefe words : & ẙ fence ẙ may be
gathered of it is very falfe. And yet is ẙ greek boke both vn

corrupt (except it be in a worde or two) and full of perfect,
senfible and pleafaunt fentences. this is it.

The prited book hath ἴυσαυ falſely.

The greke booke hath πγαικσ falſely.

τοισ τι ἀρκτικοῖς, ὅτι περὰ πᾶσιμ * ὄσιμ,ὅτι βῖσ αὐ ρῖσ πανταχα τίσ ἀμ
διοριζω πῶσ ἐυκράτσσ, ἀιπερ εἰσιμ ἀμιτάπʒωντι. τὸ μὲυ ὀυῶ μὴ περὰ πᾶσιμ ἀναι τὸσ ἀρκτικϑὑσ, ἐδὲμ ἀμ ἄμπϑσ τὸμ ἐλεγχομ. εἰ γὰρ περὰ βῖσ τὴμ
ἐυκραπομ ὀικῦσιμ * εἰναι πᾶσι, πρὸσ ὅσπερ κỳὶ λέγτται μόνσσ ἐυκρατσσ,τὸ δὲ
μὴ παντεχοῦ τὸμ αὐρμ τρόπομ, ἀλλὰ μεταπίπʒειμ,κỳλῶσ ἄλημʒαι.αὐρσ δὲ
διαιρῶμ εἰστὰσ ζώνασ,πέντι μὲυ φήσιμ εἰναι χϙνσίμκσσ πρὸς τὰ ὁράνια.ρὐρρμ
δὲ * περισκίεσ δ'υὸ τὰσ ὑπὸ ρῖσ πολϑῖσ μεγϣῖ ριʒ ἐχόντωμ βὑσ τροπικϑὑσ αρ-
κλικϑὑσ,ἐπερσκίεσ δὲ τὰσ ἐφεξῆσ πανταις δ'υὸμεγϣῖ ρʒ ὑπὸ βῖσ τροπικϑῖσ
ἀκϑωῖω ρ, ἀμφίσκιομ δὲ ϯ μεταξὑ ρʒ βοπικῶμ.

Whiche I doo tranſlate thus.

Arcticis vero circulis (vt qui nec apud omnes exiſtant, nec iidem
vbiqz perſeuerent) quis vnquam temperatas Zonas (quae immutabiles funt) terminaret? Ceterum illud quod non apud omnes exiſtant
Arctici circuli, nihil facit ad reprehenſionem. quum fatis fit, fi modo
fint apud omnes incolas temperatae ipfius zonae, ad quos folos temperata dicitur. quod vero adiecit, non vbiqz feruare eos eandem ratio-
nem, fed varie mutari, hoc quidem recte adfumptum eſt. At qz ipfe Po
fidonius dum Zonas deſtinguit, quinqz inquit vtiles effe ad coeleſtes
obſeruationes. quarum duae, que Polis fubiacent, vmbras circumſluas
habent, vnde Perifciae dicuntur : ibiqz finiuntur vbi tropici ipfipro arcticis circulis habentur. has fequuntur aliae totidem, eo pertingentes,
vbi Tropici verticibus incolarum imminent, atqz in his vmbrae me-
ridianae in vnam plagam porriguntur femper, hinc Heterofciae vocan
tur. quinto vero quae inter tropicos iacet, in vtrunqz latus viciffim vmbras mittit, atqz Amphifcia nuncupatur.

Which words may be englifhed thus. What man (faith Po
fidonius) wold affigne the Arctike circles to be as bounds to
the temperate zones? feing thofe circles ar not in euery Climate : nother do they continue vniforme and of one fort to
all cuntries. Thefe wordes (faith Strabo) that they be not in
euery climate, maketh nothing to the reproofe. for it is fuffi
cient that they be incident to all the inhabitants of the temperate zone, in refpect to whom alone that temperate zone
beareth his name : but thofe other woordes, that they keepe
not

not one vniforme manner in all places, but are diuerſly chan
ged : that is well alleaged. Alſo Poſidonius him ſelfe when he
diſtincteth the zones, doth ſay, that fiue zones are needefull
and ſufficient for celeſtiall obſeruations whereof two which
be vnder the poles, are caled Periſciae, or Round ſhadowed,
bicauſe their ſhaddowes run round about them. And theſe
zones extend to that place, where the tropik circles and the
Arctike circles are all one. After theſe there do follow two
other, which reache from thence vnto thoſe partes, that are
directly vnder the tropiks and theſe haue their noone ſhaddowe running one waies ſtyll, and therfore are called Hete-
roſciae, or Single ſhadowed. The fift zone lyeth betwene the
tropikes, and caſteth the noone ſhadows 2 waies, wherefore
the Greekes call it Amphiſcion, that is Double ſhadowed.

Scholar. By this tranſlation (which is worth a paraphra-
ſis) I doo not only perceaue the ſence of theſe wordes whiche before were darke, partly for the hardnes of the matter,
and partlye for the hypallage, in changinge of the ſpeakers
perſon, but alſo I eſpye the monſtrous ſhape of the old tran
ſlation. And by this I gather alſo that Strabo woulde not
haue the Temperat zones to be bounded by the Arctik and
Antarctike circles.

Maſter. His mynde appeareth more manifeſt anon after
where he blameth Polybius, for aſſiginge thoſe circles as
boundes of the zones : whereof one ſhould be incloſed with
in that circle, and the other ſhould extend from it to the next
tropike. then he concludeth thus : that thoſe vnconſtant circles, may be no boundes of certentye.

εἴρηται γάρ ὅτι τῷ μετακ᾽ηζϑύοη σημείοισ ὀυχ ὁριϛὶου τὰ ἀμετκηζ᾽ηζαντα.

Dictum enim eſt, qod per ſigna tranſmigrantia, ea quae non mu-
tantur, terminare non conuenit.

For I haue ſayde before, that chaungable limites may not
be appointed as boundes to vnchaungable places.

Sch. Thus it appeareth, that the diſtinction of zones by
                                      P.iij.   the

the Arctike and Antarctike circles were no conftant diftinction. and fo is autoritye of one forte repelled by thaucto-
ritie of an other forte.

Mafter. You maye not weighe the matter by auctoritye,
for fo fhoulde that former doctrine continue ftyll, feynge I
aleaged for it Parmenides, Ariftotle, Polybius, Cleomedes
and Proclus, & againft them only Pofidonius and Strabo,
which maye feeme the weaker in numbre : but then confidre
that the firfte fort bring only affirmation for their teftimo-
ny, and bare autoritye : the other confute theym by good
reafon and fubftantiall argumentes, whiche are farre to bee
efteemed aboue anye autoritye.

Scholar. Then credityng reafon againft autority, I muft
fay, that the Zones muft be otherwaies diuided, peraduenture as I dyd learne of you before, agreable to Iohn de Sacro bofco his mynde, whom you called the reftorer of the
Zones.

Mafter. Yea in deede : for although Pofidonius and Strabo did teache the like diftinction, yet did they not fo openly
name the true limites, howe bee it in effecte they meane the
fame : for when Strabo faith, that the Cold zone doth reach
to that place, where the Tropike is the Arctike circle, hee
dooth meane that there, where this firfte Zone endeth, and
the temperate Zone beginneth, the Pole is *66* degrees and
a halfe aboue the horizonte, and fo mufte the fame Pole bee
from the toppe of their headdes in that place 23 degrees and
a halfe : in whiche diftaunce bicaufe the Poles of the Zodiake do defcribe a circle, therfor doth Iohn de Sacro bofco
call that circle the Arctike circle, in that confounding it in
name with an other circle of the Greekes : wherfore I thinke
it more reafonable for auoyding confufion, to gyue it a fe-
The Polare uerall name, and call it the Polare circle, and the other to be
circles. called ftyll the Arctike circle, as the greeks longe before did
name it. And this diftinction of the zones by the two Tropikes, and the two Polare circles doth diftinct exactly thofe
                                                                    thre

three varieties of fhaddowes before mentioned. whiche is a certaine and notable difference, not imagined by men whiche may erre, but wrought by the fonne, which can not erre. But heere mufte I admonifh you of an other erroure, gathe red not of grounded reafon, but of phantafticall imagination, by occafion of whiche, this fonde diftinction of zones was imagined. *An other erroure.*

Bicaufe the elder Grekes had no trade into the fouth parts of Afrike, nother the Ethiopians again into Grece, and farther by reafon the fonne runneth ftill ouer their headdes, that dwell betweene the tropikes, manye of the Latines as well as of the Grekes phantafied that there did dwell no inhabitantes, neither could dwell there for the vehement heat : wherfore they called it the Burned Zone. And of lyke occafion where they moued to accompt two other zones, that be nigh the poles, to be vninhabited for cold, by reafon that the fonne doth neuer come nigher to them then the Tropik circles : but how muche herein they were deceaued, it maye be declared not only by reafon, and by experience, but alfo by autority of many of their owne writers, as namely Eratofthenes, Pofidonius, Polybius, and Ptolemye. but as this is a matter more agreeable to the treatife of Geographye or Cofmography, then of the Sphere, fo will I ouerpaffe it for this time, and will returne to the refte of the circles of the fphere, amongeft which the Zodiake as principall, doth offre it felfe, as the common theatre and ftage of all the planets motion, and of the chiefe fignes and celeftiall figures. *The Zodiake.*

Scholr. Are there I pray you fuche figures in the Zodiake, as Aftronomers do defcribe?

Mafter. There are fome that affirme no leffe, and teftifye that they haue in a cleere ayre perceaued them : but for the refte of the forme, I will fay nothinge now : onlye this I doo affirme whiche I know, that all the ftarres whiche aftronomers do name to be there, maye eafily be feene there, and in lyke forme as they doo place them.

P.iiij. Schol.

Scholar. If the formes of beafts be not there, why do they call it by that name of Zodiake, whiche name is deriued as many do affirme, of ʒῴδιον; that fignifieth a beafte.

Mafter. The Signes doo beare the names of beaftes, and therfore may that cirde take the like denomination alfo : but yet I denyed not that the verye formes were there, but that they are not eafilye feene in fuch exacte fhape as they be por tured, and as fome men write that they haue feene them : but howe fo euer it bee, the certenty is, that the 12 fignes are con tained in that zodiake, and therfore doth Tullye with other latine men call it Signifer, that is, the Circle of the Signes : but whye thofe names were giuen to euerye figne rather then other, dooth not appertaine fo muche to this treatife, as to that Iudiciall arte, whiche hath more ground of reafon then many men thinke.

What is to bee in a Signe.

Scholar. When you faye that the Sonne is in anye figne, you do not meane (I am fure) that the Sonne hath lepte fo high from his owne fphere, into the fphere of the Fixed ftar-res, where the zodiake and the fignes be, but that the Sonne is directly vnder the fame figne, and in a righte line betwene that figne and the centre of the earthe.

Mafter. You faye well. That is the common vnderftan-dinge, when we fpeake of the place of the fonne : but bicaufe other Planettes doo decline from the myddle of that zo-diake, fome tymes towarde the north, and other times to-ward the fouthe, therfore haue all aftronomers appointed a conuenient breadth to the zodiake, according to the decli-nation of the Planets : howe bee it proprelye they doo call

The lati-tude of Pla-netes.
Their decli nation.
Their lon-gitude.
The fecond fignificatiō of a figne.

that the Latitude of the Planetes, when they fwarue frome the Ecliptike line : and the Dedination of them is their di-ftaunce fouthe or northe from the equinoctiall line : fo doo they call the motion of them in Longitude, theyr diftaunce by theyr naturall courfe frome the beginninge of Aries, which is the beginning of the zodiak. And now appointing the latitude of the zodiake to bee twelue degrees (although fome

some planetes may runne in latitude on the one side almoſt
8 degrees) bycauſe that quantitie is moſte receaued, then is
euerye ſigne twelue degrees broad, and thirtie degrees long.
and ſo maketh a longe ſquare : frome the corners of whiche
long ſquare you may imagin lines to be drawē to the centre
of the earth : and what ſo euer commeth within the boundes
of thoſe lines, is accompted to bee in that ſigne : and this is
the ſecond ſignification of a ſigne. A third ſignification ther *The thyrde*
is, which we vſe when we ſay that the bright ſtarre Arcturus *ſignificatio*
is in Virgine, where as in deed he is aboue 30 degrees north *of a ſigne.*
from the Ecliptike line : which is farre out of the breadth of *The Pole*
the Zodiake : and ſo we ſay that the pole ſtarre is in Taurus, *ſtarre.*
whiche is from the Ecliptike line *66* degrees. and likewayes
we name all the ſtarres in the ſkye to bee in ſome ſigne, bee
they neuer ſo farre from the Ecliptike line, and the Zodiak.
Therfore to know what is vnderſtand by the name of a ſigne
in this ſignification, you muſt imagin *6* circles to be ſo dra
wen about the Globe, that they may paſſe by the beginning
of all the ſignes (for euery circle will ſerue for two ſignes beinge contrarye one againſt the other) and ſo ſhall the whole
Zodiake and all the globe alſo be parted into twelue equall
partes, yf you haue drawen thoſe circles rightly & that they
do paſſe al by the two poles of the Zodiak. Now mark how
thoſe 2 lines that do incloſe any ſigne, ar wideſt a ſonder in ẙ
myddle of the Zodiake, and from thence toward eche pole
of the zodiake they come nearer and nearer, tyll they touch
in the Pole it ſelfe. All the ſpace betweene anye two ſuche ſe-
micircles from one Pole to the other, is named a ſign in the
thyrde ſignification : ſo that what ſo euer ſtarres bee within
that ſpace, are named to bee in that ſigne which is within the
ſame ſpace : of all theſe three diuers formes of ſignes heere
maye you ſee examples. of the fyrſte by A, where the Sonne
ſtandeth vnder the ſigne of Cancer. of the ſeconde forme
you haue an example by B, and of the thirde ſorte you haue
twoo varieties, one by, C and an other by D. So that what

ſo

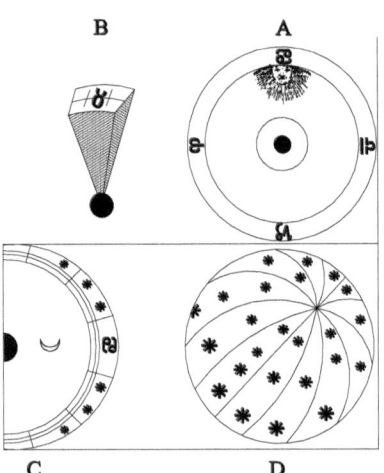

B   A

C   D

fo euer Planet doth come within y̆ boundes of that figure B, is named to be in the figne of Taurus : & what fo euer Planete or fixed ftar is within the compas of the figure C, is iudged to be in Cancer : as y̆ Moone is ther reprefented to be and all the ftarres there portured, & fo maye you iudge of anye other figne. Nowe

this may fuffife for the explication of the zodiake, after whom foloweth nexte the Colures, whiche take their names in Greeke of vnperfectnes, bycaufe they bee neuer feene all aboue the grounde in any oblique fphere : whereby it appeareth, that good Iohn de facro bofco was much deceaued in comparing them to the copaffed bowing of a wild bulles tayle, as thoughe they tooke their names thereof : but men muft bear with the ignorance of that time, for lack of know ledge in the Greeke tonge. Thefe Colures ferue principally for the diftinction of the four chiefe pointes in the zodiake, as before is declared. and bycaufe the pointe of the interfection or croffinge of the ecliptike line and the equinoctiall, doothe fufficiently expreffe two of thofe pointes in the beginning of Aries and Libra, therfore the greekes do affigne comonly but one Colure, for the other two tropike pointes, and none for thefe equinoctiall pointes. How be it, bycaufe they ferue alfo for the declinations and latitudes of fixed ftarres and Planetes, I thinke it better to defcribe them, then to omitte them. And thus haue I lyghtly touched all the cir
                                                                                          des

des that be fixed in the fphere, and moue with it. Nowe remaineth other two, which ftand ftyll alwaies and moue not,
of whiche the fyrfte is the Horizonte, and the nexte is the
Meridiane. The horizont is of twoo diuers fortes. the one *The Hori-*
doth extend on euery fyde vnto the firmament, and ferueth *zonte.*
as it were peculiarly for the partition of the heauens, and di- *The celefti-*
uideth the fkie iuftly into two halues, wherof of the one appea *al horizont.*
reth vnto vs aboue that Horizonte, and the other is hidde
from vs, vnder the fame horizont : this horizonte hath his
name of the fkie, and is called the Celeftial horizont, and his
diameter is as large as the diameter of the eight fpher, which
is the fartheft and higheft part of the fkye that we canne fee :
this large horizont our fight doth inforce vs to acknowledg
as a iufte horizont, although reafon canne fynde in it fome
wante of exacte precifenes. And therfore Proclus doth not
well diftincte this horizont from the other, by naminge the
other a fenfible horizont, and affirming this to be confidered only by reafon, where as in deede we neede reafons helpe
more in iudginge the other horizont, whiche I thinke mofte
aptlye to bee called the Earthly horizont, bycaufe it ferueth *The Earth-*
for fightes on the earthe and water onlye, and reacheth not *ly horizont.*
vnto the fkie : no, his femidiameter excedeth not (as Macro
bius faith) 180 furlongs, that is 22 myles and a halfe : and his
whole diameter comprehendeth but only 45 myles in length.
So that if any man do ftande on a plaine grounde or on the
fea, he maye fee rounde about him euery waies 22 myles

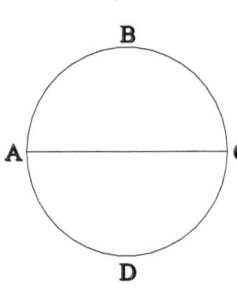

and a halfe : that is in round compas
of the whole horizonte 141 miles &$\frac{3}{7}$.
I meane that feing the right line A,
C, is 45 miles, the whole circle A
B C D, muft bee accompted 141 $\frac{3}{7}$
myles in compas. This faynge of
Macrobius is more nygher to the
truth then Proclus affertion, which
is that the diameter fhuld be in this
                                    hori-

horizont, 2000 furlonges, that is 250 myles, wherby he mea
neth that a manne may fee euery waye in a playne 125 myles
from him : whiche affertion euery maryner dooth knowe to
be falfe : for it is well knowen by often and good obferuation, that in plaine ground, or on the fea, they can not difcern
well aboue 20 myles, and therefore do all mariners call that
A kenning. diftaunce commonly a Kenninge : whiche is as muche as a
manne maye well fee : yet from a hill or highe grounde men
maye fee farther, and efpecially they maye fee other hilles or
clyffes, but that is no certaine vewe, nor iufte kenninge : yet
in that fort men may fee 60 miles, or at the mofte 90 miles :
but 125 myles is to greate a diftaunce, for to vewe any thing
from a high place, and therfore of more force it is to excef

A demon-
ftration a-
gainft pro
clus.

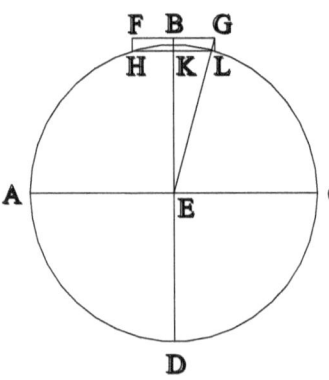

fiue a diftaunce to vewe
any thinge in an equall
plaine, as the horizont
muft needes be, for de-
claration wherof, I fup-
pofe this figure to re-
prefent the whole globe
of the earthe, and the
earthly horizon to be
expreffed by the ryghte
lyne F B G : vnto which
line ther is an other dra
wen as a iufte parallele,
which is H K L. of lyke
lengthe precifely with the earthly horizonte, and two other
lines ioyninge them at the eandes, makinge a longe fquare
of all righte angles, fo that two of thofe angles do lyght on
the circumference of the circle of the earthe. Then draw I a
right line from E which is the centre of that circle, vnto B,
and an other from the fame centre E vnto G : wherby ther is
made two triangles E B G, and E K L. Now prefuppofing
that B is the place where we ftande on the earth, and H and
L, the

vnto whiche the Semidiameter of 1000 furlonges of oure earthlye Horizont, dothe extende on bothe fides : and frome the one of them is drawen a right line to the other, that line muft needes fall within the circle.

Scholar. That is true, accordinge to the 47 Theoreme of the Pathwaye.

Mafter. Then mufte the line K E, be fhorter then the lyne B E, and fo B and K, are notably diftaunte.

Scholar. That is certaine.

Mafter. And bicaufe the righte line F B G, is parallele to the righte line H K L, there muft be as muche diftaunce be- tweene G, and L, as there is betwene B and K.

Scholar. That foloweth by the definition of Paralleles.

Mafter. Then as K, is notably vnder B, fo muft L be no- tably vnder G : that is to fay vnder the Horizont, and ther- fore can not be feene.

Scholar. It is againft the definition of an horizonte, that anye thinge vnder it fhoulde be feene.

Mafter. Then if the femidiameter of the Horizonte fhall extend no farther then that a meane quantitie maye be feene on the earth, it maye not be fo longe as Proclus hath limi- ted it. Alfo by the two triangles aforefaide, whofe angles are like, and therfore their fides proportionable, & other waies diuerfly, by the former figure, it may be demonftrate, that the righte line E G is muche longer then E L, whiche is the femidiameter of the earthe, fo that the horizont in fo much diftaunce is farre hygher then the earth is there, and therfore canne not bee aptelye called a Senfible Horizonte, nor an Earthly Horizonte, as Proclus meaneth. But it appeareth that Proclus dydde rather in this doctrine followe fome other mennes opinion then hys owne reafon, as he dooth alfo in the declaration of the chaunge of the Horizontes and the Meridianes, for betweene eafte and wefte, hee faythe that the Meridianes chaunge at the eande of 300 furlonges : but betweene fouthe and northe hee doth af-

Q.i.   figne

ſigne no chaung vnto the Horizont within 400 furlongs.
In whiche woordes there are two errours included : the one
that the horizonts be not like in chaunge betwene eaſte and
weſte, and betwene ſouthe and northe.

Scholar. Nay he ſpeaketh only of the Meridianes (I trow)
betwene eaſte and weſt, and not of the Horizontes.

Maſter. As thoughe we might chaunge the one, and not
vniformely chaunge the other.

Scholar. Truthe it is, that ſeing the meridiane doth cutte
the Horizonte with right angles, they both muſt needes o-
ther ſtand bothe ſtill, other chaunge bothe alike : wherefore
this firſte erroure can not be excuſed.

Maſter. And the ſeconde errour is as manifeſt as it : for
therby he ſuppoſeth that the Climates do chaunge by equal
quantity of furlonges or miles, which errour is to manifeſt :
for nighe vnto the equinoctiall, 2150 furlonges northwarde
do cauſe increaſe but of a quarter of an hower in the longeſt
daye. And with vs in the ſouthe parte of England, 700
furlonges northwarde dooth cauſe increaſe of a quarter of an
hower in the longeſt daye, and in the north partes of Scotlande, 320 furlonges doo giue as great an increaſe : in Iſe-
lande 4 furlonges yeldeth the lyke increaſe : and ſo ſtyll the
farther northe you go, the ſmaller ſpace of ground bringeth
the like increaſe in the longeſt daye.

Scholar. Hereby I perceaue, that who ſo euer will trauaile
in theſe ſciences with profit, muſt lean rather to reaſon, then
to authoritye, els he may be deceaued.

Maſter. That rule is generall in all artes.

Scholar. And if Proclus rule be not certen, what rule may
I haue more certen? M. For the alteratiō of the Horizonte
betwene ſouth & north, bicauſe not only the climats do chãg
therwith, but alſo the quantities of ẙ daies, I wil anon before
the doctrine of the aſcenſions, giue you a table generall for
all climates in the earthe. And as for the chaunge of the horizontes or of the meridianes betweene eaſte and weſte, you
                                                                          ſhall

THE CASTLE OF KNOWLEDGE   183

A TABLE FOR THE DIFFE-
rence of howers accordinge to the diftaunce of
myles from eafte to wefte, vnder the
Equinoctiall.

The diftaunce of miles.	The minutes of an hower.	The diftaunce of miles.	Howers.	The minutes of an hower.	The diftaunce of miles.	Howers.	The minutes of an hower.	The diftaunce of miles.	Howers.	The minutes of an hower.
15	1	465	0	31	915	1	1	1365	1	31
30	2	480	0	32	930	1	2	1380	1	32
45	3	495	0	33	945	1	3	1395	1	33
60	4	510	0	34	960	1	4	1410	1	34
75	5	525	0	35	975	1	5	1425	1	35
90	6	540	0	36	990	1	6	1440	1	36
105	7	555	0	37	1005	1	7	1455	1	37
120	8	570	0	38	1020	1	8	1470	1	38
135	9	585	0	39	1035	1	9	1485	1	39
150	10	600	0	40	1050	1	10	1500	1	40
165	11	615	0	41	1065	1	11	1515	1	41
180	12	630	0	42	1080	1	12	1530	1	42
195	13	645	0	43	1095	1	13	1545	1	43
210	14	660	0	44	1110	1	14	1560	1	44
225	15	675	0	45	1125	1	15	1575	1	45
240	16	690	0	46	1140	1	16	1590	1	46
255	17	705	0	47	1155	1	17	1605	1	47
270	18	720	0	48	1170	1	18	1620	1	48
285	19	735	0	49	1185	1	19	1635	1	49
300	20	750	0	50	1200	1	20	1650	1	50
315	21	765	0	51	1215	1	21	1665	1	51
330	22	780	0	52	1230	1	22	1680	1	52
345	23	795	0	53	1245	1	23	1695	1	53
360	24	810	0	54	1260	1	24	1710	1	54
375	25	825	0	55	1275	1	25	1725	1	55
390	26	840	0	56	1290	1	26	1740	1	56
405	27	855	0	57	1305	1	27	1755	1	57
420	28	870	0	58	1320	1	28	1770	1	58
435	29	885	0	59	1335	1	29	1785	1	59
450	30	900	1	0	1350	1	30	1800	2	0

Q.ij.

ſhall vnderſtande that 15 myles difference from eaſte toward
weſt, doth make the ſonne riſinge, the none ſteed, and Sonne
ſetting, to be later by one minut of an houre. and ſo 30 miles
2 mynutes : 120 myles 8 minutes : 225 myles 15 minutes. which
is a quarter of an hower. And for exaumples ſake more then
for any other cauſe I giue you here this table, which you may
eaſylye increaſe by the lyke fourme, vntyll you haue accom-
plyſſhed the whole 24 howers, yf you lyſte. howe be it
hee that is readye in accompte of Arithmetike, needeth
not anye ſuche tables of ayde. This table is calculate onlye for ſuche places as dyffer not aboue 1800 myles bee-
tweene eaſte and weſte, hauynge no difference or verye lyttle
in their diſtaunces betweene ſouthe and north, as touching
this conſideration. And it ſerueth onlye for the middle climate of the worlde vnder the equinoctiall circle. for euerye
other climate, yea and euerye degree in latitude of eche climate, muſt haue a ſeuerall table, whiche maye not well be ſet
forth in this brief introductiõ, but an other time ſhall ſerue
herafer for it, yf you call on me and put me in mynde therof, elſe the neceſſitye of prouiſion for my familye will make
me forget ſuche promiſes : how be it bycauſe you ſhall not
thinke that I haue done more for them that dwell vnder the
equinoctiall (or nygh vnto it in Guynea or in Calecut) then
for our own cuntrie, I haue drawen the like table for the eleuation of 52 degrees, whoſe vſe is euen one with the other
before. wherefore if I knowe the diſtaunce of myles bee-
tweene anye twoo places vnder this latitude of 52 degrees,
or nyghe thereto, as ſoone as I haue found out that numbre of myles in the table vnder that title, in the nexte columpne on the righte hande, I maye ſee howe manye minutes they do differ in theyr howers.

   Scholar. So that the miles exceede not 1110, for this
table hathe no greater numbre.

   Maſter. If you lyſte by this preſident, you may increaſe
the table as muche as you wyll.

                                          Scho-

THE CASTLE OF KNOWLEDGE   185

A TABLE OF THE DIFFERENCE

of howers, accordinge to the distaunce of miles
from easte to weste, for the eleuation of 51
degres, 55 minutes.

The distaunce of miles.	The minutes of an hower.	The distaunce of miles.	Howers.	The minutes of an hower.	The distaunce of miles.	Howers.	The minutes of an hower.	The distaunce of miles.	Howers.	The minutes of an hower.
9 1/4	1	286 3/4	0	31	564 1/4	1	1	841 1/4	1	31
18 1/2	2	296	0	32	573 1/2	1	2	851	1	32
27 3/4	3	305 1/4	0	33	582 3/4	1	3	860 1/4	1	33
37	4	314 1/2	0	34	592	1	4	869 1/2	1	34
46 1/4	5	323 3/4	0	35	601 1/4	1	5	878 3/4	1	35
55 1/2	6	333	0	36	610 1/2	1	6	888	1	36
64 3/4	7	342 1/4	0	37	619 3/4	1	7	897 1/4	1	37
74	8	351 1/2	0	38	629	1	8	906 1/2	1	38
83 1/4	9	360 3/4	0	39	638 1/4	1	9	915 3/4	1	39
92 1/2	10	370	0	40	647 1/2	1	10	925	1	40
101 3/4	11	379 1/4	0	41	656 3/4	1	11	934 1/4	1	41
111	12	388 1/2	0	42	666	1	12	943 1/2	1	42
120 1/4	13	397 3/4	0	43	675 1/4	1	13	952 3/4	1	43
129 1/2	14	407	0	44	684 1/2	1	14	962	1	44
138 3/4	15	416 1/4	0	45	693 3/4	1	15	971 1/4	1	45
148	16	425 1/2	0	46	703	1	16	980 1/2	1	46
157 1/4	17	434 3/4	0	47	712 1/4	1	17	989 3/4	1	47
166 1/2	18	444	0	48	721 1/2	1	18	999	1	48
175 3/4	19	453 1/4	0	49	730 3/4	1	19	1008 1/4	1	49
185	20	462 1/2	0	50	740	1	20	1017 1/2	1	50
194 1/4	21	471 3/4	0	51	749 1/4	1	21	1026 3/4	1	51
203 1/2	22	481	0	52	758 1/2	1	22	1036	1	52
212 3/4	23	490 1/4	0	53	767 3/4	1	23	1045 1/4	1	53
222	24	499 1/2	0	54	777	1	24	1054 1/2	1	54
231 1/4	25	508 3/4	0	55	786 1/4	1	25	1063 3/4	1	55
240 1/2	26	518	0	56	795 1/2	1	26	1073	1	56
249 3/4	27	527 1/4	0	57	804 3/4	1	27	1082 1/4	1	57
259	28	536 1/2	0	58	814	1	28	1091 1/2	1	58
268 1/4	29	545 3/4	0	59	823 1/4	1	29	1100 3/4	1	59
277 1/2	30	555	1	0	832 1/2	1	30	1110	2	0

186   THE FOVRTHE TREATISE OF

Scholar. Bicaufe examples do make rules manifeft, I pray you let me proue one example. London and Briftow are 94 myles a fonder, and as I haue hearde you faye, they are not muche different in latitude : I defire to know their difference in howers, therfore I feeke for 94 vnder the title of diftaunce of myles, and I can not find it there, for 92 and a halfe is to lyttle, and 101 $\frac{3}{4}$ is to greate.

Mafter. And in lyke rate is there difference of minutes : for 10 minutes is to lytle, and 11 minutes is to greate. but to geffe mofte nearest : as 92 and a halfe is nigher to 94 then 101 $\frac{3}{4}$: fo is 10 minutes more nearer their true difference then 11. And for this time this maye fuffife, althoughe I can giue you a precife rule by the part proportionable to fynde oute the iufte parte of euery minute, but that were more curious then profitable in this place : Therfore will I leaue it, and declare vnto you, how you may make the lyke table for any latitude of euen degrees.

Scholar. I do perceaue by thefe two tables, that if I haue ones the fyrft numbre which muft be fet againft one minute of tyme, then muft I double it for two minutes, and triple it for thre minutes, and fo forth, ftyll multiplying the fyrfte numbre of myles by the numbre of minutes againft which it fhall ftende.

Mafter. You take it well, and therfore feyng you doubte only of the fyrft numbre, I will giue you a table by whiche you may eafily find out that firfte numbre for all degrees of latitude of any region. And this is it. where in the firft columne you fee placed the degrees of latitude, and in the feconde columne are fet the myles with their fractions, which ferue for one degree of longitude, in eche of thofe dyuers latitudes. By this table may you make any table for any eleuation of hole degrees, accordinge to the example of the former two tables.

Scholar. That do I perceaue nowe very well, and can do it, I doubt not, fufficiently for anye Climate, yf I were as

certen

A TABLE DECLARINGE

how many myles do anſwere to one minute of
tyme, in euery ſeuerall latitude.

Degrees of latitude	Miles agrig to i minute of time	Degrees of latitude	Miles agrig to i minute of time	Degrees of latitude	Miles agrig to i minute of time
0	15				
1	14 $\frac{230}{240}$	31	12 $\frac{103}{220}$	61	7 $\frac{13}{48}$
2	14 $\frac{79}{80}$	32	12 $\frac{173}{240}$	62	7 $\frac{1}{24}$
3	14 $\frac{47}{48}$	33	12 $\frac{139}{240}$	63	6 $\frac{97}{120}$
4	14 $\frac{77}{80}$	34	12 $\frac{21}{48}$	64	6 $\frac{69}{120}$
5	14 $\frac{113}{120}$	35	12 $\frac{69}{240}$	65	6 $\frac{81}{240}$
6	14 $\frac{11}{12}$	36	12 $\frac{2}{15}$	66	6 $\frac{1}{10}$
7	14 $\frac{71}{80}$	37	11 $\frac{47}{48}$	67	5 $\frac{207}{240}$
8	14 $\frac{41}{48}$	38	11 $\frac{197}{240}$	68	5 $\frac{149}{240}$
9	14 $\frac{44}{80}$	39	11 $\frac{79}{120}$	69	5 $\frac{3}{8}$
10	14 $\frac{37}{80}$	40	11 $\frac{59}{120}$	70	5 $\frac{31}{240}$
11	14 $\frac{87}{120}$	41	11 $\frac{77}{240}$	71	4 $\frac{53}{60}$
12	14 $\frac{161}{240}$	42	11 $\frac{7}{48}$	72	4 $\frac{19}{30}$
13	14 $\frac{37}{60}$	43	10 $\frac{233}{240}$	73	4 $\frac{31}{80}$
14	14 $\frac{133}{240}$	44	10 $\frac{19}{24}$	74	4 $\frac{2}{15}$
15	14 $\frac{117}{240}$	45	10 $\frac{73}{120}$	75	3 $\frac{53}{60}$
16	14 $\frac{101}{240}$	46	10 $\frac{101}{240}$	76	3 $\frac{102}{240}$
17	14 $\frac{83}{240}$	47	10 $\frac{11}{48}$	77	3 $\frac{3}{8}$
18	14 $\frac{4}{15}$	48	10 $\frac{9}{240}$	78	3 $\frac{7}{60}$
19	14 $\frac{11}{60}$	49	9 $\frac{101}{120}$	79	2 $\frac{207}{240}$
20	14 $\frac{23}{240}$	50	9 $\frac{77}{120}$	80	2 $\frac{29}{60}$
21	14 $\frac{19}{240}$	51	9 $\frac{53}{120}$	81	2 $\frac{83}{240}$
22	13 $\frac{100}{120}$	52	9 $\frac{7}{30}$	82	2 $\frac{7}{80}$
23	13 $\frac{97}{120}$	53	9 $\frac{1}{240}$	83	1 $\frac{109}{240}$
24	13 $\frac{160}{240}$	54	8 $\frac{44}{60}$	84	1 $\frac{17}{30}$
25	13 $\frac{143}{240}$	55	8 $\frac{29}{48}$	85	1 $\frac{37}{120}$
26	13 $\frac{29}{60}$	56	8 $\frac{91}{240}$	86	1 $\frac{11}{240}$
27	13 $\frac{11}{30}$	57	8 $\frac{41}{240}$	87	$\frac{47}{240}$
28	13 $\frac{50}{240}$	58	7 $\frac{19}{30}$	88	$\frac{21}{40}$
29	13 $\frac{29}{240}$	59	7 $\frac{87}{120}$	89	$\frac{21}{80}$
30	12 $\frac{119}{120}$	60	7 $\frac{1}{2}$	90	0

Q.iiij.

Of the climates. certaine of their boundes. but that maye I learne by fuche table as Orontius and dyuers other haue fette forthe all readye.

Mafter. In deede bothe Orontius and other haue fet forth fuche tables, whiche maye fuffice for an Introduction, but Orontius extendeth not his table aboue the latitude of *66.* degrees and a halfe, fo there refteth vnto the northe Pole 23 *The famous* degrees and a halfe, whiche coafte hytherto hath been kno- *aduenture* wen to very fewe men, but nowe of late by the famous ad- *vnto Mof-* uenture of that woorthye companye of our Englifhe mar- *couia by* chaunts for Mofcouia, that coaft is difcouered vnto 75 de- *the northe* *Ocean.* grees of latitude nighe hande : and our hope is that if they doo continue as they haue valiantlye begonne, they fhall difclofe thofe vnknowen people whiche dwell directlye vn- der the Pole, or at the leafte waies difcouer that climate, fuche as it is, to the full fatifffaction of that importune defire, whiche hathe forced manye thoufandes to wiffhe, that whiche not one yet (that we knowe) coulde attayne : whereby they fhall not onlye profite their countrie, but fhall procure to theim felues greate ryches and treafure : and that whiche is mofte to bee defired, immortall fame. Wherefore for my parte to further their knoweledge in the atchiuinge of their woorthye attempte, as I haue all readye in this booke giuen fome lighte, fo wyll I (God wyllinge) hereafer gyue more lighte : and for an earnefte thereof I will nowe exhibyte to you a table of the Climates extended to the verie Pole, whereby you maye learne not onlye the beginninge and eande of euerye climate, but alfo the iufte quantitie of the longeft and fhor- tefte daye in eche of theim, and in all other places to the Pole felfe : the reafon whereof you fhall better vn- derftande by the diuerfities of the afcenfions. But bicaufe (as I faide beefore) that euerye Climate dif- fereth frome other, by the fpace of halfe an hower in the quantitye of their longeft daye, therfore did the greekes

and

THE CASTLE OF KNOWLEDGE   190

and namely Ptolemye, for a more precifenes make a certain
diftinction for euery quarter of an howers difference, whiche he calleth only by the generall names of paralleles, as it
doth at large appear in the fixte chapter of the fecond boke
of his Almageftes, whereof at anye other tyme I will more
largelye intreate. And for this prefent time will onlye fette
forthe the fumme of that matter in a table, whofe firfte columpne doth containe the numbre of the paralleles as Ptolemye did diftincte them. The feconde columpne contayneth a more exacte partition of thofe paralleles accordinge
vnto the increafe of the longeft daye, by a quarter of an
hower, whiche Ptolemye obferued not, after hee came to 18
howers of lengthe : but I obferue ftyll, vntill 24 howers of
length. after which time and place, bicaufe the increafe of the
longeft daye is greater and greater continuallye, I thinke it
not good to make fo curious a table for euery quarter of an
hower, but (as Erafmus Reynhold doth) to make the
diftinction thence forthe by halfe a degree of difference in
eleu-ation of the Pole, as by the table you maye fee.

In this table are fette forthe 96 paralleles iuftlye : and but
38 by Ptolomies partition : the caufe whereof, I will fhewe
you an other time. Of thefe paralleles are made 24 Climats
betweene the Equinoctiall circle & the Tropike of Cancer.
eche differinge frome the other by halfe an hower, as the lafte
columpne of the table declareth. but the elder Greekes dyd
not knowe verye well thofe North cuntries, and therefore
did they affigne only 7 climats according as I haue fet them
annexed to the firfte columne of this table.

T H E   F O V R T H E   T R E A T I S E OF

A TABLE FOR THE IVSTE
diftinction of Climates.

The numbre of the 7 climates accordinge to the elder Greekes.	Parallels after Ptol.	Parallels more exact.	De	Mi	H.	M.	The Climates.	The names	Parallels after Ptol.	Parallels more exact.	De	Mi	H.	M.	The Climates.
	1	1	0	0	12	0	1	of the 7 cli-	25	25	58	27	18	0	13
	2	2	4	18	12	15		mates after		26	59	15	18	15	
	3	3	8	34	12	30	2	fome chiefe	26	27	59	59	18	30	14
	4	4	12	43	12	45		place in thē		28	60	40	18	45	
1	5	5	16	44	13	0	3	by Meroe.	27	29	61	18	19	0	15
	6	6	20	34	13	15				30	61	53	19	15	
2	7	7	24	11	13	30	4	by Siene.	28	31	62	25	19	30	16
	8	8	27	36	13	45				32	62	55	19	45	
3	9	9	30	48	14	0	5	by Alex-	29	33	63	22	20	0	17
	10	10	33	46	14	15		andria.		34	63	47	20	15	
4	11	11	36	30	14	30	6	by the		35	64	10	20	30	18
	12	12	39	3	14	45		Rodes.		36	64	31	20	45	
5	13	13	41	23	15	0	7	by Rome.	30	37	64	49	21	0	19
	14	14	43	32	15	15				38	65	6	21	15	
6	15	15	45	31	15	30	8	by Ponte		39	65	22	21	30	20
	16	16	47	21	15	45		Euxine.		40	65	35	21	45	
7	17	17	49	1	16	0	9	by Boris-	31	41	65	47	22	0	21
	18	18	50	34	16	15		thenes.		42	66	58	22	15	
	19	19	51	59	16	30	10	by En-		43	66	7	22	30	22
	20	20	53	17	16	45		glande.		44	66	15	22	45	
	21	21	54	30	17	0	11		32	45	66	21	23	0	23
	22	22	55	36	17	15				46	66	25	23	15	
	23	23	56	38	17	30	12			47	66	29	23	30	24
	24	24	57	34	17	45				48	66	31	23	45	
									33	49	66	31½	24	0	

with

THE CASTLE OF KNOWLEDGE   191

with the quantities of their longeſt dayes, and the Eleuation of the Pole.

Parallels after Ptol.	Parallels more exact.	Elevation of the Pole. Deg	Mi.	Quantitye of the longeſt daye. Dai.	Ho.		Parallels after Ptol.	Parallels more exact.	Elevation of the Pole. De.	Mi.	Quantitye of the longeſt daye. Dai	Ho.
	50	67	0	23	11			74	79	0	127	19
34	51	67	30	33	17			75	79	30	130	17
	52	68	0	41	14			76	80	0	133	13
	53	68	30	48	6			77	80	30	136	8
	54	69	0	54	3			78	81	0	139	3
	55	69	30	59	12			79	81	30	141	11
35	56	70	0	64	11			80	82	0	144	14
	57	70	30	69	4			81	82	30	147	7
	58	71	0	73	13			82	83	0	150	0
	59	71	30	77	17			83	83	30	152	16
	60	72	0	81	17		38	84	84	0	155	8
	61	72	30	85	14			85	84	30	158	0
	62	73	0	89	8			86	85	0	160	15
36	63	73	30	92	22			87	85	30	163	5
	64	74	0	96	20			88	86	0	165	19
	65	74	30	99	21			89	86	30	168	9
	66	75	0	103	5			90	87	0	170	23
	67	75	30	106	11			91	87	30	173	13
	68	76	0	109	16			92	88	0	176	2
	69	76	30	112	20			93	88	30	178	16
	70	77	0	115	22			94	89	0	181	5
	71	77	30	118	22			95	89	30	183	19
	72	78	0	121	22			96	90	0	186	7
37	73	78	30	124	21							

Howe be it bicaufe you fhall know what names thelder grekes dyd giue them (whyche names hath beene retayned euer
fith that time) I haue here drawen a lyke table as your other
authors haue fette forthe, that you may the better conferre
the figure with the table, and the more eafilye vnderftande
the one by the other. in whiche figure the circle A, B, C, D,

*The names
and ordre
of the Cli-
mates.*

reprefetēth
the Hori-
zont, & the
righte line
A C, ftan-
deth for the
Meridiane
line. A is ẙ
north pole
and C, the
fouth pole.
B the eafte,
& D ẙ weft.
B D beto-
kening the
Equinoᶜti-
all, and E F
the tropike

of Cancer, GH, the tropike of Capricorne. and al the other
lines are the boundes of the Climats eche in his order. The
firft Climat taketh name of Meroe, a famous Iland in Ethio-
pia vnder Egypt, inclofed by the riuer Nilus. the fecōd Cli
mat is named of Syene, a city of Egypt, lying direᶜtli vnder
ẙ tropik of Cancer. The third Climate is called after Alexā-
dria, a notable city & an anciēt vniuerfity in egypt alfo, lying
on the north fhore of it. The fourth climate beareth ẙ name
of ẙ Rodes, an iſland better knowē then kept, and yet better
lofte then kepte fo derely. The fifte Climate is expreffed by
the name of Rome, a citye in Italye well ynoughe knowen.

                                                    The

The fixte climate is called after the Euxine fea, commonlye
called Ponte. The feuenth Climate reacheth from the parallele that paffeth by the mouthe of the riuer Borifthenes,
and extendeth to the parallele that runneth by the fouth par
tes of Englande, as Ptolemy witneffeth in the fecond booke
of his Almageftes. And although more maye bee faide of
the Climates, yet I will referue it to the treatife of Cofmographye, and at this time will faye no more, but that on the
other fide of the Equinoctiall towarde the Southe, there
are the like Paralleles, and the like Climates, with the fame   *The fouthe*
quantities of diftaunce from the Equinoctiall, and the like   *Climates.*
increafe of daies.

Scholar. The diftaunce of anye Climate or Parallele
frome the Equinoctiall is equall all wayes with the eleuation of the Pole aboue the Horizonte, as I maye eafilye
coniecture : fo that when I knowe the one, I mufte needes knowe the other : and that maketh me nowe to thinke
that yf I knowe anye eleuation of the Pole, I maye by
thys table eafilye declare howe farre that Parallele whi-   *The vfe of*
che ferueth for that eleuation, is frome the Equinocti-   *the table of*
alle circle : and howe longe the longeft daye is in that   *Climates.*
place : and if it chaunce that the latitude of anye region
whyche I doo feeke for, bee not in thys table iuftelye expreffed, I mufte then geffe by the proportion of thofe
twoo numbres, betweene whyche it ftandeth, what the precife lengthe of the longeft daye is.

Mafter. Thys table it felfe fuffifeth for eche quarter
of an hower betweene the longeft nighte of 24 howers,
and the longeft daye of 24 howers : but for more exacter
partes of tyme I woulde not wiffhe you to trauaile yet, tyll
I maye hereafter gyue you full rules for it : efpeciallye
feeynge thys quarter of the hower is the difference of
the whole daye, whiche mufte be parted into twoo partes, and the one halfe quarter to bee affygned to the

R.i.   difference

difference of the Sonne rifinge, and the other halfe quarter
the difference of the fonne fetting.

Scholar. That difference is more precife then our clocks
or dials do ferue vnto, and therfore I may well ynoughe bee
fatiffied with it for this time : wherefore I pray you now proceede to the Afcenfions.

Of the Af-
centions. Mafter. The vfe of the name of the Afcenfions, hathe
greate diuerfitye in it, therfore I mufte by diuifion and definition diftincte fo thofe diuers varieties, that you may iuftly knowe them eche in his kinde. And fyrft, for the name of
Afcenfion in generall, it doothe betoken the rifinge of anye
ftarres or fignes (what fo euer they be) aboue the Horizont.
But now is there dyuers obferuations of feuerall perfons
touching the rifinge of the ftarres, for Aftronomers vfe to
obferue theyr ryfinge in fourme, that is to faye, whether
they ryfe ryghte or obliquely, not regardynge (in that
confideration) the difference in the time of the daye : where
as the conninge Maryners, and authors of hufbandrye,
yea and good Phyficians alfo as well as Aftronomers do
marke their rifinge at twoo times principallye, that is when
they rife iufte at the Sonne fettinge, or els iufte at the Sonne
ryfynge.

Scholar. If Aftronomers doo nonfider onlye the fyrfte
forme, then thefe other formes do not appertaine to thys
treatife, whiche is of Aftronomye peculiarly.

Mafter. Althoughe thofe rifinges and fettinges of the
ftarres which Phyficions and other good writers of hufban
drye and writers alfo of nauigation, doo ofte times fpeake
of in their writinges, as beynge fuche, whiche in aunciente
Kalendars haue beene fette forth plainlye for all menne to
vnderftande, and fo myghte bee at this tyme alfo, yet he
that fhoulde well fette theym fo forthe, oughte to bee fkyl-
full in Aftronomye, els canne hee not doo it woorthy the
readynge, and therefore it belongeth to Aftronomers to
                                                    determine

THE CASTLE OF KNOWLEDGE 195

determine their true times. Howe bee it bycaufe Poetes haue oftener made mention of fuche ryfinges, then Aftronomers haue doone, therefore doothe Ioannes de Sacro Bofco and others alfo call them Afcenfions Poeticall : not as fayned matters, but as thinges often remembred in Poetes bookes. And as I fayde, they putte difference betweene the ryfynge of thofe ftarres in the mornyng wyth the Sonne, and the rifynge of the fame at the Sonne fettynge. The fyrfte manner of rifinge with the Sonne, they call in Latine, Ortus Cofmicus, Mundanus and Matutinus : whiche maye well bee named in Englyfhe the Mornynge ryfinge : the other forte whiche in Englifh ought to bee called the Euenynge rifinge, is named truely in Latine ortus Vefpertinus or Acronychus, and not Temporalis or Chronicus.

Scholar. Yet manye doo call it fo, and Ioannes de Sacro bofco fheweth a reafon of that name, bicaufe (fayth he) that Aftronomers vfe that time after the Sonne fettinge beft for markinge the courfe of the ftarres.

Mafter. Ignorance of the Greeke tongue hathe hindred muche manye good wittes : whiche maye often appeare not only in good Iohn de facro bofco, but alfo in many writers within thefe 300. yeares efpeciallye : but wee mufte wynke at fuche faultes, whiche rather were the faultes of the time, then of the perfons. and for this name Acronychus, is eafilye tourned into Chronicus. The fyrfte name is often readde in Ptolemye and other Greeke wryters, and is named of the begynnynge of the nyghte, whiche name by ignoraunce was tourned into Chronicos in Greeke, and fo accordinglye was called Temporalis in Latine, and then an ymagined reafon clouted thereto : lykewaies alfo in the thyrde kinde of ryfynge and fettinge, whereof the fame author doothe make mention, hit appeareth that hee was fomewhat deceaued, for that owghte not to bee called proprelye ryfynge of anye Starre

R.ij.   when

The thyrde kinde of settynge.

when it getteth oute of the Sonne beames, and maye ſhewe or ſhine at eueninge or mornynge. but it oughte rather to be called Apparition or appearynge of that ſtarre. And contrarye wayes when anye ſtarre is ſo nyghe vnto the Son that the Sonne doothe take awaye or hyde the lyghte of it, it oughte to bee called the Hydynge or occultation of that ſtarre, and not the ſettynge of that ſtarre, ſyth ſettynge and ryſynge haue propre relation to the Horizonte, and yet doothe hee and mennye other contrarye to the lear- ned Greekes call the fyrſte, the Sonnelye ryſynge of the ſtarre, and the other, the Sonnelye ſettyng of him. where as Ptolemye and the learned Greekes call the one φάσις, that is in Latine Apparitio, the ſhewynge of the ſtarre. and the contrarye is called in Greeke κρύψις, and in Lat- ine Occultatio, the darkenynge or hidynge of the ſtarre. whiche chaunce happeneth commonly to any ſtarre being within 15 degrees of the Sonne. this paſſion is called of ma-

Combuſtiō. Oppreſſion.

ny men Combuſtion : Other contract the name of combu- ſtion to ſyxe degrees, and call this Oppreſſion. but of all theſe, I will an other time declare my full mynde, for the iuſte knowledge hereof appertaineth to a higher Arte. And ſo will I hereafter giue you a table declaringe the mor- ninge and euenynge ryſynge and ſettinge of all the moſte notable ſtarres, for the matter is not ſo eaſye as it ſeemeth to bee.

Scholar. I vnderſtande it thus : that when the Sonne is in anye parte of a Signe, thoſe ſtarres whiche be in the ſame parte of that Signe, doo riſe with the Sonne, and thoſe whi- che be in the like degree of the contrarye ſigne, they riſe at the Sonne ſettynge.

Maſter. Your taking is true, for ſuche ſtarres as are nigh vnto the Ecliptike line : but yet ſuch ſtarres as be farre from the ecliptike line, may riſe or ſet with the Son, although thei be in an other Signe then the Sonne is, & ſo may they ryſe or ſet before or after ẙ ſon, although thei be in one degre of any
signe

Signe with the Sonne. And here maye you not forgette that *The eue-*
the ſtarre that ſetteth with the Sonne, is named to haue an *ning ſettíg.*
euening ſetting : and the ſtarre that ſetteth in the weſte at the *The mor-*
Sonne riſing, is iudged to haue the morning ſetting : wher- *ning ſettíg.*
by it foloweth, that the ſtarre that hath the morning riſing,
hath alſo the euening ſetting : and he that hath the eueninge
riſinge, hath the morning ſetting : thus haue I ſpoken rude-
ly and lyghtly for this time, but in the table of theſe riſinges
and ſettinges, you ſhall haue a more exacte forme of knowledge ſet out for you, touching this matter. And nowe to re
tourne to thoſe aſcenſions which be peculiarly called Aſtro
nomicall, fyrſte, for the definition you muſte vnderſtand, *Aſcentiō a-*
that Aſcenſion aſtronomicall is the certaine limitation of *ſtronomical.*
ſom pointe of the equinoctiall circle, whiche riſeth iuſtelye
with anye ſtarre, and largely taking the vſe of that name. It
betokeneth alſo the arke of the Equinoctiall circle, whiche
lyeth betweene the beginninge of the ſame Equinoctiall at
Aries, and extendeth to the iuſte degree that riſeth with any
ſtarre or ſigne. Thirdly the aſcenſion of a ſigne or conſtellation (whiche includeth a certaine meaſure in lengthe,) is
that iuſte arke of the equinoctiall, which doth paſſe the Horizont with that whole ſigne or conſtellation.

 This aſcenſion is commonly dyuided into twoo kyndes,
the one is called Ryghte aſcenſion, and the other Oblique or Crooked aſcenſion. Ryghte aſcenſion, is defined *Ryght aſ-*
to bee that, with whiche a greater portion of the Equino- *cention.*
ctiall dooth aſcende. And that is called Crooked or Ob- *Croked aſ-*
lique aſcenſion, with whiche a leſſer portion of the Equino- *cention.*
ctiall doth aſcende.

 Scholar. I heare you ſpeake of a leſſer portion and a greater portion, but where vnto thoſe compariſons ought to be
referred, I can not tell, excepte I ſhall referre the one to the
other.

 Maſter. That maye you not doo, for ſo one aſcenſion
     R.iij.   myghte

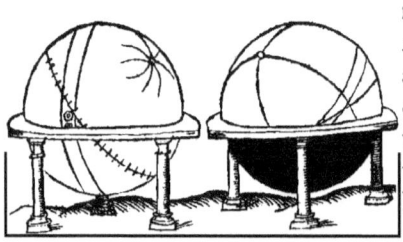

mighte bee called right & croked alſo, at the leaſt in diuers comparifons : but that can not be, no-ther is it permitted by any aſtronomers.
Scholar. How may it appeare that ſuche abſurditie woulde follow?

Maſter. To the intente that I maye alleage nothinge, but that whiche ſhall not only be certaine and true, but alſo ſhall be manifeſt to you, I will firſte inſtructe you in the vnder-ſtanding of thoſe Aſcenſions, and after that I will return to the proof of theſe my woordes. And for the better vnder-ſtandinge of both definitions, I will name vnto you a thirde Aſcenſion, which muſt be as the rule of thoſe other 2, and that is the Meane aſcenſion, for ſeyng you can not well refer greater and leſſer but other to one common meane, or els eche to other : and I haue ſaid before (and wil proue it anon) that they can not be compared togither, therfore muſt they bee referred and compared to one common meane, whiche I call the Mean aſcenſion, bicauſe that with it ther aſcendeth not ſo muche of the Equinoctiall, as with the right aſcenſiō, nor ſo lytle as doth aſcende with the crooked aſcenſion. and for this cauſe may it well be called a Mean aſcenſion. Again it maye be called a Meane aſcenſion, bicauſe it is without all exceſſe : for the portion of the Equinoctiall whiche aſcen-deth with it, is equall to it in preciſenes of degrees, ſo that neither of them excedeth other.

The meane Aſcenſion.

Scholar. It ſeemeth reaſonable that all exceſſes beinge re-ferred to anye one thinge, do approue that one as a meane betwene them, namely when the exceſſes decline to both ex-tremities, as more and fewer, greater & leſſer do. but in al this kinde of doctrine, the wordes are more eaſye to bee vnder-ſtande, then the matter. Therfore excepte ye do with exam-ples

THE CASTLE OF KNOWLEDGE   199

ples declare thefe varieties of Afcenfions, I doubte it wyll
be longe before I fhall well conceaue them and rightlye diftincte them.

Mafter. You haue learned before, that there is two varieties of Spheres, a Righte Sphere, and a Bowing fphere : and
as in eche of thefe the Equinoctiall doth kepe one vniforme
afcenfion, that is to fay, ẏ in 24 houres iuftlye all the equinoctiall doth afcende, and fo confequently in euerye hower of
the daye 15 degrees of the Equinoctiall doo paffe the righte
horizont, fo the Zodiake whiche is the circle of the fignes,
by meanes of his obliquitie, dooth not keepe vniforme afcenfion anye where in any pofition of Sphere. for although
the whole Zodiake do afcend iuftly in 24 howers, yet in euery hower, vnequal portions of it do afcend, and that diuerfly, according to the diuerfities of the Climates. But in a ge- *Certain ge-*
neraltye of differences, you may take thefe generall rules. *nerall rules
in a righte*

In the right fphere, euerye quarter of the Zodiak hath an *Sphere.*
equall or Meane afcenfion, with euery quarter of the equi- I.
noctiall, beginning the quarters at the 4 principall points,
whiche I haue before fet forthe : for if you fhoulde take three
fignes in other partes of the Zodiak, their afcenfions wyll
not agree with a quarter of the Equinoctiall, fith there is no
one figne that doth equally agree in afcenfion with the lyke
portiõ of the Equinoctial, that is to fay, with 30 degres in it.

Scholar. This rule is in Ioannes de Sacro bofco, and in
Orontius alfo.

Mafter. Then you beleue it the better.

Scholar. Yea in deede.

Mafter. Then tell me whether the afcenfion of one of thofe
quarters of the Zodiake, ought to be called a Right afcenfion, or a Crooked afcenfion.

Scholar. Neither of bothe, as I do vnderftande their definitions, feeyng the arke of the Equinoctiall that afcendeth
with them, is nother greater nother yet leffer then they,
as thefe definitions do importe, but is equal with them, and
                                                        ther-

therfore it feemeth to me more apte to call it a Meane afcenfion after your definition.

Mafter. You faye truthe, and therefore is their doctrine
imperfecte, that make but two afcenfions, where thre ought
to be diftincte, (and them felues name thre in vfe, and but 2
in diftinction and definition) namely feyng (as Tullye hath
fayd) it is the greateft faulte that can be, to omitte any mem
bre in diuifion : but to omitte their faultes in omiffion, and
to retourne to their better declaration. This fecond rule do
they alfo approoue, yea and natures ordre doth neceffarily
inferre the fame, that euerye twoo fignes or partes of Signes
equall in quantitie, and lyke diftaunte from anye one of the
4 principall pointes, haue equall afcenfions eche to other.

Scholar. That is to meane, that Taurus, and Aquarius
haue equall afcenfion, bicaufe they are equally diftaunt from
the Equinoctiall pointe of Aries.

Mafter. And fo haue Taurus and Leo, bicaufe they
differre equallye frome the Tropicall pointe of Cancer,
and fo of all the other. But to the intente that you maye
the better vnderftande all this that is faide, and the refte that
is to be faide, I haue here fet forthe in a table the iufte numbres of degrees of the Equinoctiall circle, which do anfwer
to the degrees of euery figne in their afcenfions in the right
Sphere. So that if you defire to knowe the afcenfion of any
degree of anye figne, firfte feeke out the figne, and then in
the firfte columne looke for the noumbre of the degree,
againft whiche in the common corner vnderneth the Signe
you may fee the numbre of the degrees and Minutes of the
Equinoctiall, that do afcende with that degree of the figne.
And thofe degrees be accompted frō the beginning of the
Equinoctiall at Aries, and fo orderly after ẙ naturall courfe
of the fignes. wherby you maye perceaue, that Aries, Taurus and Gemini all three togither haue for their afcenfion
90 degrees, whiche numbre agreeth with the quantitie of 3.
fignes, and therfor is their afcenfion Meane. Alfo I maye
                                                        faye

II.

THE CASTLE OF KNOWLEDGE   201

A TABLE FOR THE ASCENSIONS
of the twelue Signes in the Righte
Sphere.

Degrees of signes.	Aries		Taurus		Gemini		Cancer		Leo		Virgo	
	Deg.	Min.	Deg.	Min.	Deg.	Min.	Deg.	Min.	Deg.	Min.	Deg.	Min.
1	0	55	28	52	58	51	91	5	123	14	153	3
2	1	50	29	49	59	54	92	11	124	16	154	0
3	2	45	30	47	60	57	93	16	125	18	154	57
4	3	40	31	45	62	0	94	22	126	20	155	54
5	4	35	32	43	63	3	95	27	127	21	156	50
6	5	30	33	41	64	7	96	32	128	23	157	47
7	6	26	34	39	65	10	97	37	129	24	158	44
8	7	21	35	38	66	14	98	43	130	25	159	40
9	8	16	36	36	67	18	99	48	131	26	160	36
10	9	11	37	35	68	21	100	53	132	27	161	32
11	10	7	38	34	69	26	101	59	133	28	162	28
12	11	2	39	33	70	30	103	3	134	28	163	24
13	11	57	40	33	71	34	104	8	135	28	164	20
14	12	53	41	32	72	38	105	12	136	28	165	16
15	13	49	42	32	73	43	106	17	137	28	166	11
16	14	44	43	32	74	48	107	22	138	28	167	7
17	15	40	44	32	75	52	108	26	139	27	168	3
18	16	36	45	32	76	57	109	30	140	27	168	58
19	17	32	46	32	78	2	110	34	141	26	169	53
20	18	28	47	33	79	7	111	39	142	25	170	49
21	19	24	48	34	80	12	112	42	143	24	171	44
22	20	20	49	35	81	17	113	46	144	22	172	39
23	21	16	50	36	82	23	114	50	145	21	173	34
24	22	13	51	37	83	28	115	58	146	19	174	30
25	23	10	52	39	84	33	116	57	147	17	175	25
26	24	6	53	40	85	38	118	0	148	15	176	20
27	25	3	54	41	86	44	119	3	149	13	177	15
28	26	0	55	44	87	49	120	6	150	11	178	10
29	26	57	56	46	88	53	121	9	151	8	179	5
30	27	54	57	49	90	0	122	11	152	6	180	0

THE SECOND TABLE OF THE
Afcenfions of the twelue Signes in the Righte
Sphere.

Degrees of fignes.	Libra		Scorpius		Sagittari		Capricor		Aquarius		Pifces	
	Deg.	Min.	Deg.	Min.	Deg.	Min.	Deg.	Min.	Deg.	Min.	Deg.	Min.
1	180	55	208	52	238	52	271	5	303	14	333	3
2	181	50	209	49	239	54	272	11	304	16	334	0
3	182	45	210	47	240	57	273	16	305	18	334	57
4	183	40	211	45	242	0	274	22	306	20	335	54
5	184	35	212	43	243	3	275	27	307	21	336	50
6	185	30	213	41	244	7	276	32	308	23	337	47
7	186	26	214	39	245	10	277	37	309	24	338	44
8	187	21	215	38	246	14	278	43	310	25	339	40
9	188	16	216	36	247	18	279	48	311	26	340	36
10	189	11	217	35	248	22	280	53	312	27	341	32
11	190	7	218	34	249	26	281	58	313	28	342	28
12	191	2	219	33	250	30	283	3	314	28	343	24
13	191	57	220	33	251	34	284	8	315	28	344	20
14	192	53	221	32	252	38	285	12	316	28	345	16
15	193	49	222	32	253	43	286	17	317	28	346	11
16	194	44	223	32	254	48	287	22	317	28	347	7
17	195	40	224	32	255	51	288	26	319	27	348	3
18	196	36	225	32	256	57	289	30	320	27	348	58
19	197	32	226	32	258	2	290	34	321	26	349	53
20	198	28	227	33	259	7	291	39	322	25	350	49
21	199	24	228	34	260	12	292	42	323	24	351	44
22	200	20	229	35	261	17	293	46	324	22	352	39
23	201	16	230	36	262	23	294	50	325	21	353	34
24	202	13	231	37	263	28	295	53	326	19	354	30
25	203	9	232	39	264	33	296	58	327	17	355	25
26	204	6	233	40	265	38	298	0	328	15	356	20
27	205	3	234	42	266	44	299	3	329	13	357	15
28	206	0	235	44	267	49	300	6	330	11	358	10
29	206	57	236	46	268	55	301	9	331	8	359	5
30	207	54	237	49	270	0	302	11	332	6	360	0

faye, that the lafte degree of Gemini, or anye ftarre in that degree, or in the lafte degree of Virgo, Sagittarius or Pifces, haue a Meane Afcenfion, fo that the fame ftarre haue no latitude : how be it in the eande of Gemini and Sagittarye, althoughe they haue neuer fo muche latitude, yet is their afcenfion meane. whiche prerogatiue thofe two points haue, bicaufe the lynes or circles of their longitudes doo touche bothe the Poles of the Zodiake and of the Equinoctiall, and fo dothe no other circle of longitude : wherefore all ftarres out of thofe places limited where fo euer they be, they haue no Meane afcenfion, but other Ryghte afcenfion, or els Crooked.

Scholar. Thus I perceaue that the twoo tropike pointes haue a priuiledge aboue the two equinoctiall pointes in the afcenfions.

Mafter. It feemeth fo in the righte fphere, but in the Oblique fphere the Equinoctiall pointes haue the greater priuilege : for alwaies in all places where they doo afcende, they keepe their meane afcenfion, but fo dooth not the tropike pointes in anye oblique fphere. no nother anye ftarres of their longitude, that is to faye in their Colure. for although twoo pointes in the fkie, where their Colure dooth cutte the Equinoctiall circle, haue a meane afcenfion, yet in thofe 2. places is there no ftarre that hath beene noted, as hereafter you fhall better vnderftand. But that you maye in the mean feafon knowe what fignes doo afcende righte, and which do afcende crokedlye in the righte fphere, you fhall marke this lytle table whiche I haue drawen out of the former great table, where you fee that 4 fignes agree ftyll in their afcenfion, and the firfte 4 haue but 27 degrees and 54 minutes of the Equinoctial anfwering to eche of their afcenfions : the other 4 fignes haue 29 degrees, 55 minutes for their afcenfion : and the lafte 4 haue 32 degrees and 11 minutes agreeing to theyr rifing , which degrees and minutes added togither, do make iufte 90 degrees that is exactlye one quarter of the equinoctiall

A briefe table for the righte Sphere.

Afcenfion	The twelue Signes.				Partes of the Equinoctiall		Partes of tyme	
					Deg.	Min.	Ho.	Min.
Crooked	Aries	Virgo	Libra	Pifces	27	54	I	$51\frac{3}{5}$
Crooked	Taurus	Leo	Scorpius	Aquarius	29	55	I	$59\frac{2}{3}$
Ryghte	Gemini	Cancer	Sagittarius	Capricornus	32	II	2	$8\frac{11}{15}$
The addition of thofe partes eche to his own kinde					90	0	6	0

ctialll and fo are eche ternary of thofe Signes one iufte quarter of the Zodiake.

Scholar. And in like cafe I perceaue, the 6 howers of time
that anfwereth to thofe whole quarters, is alfo the iufte quar
ter of the naturall day, which amounteth by the addition of
the three feuerall times agreing to thofe 3 feuerall afcenfions.
And as I vnderftand it, the quantitye of tyme is gathered
after the rate of 15 degrees afcendinge euerye hower, as you
faide before. fo that euerye degree afketh 4 minutes of an
hower : and 15 minutes of a degree in the Equinoctiall doo
ryfe in one minute of an hower : for this is alwaies to bee re-
mēbred, that a minute is euermore the 60 part of that thyng
whervnto it is referred. But now ther commeth to my mind
the fayinge of Ioannes de Sacro Bofco, whiche longe hathe
troubled my minde, and I can not learne of anye man howe
to vnderftande him well : for in mine opinion his woordes
import an impoffibilitie. he blameth this argument as euel :
Thefe two arkes are equall, and they begin to rife togither,
and continually ther rifeth a greater portion of the one arke
then of the other : ergo that arke will bee full rifen fooneft,
whofe greater portion did alwaies rife. This argumente feemeth inuincible in mine opinion, and yet Iohn de Sacro bo
fco for improuing of it alleageth an example, wherby as he
feemeth to intend, the antecedent maye be true, and the con-
fequente falfe : and therefore the argumente mufte needes be
naught.

Mafter. Repeat you his example, that we may examine it.
                                                      Scholar.

Scholar. He willeth to take any quarter of the Zodiake, compared with his like quarter of the Equinoctiall, and to begin with that quarter from the fyrste pointe of Aries, to the latter ende of Gemini, alwaies the greater portion riseth of the Zodiake, and the lesser of the equinoctiall, and yet those two quarters ascend fully togither : and the lyke muste you vnderstande of the thirde quarter, from the beginning of Libra to the eande of Sagittarye. but contrarye waies, in the quarter that lyeth frome the fyrste parte of Cancer, to the laste of Virgo, the portion of the Equinoctiall in ry-synge, is styll greater then the parte of the Zodiake that ri-seth with it : and yet those bothe arkes doo rise iustly to gi-ther at the eande.

Master. Here is a greate fallation by Amphibologye, as Logitians do call it, so that in one sence it maye be true, and in an other it is false. And fyrste for declaration of Iohn his meaning (as I thinke) marke as many partes of those 2 firste quarters as you lyste, and still by the former table, as well as by tournynge the Sphere it selfe, it wyll appeare manyfestly, that the portion of the Zodiake is euer greater then the matche portion of the Equinoctiall.

Scholar. That is moste true. for with 12 degrees of Aries there ascendeth of the equinoctiall 11 degrees and twoo mi-nutes only of the Equinoctiall, that is 58 minutes lesse : with 30 degrees of Aries there riseth but 27 degrees and 54 mi-nutes, whiche is lesse by two degrees and syxe minutes : also in Taurus, 15 degrees hath for their ascension 42 degres and 32 minutes, that is twoo degrees and 28 minutes to lytle : the laste of Taurus ascendeth with 57 degrees and 49 minutes, whiche shoulde be 60 if it were equall with the degrees of the Zodiake. Againe the 16 degree of Gemini answereth to the ascensiō of the 74 degree and 48 minutes of the equinoctial, whiche in equalitye would be 76 : and the 29 degree of Ge-mini should haue by ordre of equalitie the 89 degree of the equinoctial, & hath but 88 degrees & 55 minuts, which is lesser

by 5 minutes then equalitye requireth, and fo doth it appear
in all the refte, faue in the verye lafte degree of Gemini, wher
bothe numbres appeare euen.

Maft. Then are the wordes of Iohn de facro bofco true.

Scholar. This matter troubleth me to muche : for of this
am I affured, that if anye two qnantities be equall togyther,
and a leffer portiō of the fyrfte matched with a greater part
of the fecond, then of neceffitye that parte that remaineth of
the fyrfte quantitie, muft needes be greater then that that
refteth of the feconde.

Mafter. That is true alfo : for if you abate vnequall partes
from 2 equall quantities, the portions that remaine will be
vnequall, and that parte will bee leafte, frome whiche the
greater portion was abated.

Scholar. As that can not be falfe, fo it feemeth to me, that
feyng there doth afcende with the whole figne of Aries but
27 degrees, and 54 minutes, there muft needes remain 62 de
grees and 6 minutes of that quarter, and that is more then
the 60 degrees which refteth of the like quarter of the Zodiake. Now thofe 62 degrees and 6 minutes will afcend with
the 60 degrees of the Zodiake, fo that then there dooth not
ftyll afcende a leffer portion of the Equinoctiall : for as the
fyrfte portion was leffer, fo this feconde parte is greater.

Mafter. Your coniecture is good : and to approue it the
better, you may conferre fome leffer partes of thofe 2 quarters togither, as from the 20 degree of Taurus, to the 10 de
gree of Gemini, the degrees betweene them are 20 : & to know
the arke of the equinoctiall that afcendeth with thofe 20 degrees, fubtracte the leffer from the greater, and the afcenfion
of thofe 20 degrees wyll remayne.

Deg.	Min.	
68	21	Scholar. The afcenfion of the 20 degree of Taurus is 47 degrees and 33 minutes : the afcenfion of the 10 degree of
47	33	Gemini is 68 degrees, and 21 minutes. wherfore fetting thofe
20	48	numbres in conuenient ordre, and making fubtractiō duly,

ther refteth 20 degres & 48 minuts, fo is this portiō of ẙ equi
noctiall

noctiall the greater by 48 minutes.

Mafter. Proue again from the 28 degree of Taurus, to the 28 degree of Gemini : whiche difference is 30 degrees.

Scholar. With the 28 degree of Taurus there dooth af-cende 55 degrees, and 44 minutes : and with the 28 of Gemini, 87 and 49. and by Subtraction the difference appeareth to bee 32 degrees, and 5 minutes. fo is the arke of that Equinoctiall greater by two de-grees and 5 minutes, then the matche arke of the Zodiake. And therefore are not Iohn de Sacro bofco his woordes true.

$$\begin{array}{cc} 87 & 49 \\ 55 & 44 \\ 32 & 5 \end{array}$$

Mafter. Prooue yet more before you condemne him. try the arke from the tenth degree of Taurus, to the 22 degre of the fame figne, whiche arke includeth 12 degrees of the Zo-diak.

Schol. The 10 degre of Taurus, afcedeth with 37 degrees & 35 minuts of the equinoctial : y̆ 22 degre of y̆ fame fign hath for his afcenfiõ 49 degrees & 35 minuts, y̆ difference between them by fubtractiõ is found to be 12 degres iuft : and fo that arke of the Equinoctiall is equall with his matche arke in the Zodiake.

$$\begin{array}{cc} 49 & 35 \\ 37 & 35 \\ 12 & 0 \end{array}$$

Mafter. Yet ones more proue the arke frõ the laft degre of Aries to y̆ fecond degre of Gemini, which ark is 32 degrees.

Scholar. The laft degree of Aries rifeth with 27 degrees, and 54 minutes : and the 2 of Gemini hath 59 degrees and 54 minutes in his afcenfion. betwene which 2 numbres, the diftaunce is 32 degrees exactly, and fo are thofe arkes equall alfo, and neither of thofe 2 examples do make the arke of the Equinoctiall leffer then the matche arke in the Zodiake : fo that they make agaynft Iohn de Sacro bofco.

$$\begin{array}{cc} 59 & 54 \\ 27 & 54 \\ 32 & 0 \end{array}$$

Mafter. In deede as his woordes be placed in the Prefent time, they can not be true, but his meaning may be more fa-uourably gathered, by turning the Prefent time into y̆ Per-fect time, & referring the name of afcenfion to the whole arke

S.ij.   that

that is fully ryſen in that quarter, as I dyd in the explication
of his wordes occaſion you to make proofe : wherfore take
anye parte of the fyrſte quarter, and accompt from the beginninge of Aries : or lykewaies any part of the thyrd quar
ter, and recken from the beginning of Libra, and ſo ſhall
you ſee alwaies that the portion of the Zodiake whiche is
aſcended, ſhall be greater then the parte of the Equinoctiall
that is riſen with it : & ſo ſhall it continue euen to the very laſte
degre of them bothe, and then at length doth both the quar
ters end their aſcenſions exactly togither.

Scholar. As you ſaye. nowe doo I perceaue it, ſo that the
faulte is rather in his woordes then in his meanynge.

Maſter. Such meane matters muſt be winked at in other,
but not folowed. And nowe for the ordre of Aſcenſion of ẙ
other 2 quarters which begin at Cancer & Capricorne, you
ſhall vnderſtand the lyke : but that the greater portion ẙ aſcēdeth is referred to ẙ Equinoctial circle & not to ẙ Zodiak

Scholar. So I vnderſtand by this former table that with ẙ
28 degree of Cancer there aſcendeth 120 degrees and 6 minutes of the Equinoctiall, which is two degrees and 6 minutes more then equality woulde yelde : and with the 26 degree of Virgo, there riſeth the 176 and 20 minutes of the e-
quinoctiall, whiche is alſo more then equallenes by 20 minutes : and ſo if I take anye degre of any ſigne in that ſecond
quarter, or in the fourth quarter, beginning at Capricorn,
I may lyghtly ſee by the table that the portion of the Equinoctiall in his aſcenſion is greater then the matche arke of
the Zodiake from the beginninge of Aries to that degree.
wherby it appeareth that al thoſe 6 ſignes do aſcend right, bi
cauſe a grater portiō of the equinoctiall aſcēdeth with thē.

Maſter. Then by the like reaſon, the other 6 ſignes Aries,
Taurus, Gemini, Libra, Scorpius and Sagittarius do aſcēd
crokedly, bicauſe ẙ leſſer portiō of ẙ Equinoctial doth aſcēd
with thē : after ẙ ſort of conferēce, which is cōtrary to ẙ I ſaid
before, ẙ 4 ſignes only do aſcend ryght in the Ryght ſphere
wher-

wherefore you mufte vnderftande, that for to knowe the afcenfion of euerye figne, you muft confider that figne alone,
and the arke of the Equinoctiall that dooth afcend with it,
and fo fhall you fee exactly the afcenfion of euerye figne feuerally. And here you fhall vnderftande, that all Aftronomers commonly do call the Right afcenfion fo largely, that *An other fi-*
it extēdeth to the afcenfiō of all the fignes in a Right fphere : *nification*
and fo they name the Oblique afcenfion the rifing of all the *of right af-*
cenfion.
Signes in anye Oblique Sphere, whereby it appeareth that
they giue the name of Ryghte and Crooked afcenfions, accordinge to the Horizontes or pofitions of the Sphere, and
not after the quantities of time in their afcenfion. And this
fhall fuffice at this time touchinge afcenfions in the Righte
Sphere : in which alfo the defcenfions or fettinges vnder the
Horizont, are equall with the Afcenfions, fo that they need *Of the def-*
not to haue anye peculiare declaration : but in the Oblique *cention of*
Signes.
Spheres it is not fo, but contrary waies. thofe fignes that do
afcende righte, doo defcende crooked : and they that afcende
crooked, doo defcend righte : fo that the defcenfion of anye
figne in an Oblique fphere, is equall precifely to the afcenfiō
of the contrarye figne.

    Schollar. You meane that the defcending of Aries is equal
to the afcendinge of Libra, and the defcendinge of Taurus
is one in quantity of time with the afcenfion of Scorpius.

    Mafter. So is it in deed. And in this greate varietie you
fhall marke one conftaunte vniformitie, that the afcenfion
and defcenfion of any figne in any croked fphere ioyned by
addition togither, doo make an equall fumme of time with
the afcenfion and defcenfion of the fame figne in a righte
fphere, in lyke forte ioyned togither : but to the intente that
you maye vnderftande all thefe thinges the better, and alfo
knowe the iufte afcenfion of euerye figne in this our Climat
where the eleuation of the pole is 52 degrees, I haue drawen
heere a fpeciall table for that latitude. in whiche you fhall vfe
the like manner of entringe, as you did in the other, fo that

             S.iij.   although

A TABLE OF ASCENSION OF
the Signes in 52 degrees of Latitude.

Degrees of signes	Aries		Taurus		Gemini		Cancer		Leo		Virgo	
	Deg.	Min.	Deg.	Min.	Deg.	Min.	Deg.	Min.	Deg.	Min.	Deg.	Min.
0	0	0	12	48	29	42	56	11	94	6	137	0
1	0	24	13	26	30	24	57	17	95	30	138	37
2	0	48	13	45	31	7	58	24	96	54	139	54
3	1	13	14	14	31	50	59	31	98	18	141	20
4	1	37	14	43	32	34	60	39	99	42	142	47
5	2	2	15	12	33	18	61	48	101	7	144	13
6	2	16	15	42	34	3	62	58	102	32	145	40
7	2	51	16	13	34	49	64	9	103	57	147	6
8	3	15	16	43	35	36	65	20	105	22	148	32
9	3	40	17	14	36	24	66	32	106	47	149	58
10	4	5	17	45	37	12	67	45	108	12	151	24
11	4	30	18	16	38	1	68	59	109	38	152	50
12	4	55	18	48	38	51	70	13	111	4	154	16
13	5	20	19	20	39	42	71	28	112	30	155	42
14	5	45	19	52	40	34	72	44	113	56	157	8
15	6	10	20	25	41	26	74	0	115	23	158	39
16	6	15	20	59	42	19	75	17	116	49	160	0
17	7	1	21	34	43	13	76	34	118	15	161	26
18	7	26	22	8	44	8	77	52	119	42	162	52
19	7	52	22	43	45	3	79	11	121	8	164	18
20	8	18	23	18	45	59	80	30	122	35	165	43
21	8	44	23	54	46	56	81	50	124	2	167	9
22	9	11	24	31	47	54	83	10	125	28	168	35
23	9	37	25	8	48	53	84	32	126	55	170	1
24	10	4	25	45	49	53	85	51	128	22	171	27
25	10	31	26	23	50	54	87	12	129	48	172	52
26	10	58	27	2	51	56	88	34	131	15	174	18
27	11	25	27	41	52	59	89	57	132	43	175	44
28	11	53	28	21	54	2	91	20	134	8	177	9
29	12	20	29	2	55	6	92	43	135	34	178	35
30	12	48	29	42	56	11	94	6	137	0	180	0

THE CASTLE OF KNOWLEDGE   211

Degrees of signes.	Libra		Scorpius		Sagittari.		Capricor.		Aquarius		Pisces	
	Deg.	Min.	Deg.	Min.	Deg.	Min.	Deg.	Min.	Deg.	Min.	Deg.	Min.
0	181	0	223	0	265	54	303	49	330	18	347	12
1	181	25	224	26	267	17	304	54	330	59	347	40
2	182	52	225	52	268	40	305	58	331	39	348	7
3	184	16	227	19	270	3	307	1	332	19	348	35
4	185	42	228	45	271	26	308	4	332	58	349	2
5	187	8	230	12	272	48	309	6	333	37	349	29
6	188	33	231	38	274	9	310	7	334	15	349	56
7	189	59	133	5	275	29	311	7	334	52	350	23
8	191	25	234	32	276	50	312	6	335	29	350	49
9	192	51	235	58	278	10	313	4	336	6	351	16
10	194	17	237	25	279	30	314	1	336	42	351	42
11	195	42	238	52	280	49	314	57	337	17	352	8
12	197	8	240	18	282	8	315	52	337	52	352	34
13	198	34	241	45	283	20	316	47	338	26	352	59
14	200	0	243	11	284	43	317	41	339	1	353	25
15	201	26	244	37	286	9	318	34	339	35	353	50
16	202	52	246	4	287	16	319	26	340	8	354	15
17	204	18	247	30	288	32	320	18	340	40	354	40
18	205	44	248	56	289	47	321	9	341	12	355	5
19	207	10	250	22	291	1	321	59	341	44	355	30
20	208	36	252	48	292	15	322	48	342	13	355	55
21	210	2	253	13	293	28	323	36	342	46	356	20
22	211	28	254	38	294	40	324	24	343	17	356	45
23	212	54	256	3	295	51	325	11	343	47	357	9
24	214	20	257	28	297	2	325	57	344	18	357	34
25	215	47	258	53	298	12	326	42	344	48	357	58
26	217	13	260	18	299	21	327	26	345	17	358	23
27	218	40	261	42	300	29	328	10	345	46	358	47
28	220	6	263	6	301	36	328	53	346	15	359	12
29	222	33	264	30	302	43	329	36	346	44	359	36
30	223	0	265	54	303	49	330	18	347	12	360	0

S.iij.   although

althoughe the numbres differ, yet the woorke differeth not in this table. the fyrſt columne containeth the degrees of the Signes, and the other columnes doo containe the degrees & minutes of the Equinoctiall vnder eche ſigne, accordingly as they doo anſwere to the Aſcenſion of the degrees of the ſame Signes. By this table may you ſee a great diuerſitie in the Aſcenſions from thoſe in the Righte Sphere : And yet this maye you certainly obſerue : that euerye two ſignes beinge contrarye togither, the one lyinge againſt the other, as they haue farre vnlyke aſcenſions, ſo yet if you adde their bothe aſcenſions togither, they will be equall to the aſcenſions of the ſame twoo ſignes in the Right ſphere.

Scholar. Then in as muche as the aſcenſion of Aries is in this latitude 12 degrees and 48 minutes, & the aſcenſion of Libra, 43 degrees iuſt, (abating as I ought 108 degrees) and ſo they bothe by addition do make 55 degrees, and 48 minutes. And

$$\begin{array}{cc} 12 & 48 \\ 43 & 0 \\ \hline 55 & 48 \end{array}$$

in the right ſphere eche of theſe ſignes hath for his aſcenſion 27 degrees and 54 minutes (for the contrarye ſignes there are equall in their aſcenſion) wherfore by addition there will amounte the ſame ſumme preciſely that was gathered before : and ſo like

$$\begin{array}{cc} 27 & 54 \\ 27 & 54 \\ \hline 55 & 48 \end{array}$$

waies of Taurus and Scorpius : their aſcenſions ioyned togyther maketh 59 degrees and 48 minutes : but in the righte ſphere, thoſe two aſcenſions maketh 59, 50. that is twoo minutes only difference in two ſignes, ſo is it but one minute in one ſigne, that is not to be regarded.

Maſter. Not greately, and eſpecially in an Introduction. But doo you marke here the Signes that aſcende ryght, and them that aſcende crooked?

Schollar. Although I ſee a difference by this table frome the other : I had thoughte that the more croked Sphere had made the more croked aſcenſion onlye : but yet that they alwaies had kepte one name in generall, and not haue chaunged it. but by your queſtion only I am admoniſhed of mine
erroure

THE CASTLE OF KNOWLEDGE   212

errour : for I fee that Libra (as it is eafilye vewed) dooth afcend here righte, and hath for his afcenfion 43 degrees, and
in the Righte fphere it dyd afcende crookedlye, and had but
27 degrees and 54 minutes for his afcention, and therefore
maye I doubte of all the refte, tyll I haue examined theyr afcenfions better.

Mafter. To eafe you of payne, lo here is a table of theyr
iufte afcenfions, which you maye examine at leafure.

A BRIEFE TABLE FOR
52. degrees of latitude.

Afcention	The 12 Signes.	*Partes of the Equin.*		*Partes of tyme.*	
		Degrees.	*Minutes.*	*Howers.*	*Minutes.*
Crooked	Aries, Pifces,	12	48	0	51 $\frac{3}{15}$
Crooked	Taurus, Aquarius,	16	54	1	7 $\frac{9}{15}$
Crooked	Gemini, Capricornus,	26	29	1	45 $\frac{14}{15}$
Ryghte	Cancer, Sagittarius,	37	55	2	31 $\frac{10}{15}$
Ryghte	Leo Scorpius,	42	54	2	51 $\frac{9}{15}$
Ryghte	Virgo, Libra.	43	0	2	52
The addition of those partes.		180	0	12	0

By this table you maye perceaue what fignes doo rife cro
kedlye, and whiche doo afcend righte, and that there bee of
eche forte 6. fo that from Cancer vnto Capricorne all the
fignes in direct ordre do afcende ryghte, and frome Capricorne to Cancer, in naturall ordre of the Signes, all thofe 6
fignes do ryfe crokedly. And this rule is generall in all thefe
northe dimates, that lye from 30 degrees of latitude (vnder
which Memphis and Alcayre are and mounte Sinay : alfo
the yfle of Madera, and the parte of the wefte Indies, called Terra florida) vnto 66 degrees and a halfe of latitude, in
that Climate wher Ifland lyeth and the north partes of Nor
waye, and namelye Halgoland, where Ohthere dwelte, that
was the fyrfte difcouerer of the north viage towarde Mofcouia.

Scholar. That viage I defire muche to vnderftande, and
                                                          fo

ſo do manye other.

Maſter. An other time ſhall ſerue for it, for now we haue an other matter in hande.

Scholar. Then for this preſent matter : Is there anye other varietie of aſcention betweene the Equinoctiall circle and the Latitude of 30 degrees?

Varietes of
Aſcenſions.

Maſter. Yea, muche diuerſitye : for (as you haue hearde) vnder the equinoctiall 8 ſignes do aſcend crokedly, and but 4 ryght : but from the Equinoctiall vnto 10 degrees of lati-tude, 6 ſignes aſcende ryght, (Gemini, Cancer, Leo, Scor-pius, Sagittarius, Capricornus) and other ſyxe croked, that is Aries, Tarurus, Virgo, Libra, Aquarius & Piſces. And from 10 degrees vnto 30 there are 8 ſignes that riſe right, as Gemini, Cancer, Leo, Virgo, Libra, Scorpio, Sagittarius, and Capricornus : and the other four, Aries, Taurus, Aqua rius and Piſces, riſe crokedly. but to the intent that you may haue the better habilitie to iudge of ſuche varieites, I haue here ſette forthe diuers tables for examples ſake : and namely ſuche, whiche importe anye varietie of alteration, or helpe to the apte vnderſtandinge of the ſame.

A TABLE FOR THE LA -
titude of .1. degree.

Aſcention	The 12 Signes.		Partes of the Equin.		Partes of tyme.	
			Degrees.	Minutes.	Howers.	Minutes.
Crooked	Aries,	Piſces,	27	42	1	$50\frac{12}{15}$
Crooked	Taurus,	Aquarius,	29	44	1	$58\frac{14}{15}$
Ryghte	Gemini,	Capricornus,	32	8	2	$8\frac{8}{15}$
Ryghte	Cancer,	Sagittarius,	32	16	2	$9\frac{1}{15}$
Ryghte	Leo	Scorpius,	30	4	2	$0\frac{4}{15}$
Crooked	Virgo,	Libra.	28	6	1	$52\frac{6}{15}$
The ſumme of those partes.			180	0	12	0

THE CASTLE OF KNOWLEDGE 215

A table for 10. degrees of latitude.

Afcention	The 12 Signes.	Partes of the Equin.		Partes of tyme.	
		Degrees.	Minutes.	Howers.	Minutes.
Crooked	Aries, Pifces,	25	51	1	43 $\frac{6}{15}$
Crooked	Taurus, Aquarius,	28	14	1	52 $\frac{14}{15}$
Ryghte	Gemini, Capricornus,	31	31	2	6 $\frac{1}{15}$
Ryghte	Cancer, Sagittarius,	32	53	2	11 $\frac{8}{15}$
Ryghte	Leo Scorpius,	31	34	2	6 $\frac{4}{15}$
Crooked	Virgo, Libra.	29	57	1	59 $\frac{12}{15}$
The fumme of those partes.		180	0	12	0

A table for 11 degrees of latitude.

Afcention	The 12 Signes.	Partes of the Equin.		Partes of tyme.	
		Degrees.	Minutes.	Howers.	Minutes.
Crooked	Aries, Pifces,	25	38	1	42 $\frac{8}{15}$
Crooked	Taurus, Aquarius,	28	4	1	52 $\frac{4}{15}$
Ryghte	Gemini, Capricornus,	31	27	2	5 $\frac{12}{15}$
Ryghte	Cancer, Sagittarius,	32	57	2	11 $\frac{12}{15}$
Ryghte	Leo Scorpius,	31	44	2	6 $\frac{14}{15}$
Ryghte	Virgo, Libra.	30	10	2	0 $\frac{10}{15}$
The fumme of the partes.		180	0	12	0

A table for 20. degrees of latitude.

Afcenfion	The 12 Signes.	Partes of the Equin.		Partes of tyme.	
		Degrees.	Minutes.	Howers.	Minutes.
Crooked	Aries, Pifces,	23	39	1	34 $\frac{9}{15}$
Crooked	Taurus, Aquarius,	26	27	1	45 $\frac{12}{15}$
Ryghte	Gemini, Capricornus,	30	48	2	3 $\frac{3}{15}$
Ryghte	Cancer, Sagittarius,	33	36	2	14 $\frac{6}{15}$
Ryghte	Leo Scorpius,	33	21	2	13 $\frac{6}{15}$
Ryghte	Virgo, Libra.	32	9	2	8 $\frac{9}{15}$
The fumme of the partes.		180	0	12	0

A table for 29. degrees of latitude.

Afcenſion	The 12 Signes.	Partes of the Equin.		Partes of tyme.	
		Degrees.	Minutes.	Howers.	Minutes.
Crooked	Aries, Piſces,	25	51	1	43 $\frac{20}{15}$
Crooked	Taurus, Aquarius,	28	14	1	52 $\frac{1}{15}$
Ryghte	Gemini, Capricornus,	31	31	2	6 $\frac{1}{15}$
Ryghte	Cancer, Sagittarius,	32	53	2	11 $\frac{8}{15}$
Ryghte	Leo Scorpius,	31	34	2	6 $\frac{12}{15}$
Ryghte	Virgo, Libra.	29	57	1	59 $\frac{3}{15}$
The ſumme of those partes.		180	0	12	0

A table for 30 degrees of latitude.

Afcenſion	The 12 Signes.	Partes of the Equin.		Partes of tyme.	
		Degrees.	Minutes.	Howers.	Minutes.
Crooked	Aries, Piſces,	21	9	1	24 $\frac{9}{15}$
Crooked	Taurus, Aquarius,	24	23	1	37 $\frac{8}{15}$
Crooked	Gemini, Capricornus,	29	56	1	59 $\frac{11}{15}$
Ryghte	Cancer, Sagittarius,	34	28	2	17 $\frac{13}{15}$
Ryghte	Leo Scorpius,	35	25	2	21 $\frac{10}{15}$
Ryghte	Virgo, Libra.	34	39	2	18 $\frac{9}{15}$
The ſumme of the partes.		180	0	12	0

A table for 50. degrees of latitude.

Afcenſion	The 12 Signes.	Partes of the Equin.		Partes of tyme.	
		Degrees.	Minutes.	Howers.	Minutes.
Crooked	Aries, Piſces,	13	52	0	55 $\frac{7}{15}$
Crooked	Taurus, Aquarius,	17	55	1	11 $\frac{10}{15}$
Crooked	Gemini, Capricornus,	27	0	1	48
Ryghte	Cancer, Sagittarius,	37	24	2	29 $\frac{9}{15}$
Ryghte	Leo Scorpius,	41	53	2	47 $\frac{8}{15}$
Ryghte	Virgo, Libra.	41	56	2	47 $\frac{11}{15}$
The ſumme of those partes.		180	0	12	0

A table for 60. degrees of latitude.

Afcenfion	The 12 Signes.	Partes of the Equin.		Partes of tyme.	
		Degrees.	Minutes.	Howers.	Minutes.
Crooked	Aries, Pifces,	7	16	0	29 $\frac{1}{15}$
Crooked	Taurus, Aquarius,	10	56	0	43 $\frac{11}{15}$
Crooked	Gemini, Capricornus,	22	56	1	31 $\frac{11}{15}$
Ryghte	Cancer, Sagittarius,	41	28	2	45 $\frac{13}{15}$
Ryghte	Leo Scorpius,	48	52	3	15 $\frac{7}{15}$
Ryghte	Virgo, Libra.	48	32	3	14 $\frac{2}{15}$
The fumme of those partes.		180	0	12	0

A table for 66 degrees and ½ of latitude.

Afcenfion	The 12 Signes.	Partes of the Equin.		Partes of tyme.	
		Degrees.	Minutes.	Howers.	Minutes.
Sudden	Aries, Pifces,	0	0	0	0
Sudden	Taurus, Aquarius,	0	0	0	0
Sudden	Gemini, Capricornus,	0	0	0	0
Ryghte	Cancer, Sagittarius,	64	22	4	17 $\frac{7}{15}$
Ryghte	Leo Scorpius,	59	49	3	59 $\frac{4}{15}$
Ryghte	Virgo, Libra.	55	49	3	43 $\frac{4}{15}$
The fumme of the partes.		180	0	12	0

Scholar. Sire I thanke you mofte hartely for thefe tables, for I haue not feene the lyke of them before : and theyr or-dre is fo eafye, that I neede no greate healpe in the vnder-ftandinge of them : For as in the tytle of eche of them is fette the degree of the latitude of the Region for whyche the table is calculate, fo in the fyrfte columne is fette the differences of the afcenfions in name, and in the fe-conde columne are the names of the Signes, whiche haue thofe diuers Afcenfions, eche rowe contayning two Signes, whereby they differ from the ryght Sphere, for in it 4 Si-gnes agree in one quantitie of afcenfion, wheras in all thefe

T.i.   Ob-

Oblique fpheres, only twoo fignes doo agree in lykenes of
afcenfion. And in eche of them are there fette in the thirde
column, the degrees of Afcenfion, and minutes after them,
whiche appertayne to euerye figne : and in the fourthe Columne are the partes of tyme, agreeynge to thofe partes of
the Equinoctiall circle : by whiche it maye appeare not onlye
howe manye degreees and minutes thofe Signes occupye
in their Afcenfion, but alfo howe manye howers or minutes doo anfwere to the fame. And in eche table is fette
the full quantitie of halfe a daye, and alfo of halfe the Zodiake, whiche is the full fumme by addition of all the other
The firfte rule of Ob lique Afcé tion. percelles ouer them : whereby I perceaue it to bee true, that
eche halfe of the Equinoctiall dooth equallye afcende wyth
eche halfe of the Zodiake.

Mafter. Beginninge the halues of them bothe at the
Equinoctiall pointes, in Aries and Libra, it is moft true :
but not fo yf you begin at the Tropike pointes, or in anye
other partes of theym : for yf you begynne at anye of
the northerlye Signes betweene Aries and Libra, and fo
recken 6 fignes togyther, thofe Sygnes fhall haue a ryghte
Afcenfion : for wyth them fhall afcende a greater portion of the Equinoctiall. But if you doo recken fyxe
Signes and begynne that accompte betweene Libra and
Aries, in the fouthe parte of the Zodiake, then doo
thofe fyxe fignes afcende crookedlye : for as muche as the
portion of the Equinoctiall that ryfeth with them, is leffe
then halfe of it.

Scholar. For proofe thereof I take the
table of tenne degrees of latitude, and I
begynne with Taurus, and fo doo I recken fyxe Signes, Taurus, Gemini, Cancer, Leo, Virgo and Libra, vnto which
Signes thefe fyxe numbrs anfwere as they
be here fet, accompting one numbre twife,
that

Degrees	Minutes.
28	14
31	31
32	53
31	34
29	57
29	57
184	6

that is fyrſt for Virgo, and then for Libra, and ſo the whole
ſumme of partes of the Equinoctiall is 184 degrees and 6
minutes : that is 4 degrees and 6 minutes more then halfe :
wherefore thoſe ſignes do aſcende right. And ſo I perceaue
it wyll be in the other lyke woorkes, if I doo begynne wyth
anye Signe in that northe halfe of the Zodiake, for ſeeynge Aries hathe the leaſte of all other Aſcenſions, if I
take anye other Signe, and omytte hym, I ſhall haue a
greatter noumbre then the halfe of the Equinoctiall circle. But nowe contrarye wayes if I begynne wyth anye
of the ſouthe Signes, and ſo recken ſyxe continuall Sygnes, theyr Aſcenſion you ſaye will bee an Oblyque aſcenſion, bycauſe theyr degrees wyll bee more in noumbre then the degrees of the Equinoctiall circle : for example I take my beginninge at Sagittarius, and ſo recken
forthe directelye ſyxe Signes, that is Sagittarius, Capricornus, Aquarius, Piſces, Aries and Taurus. and

for them I take the numbres of their Aſcen-	Deg.	Min.
ſions, and ſet them downe as here you ſe : ſo	32	53
that by addition they doo make 172 degrees,	31	31
and 34 minutes : that is leſſe then the halfe	28	14
circle by ſeuen degrees, and 26 minutes.	25	51
wherefore it muſte needes bee, that thoſe	25	51
Signes doo aſcende crookedlye.	28	14
	172	34

Maſter. And ſo muſte it followe where
ſo euer you begynne after Libra in that ſouthe halfe of
the Zodiake : for ſo muche as you omytte the aſcenſion
of Libra, beeynge 29 degrees and 57 minutes, and in ſteed
of it you take the aſcenſion of Aries, whiche is but 25 degrees and 51 minutes.

Scholar. Thys reaſon doothe appeare manyfeſte y-
noughe : and that not only in this table, but alſo in al the o-
ther, ſaue that in the laſte table I ſee a ſtraunge dyſa-
greemente frome all the other. for in theſe ſyxe Signes,

T.ij.   Aries

Aries, Taurus, Gemini, Capricornus, Aquarius & Pifces,
there is fet no numbres of degrees or minutes for their afcenfion, but only cyphers, whiche thyng is ftraunge to me,
for thereby may it be coniectured, that thofe 6 Signes haue
none Afcenfion at all : and yet I am fure that the fyrfte
three of them doo afcende not onlye in that Climate, but
alfo in all other Climates be north that latitude euen to the
northe Pole.

Mafter. A lyttle miftakinge dooth difturbe your mynde
muche, but yf you doo place the fphere in the Horizonte,
in fuche forte, that the northe Pole be 66 degrees and halfe
aboue the Horizonte, and then tourne the fyrfte degree of
Aries, to the eafte Horizonte readye to afcende, and afterwarde yf you tourne the Globe towarde the wefte, but by
the quantitie of halfe one degree in the Equinoctiall, you
fhall perceaue that all thofe fixe Signes whyche lye from
the wynter Tropyke vnto the Sommer Tropike, that is
to faye, Capricornus, Aquarius, Pifces, Aries, Taurus,
and Gemini, wyll afcende fodainlye in one momente all 6
at ones : fo that for their afcenfion there canne be affigned
no degree of the Equinoctiall, nother anye fenfible parte
of tyme, fyth it is doone in a momente of tyme. and therfore mufte I putte no degree for their Afcenfion, nother yet
anye tyme. And bycaufe I thoughte no leffe but that this
woulde feeme fome thynge ftraunge vnto you, therefore
haue I not touched anye thinge of the other Afcenfions
for thefe Climates that bee betweene the Tropike of Cancer and the Pole, beynge adfured that they woulde feeme
to you muche more ftraunge, then thys doothe. but
hereafter yf I perceaue that you trauayle well in thys firft
Introduction, I wyll inftructe you more largelye in
all that fhall bee needefulle for you : and in the meane
ceafon I wylle profecute the rules of thefe Afcenfions
in the Oblyque Spheres, as I dydde begynne.

Wher-

wherefore you fhall note, that althoughe eche halfe of the
Zodiake doo agree in afcenfion with eche halfe of the
Equinoctiall, yet the partes of thofe halues, I meane the feuerall fignes, and their diftincte portions doo not fo agree,
but are ether more or leffe.

Scholar. So I remembre doth Iohn de facro Bofco affirm :   *Iohn de fa-*
for (faithe hee) in that halfe of the Zodiake, which is be-   *cro Bosco*
tweene the beginninge of Aries, and the eande of Virgo,   *his rules ex*
                                                            amined.
alwaies the portion of the Zodiake whiche rifeth, is greater then the like halfe of the Equinoctiall : and yet thofe halues doo rife togither.

Mafter. This he fpeaketh of the Oblique fphere.

Scholar. So dooth he in deede.

Mafter. Propounde you an example, that I maye knowe
howe you do vnderftande it.

Scholar. I take an example out of the table of 50 degrees
of latitude, and for the fyrfte fyue Signes I fette   13   52
the quantities of their afcenfions, as heere is   17   55
feene, whyche by Addition doo make 138 de-   27   0
grees and foure minutes. fo dooth there wante   37   24
of 150 degrees, whiche are the fulle degrees for   41   53
fyue fignes, 11 degrees and 56 minutes, that arke   138   4
therefore of the Equinoctiall is leffer then the matche arke
of the Zodiake : but nowe there refteth in that halfe of the
equinoctiall 41 degrees and 56 minutes, whiche is the iufte
afcenfiõ of Virgo, in that latitude. and fo thofe both halues
doo afcend ioyntly togither.

Mafter. Prooue the lyke woorke in the table of 10 degrees of latitude.

Scholar. For the firfte 5 fignes Aries, Taurus   25   51
Gemini, Cancer and Leo, I fet their afcenfions   28   14
thus. And by addition I fynde that theyr whole   31   31
fumme for all that arkes afcenfion is 150 degrees   32   53
and three mynutes. that is three mynutes   31   34
more thenne the degrees of fyue Sygnes,   150   3

T .iij.   whiche

whiche is 5 times 30. And ſo is this example againſt the rule,
for here the greater portion is of the Equinoctiall.

Maſter. Proue yet againe in the table of one degree of latitude.

Scholar. The aſcenſions of the fyrſte 5 ſignes   27   42
in that latitude, are theſe : and make in one total   29   44
ſumme, 151 degrees, and 54 minutes : that is 1 de-   32   8
gree, and 54 minutes more then the like arke of   32   16
the 5 ſignes in the Zodiake, whiche contayneth   30   4
but onlye 150 degrees. And ſo is this example   151   54
alſo againſt the rule.

Maſter. So you haue two examples contrary to that rule.

Scholar. It can not be denyed.

Maſter. Then is that no certain rule.

Scholar. It ſeemeth ſo.

Maſter. In deede it is true onlye aboue 13 degrees of latitude. for in all climates and paralleles vnder 13 degrees of latitude, the equinoctiall maketh greateſt numbre of degrees
in his arke. ſo that Iohn de ſacro Boſco his woordes maye
not be accompted true generally (as they ſounde) but particularly betwene 13 degrees of latitude, and 66 and an halfe :
and ſo is it to be ſayde of diuers other of his rules.

Scholar. Is there the lyke diuerſitye beyonde 66 degrees
and a halfe northward?

Maſter. There is more diuerſitie, but ſuch and ſo ſtraung
as I will not at this time trouble your head withall, but wyll
appoint a more conuenient place for it.

Scholar. Then I beſeeke you to proſecute the reſt of Iohn
de ſacro Boſco his rules, touchinge aſcenſions.

Maſter. Repete you the rules.

Scholar. His nexte rule is : that in the other halfe of the
Zodiake, from the beginning of Libra, to the eande of Piſces euermore there riſeth a greater parte of the Equinoctial
then of the Zodiake, and yet bothe thoſe halues doo ryſe
fully togither.

Maſter

Mafter. Prooue it by fome examples.

Scholar. In the latitude of 30 degrees I take Libra onlye, and fynde againft it 34 degrees and 39 minutes : fo is there 4 degrees and 39 minutes more of the equinoctiall then of the Zodiake agreablye to the rule. Alfo in the table of 60 degrees with Libra, there doth afcende in the equinoctiall 48 degrees and 32 minutes. that is to faye 18 degrees and 32 minutes more then the 30 degrees of Libra.

Mafter. Affaye the lyke in the latitudes of one degree, and of 10 degrees.

Scholar. In the latitude of 10 degrees, the figne of Libra hath for his afcenfion 29 degrees, and 57 minutes of the Equinoctiall, that is 3 minutes leffe then the degrees of the Zodiake, and fo is that contrarye to the fayde rule.

Mafter. Nowe proue the other.

Scholar. In that parallele where the Pole is but one degree hyghe, the Signe of Libra afcendeth with 28 degrees and 6 minutes of the Equinoctiall, fo is that arke of the Equinoctiall leffer then the degrees of the fayde figne of Libra, by 1. degree and 65 minutes, and yet by the rule it fhuld be greater. wherfore I maye perceaue, that this rule dooth not ferue for all Latitudes, but for certaine of them. And as I thinke, not for anye aboue 10 degrees, althoughe (as you fayd) the other exception did extend to 13 degrees of latitude.

Mafter. What caufeth you to thinke fo?

Scholar. The table calculate by you for 11 degrees of latitude, where I fee 30 degrees, and 10 minutes of the Equinoctiall, affigned for the afcenfion of the figne of Libra, and there is the portion of the Equinoctiall greater by 10 minutes then the portion of the Zodiake.

Mafter. In deede for whole fignes this exception extendeth not aboue 10 degees of latitude, and no more doothe the other former exception, but yet in partes of fignes it extendeth in them both to 13 degrees, as herafter you fhall perceaue more at large. but now go forth to the nexte rule.

                                      T.iiij.   Scholar

The fourth
rule.

Scholar. The fourthe rule is this : that thofe arkes which fuccede after Aries vnto the eande of Virgo in the Oblique fphere, do abate their afcenfions in comparifon to the afcenfions that they haue in the Right fphere : namely feeyng leffe dooth rife of the Equinoctiall.

A TABLE OF ASCENSIONS

fhowinge all diuerfities of them, vnto the Polare circle, peculiare for euery feuerall Signe.

Degrees of latitude.	Aries Pifces		Taurus Aquarius		Gemini Capricor.		Cancer Sagittari.		Leo Scorpius		Virgo Libra	
	Deg.	Min.	Deg.	Min.	Deg.	Min.	Deg.	Min.	Deg.	Min.	Deg.	Min.
0	27	54	29	54	32	12	32	12	29	54	27	54
1	27	42	29	44	32	8	32	16	30	4	28	6
2	27	30	29	34	32	4	32	20	30	14	28	18
3	27	17	29	25	32	0	32	24	30	23	28	31
5	26	53	29	4	31	52	32	32	30	44	28	55
8	26	16	28	34	31	40	32	44	31	14	29	32
10	25	51	28	14	31	31	32	53	31	34	29	57
11	25	38	28	4	31	27	32	57	31	44	30	10
15	24	46	27	23	31	10	33	14	32	25	31	2
20	23	39	26	27	30	48	33	36	33	21	32	9
25	22	17	25	27	30	24	34	0	34	21	33	21
30	21	9	24	23	29	56	34	28	35	25	34	39
35	19	43	23	9	29	24	35	0	36	39	36	5
40	18	4	22	45	28	47	35	37	38	3	37	44
45	16	10	20	3	28	1	36	23	39	45	39	38
50	13	52	17	55	27	0	37	24	41	53	41	56
55	11	1	15	5	25	32	38	53	44	43	44	47
60	7	16	10	56	22	56	41	28	48	52	48	32
65	2	4	3	44	15	20	49	2	56	5	53	45
66½	0	0	0	0	0	0	64	22	59	49	55	49

Ma-

Maſter. For tryall of this rule I haue ſette forth here a table contayninge all the diuerſities (though not all the ſeuerall degrees of latitude) that happen in anye Climate vnder
67 degrees of latitude, that is vnto the Polare circle. So
that by thys table you maye examine all the rules bothe
of Iohn de Sacro Boſco, and alſo of others. Nowe therefore examine thoſe arkes that followe Aries, and ſo abate
their aſcenſions, as your rule ſaythe, frome Aries, vnto the
eande of Virgo.

Scholar. Firſte for Aries it ſelfe : I ſee that it abateth in this
table from 27 degrees and 54 minutes vnto nothinge. And
Taurus abateth alſo frome 29 degrees and 54 minutes vnto
nothinge. Lykewiſe Gemini abateth from 32 degrees and 12
minutes vnto nothinge. But contrary waies, Cancer, Leo,
and Virgo, do not abate, but increaſe the quantities of their
Aſcenſions, ſo that in the three firſte Signes onlye (that is
Aries, Taurus and Gemini) that rule is true, and in the o-
ther three Signes, Cancer, Leo and Virgo, it appeareth vtterly to be falſe.

Maſter. Yet in one manner of conſideration thoſe words
maye be true as he hath ſpoken them, though not ſo largelye as the woordes do ſound : for it appeareth that your author doth accompt the beginning of thoſe arkes (whereof
he ſpeaketh) not from diuers and ſeuerall pointes but from
one common beginning, which is the fyrſt degree of Aries,
and in that ſence his rule is true. for proofe whereof here is
two other tables ſette forthe, in whiche is declared the quan
tities of the Aſcenſions of the twelue Signes, but not in ſuch
forte as it was in the table nexte before, for there euerye arke
of the ſeuerall Signes did take his beginninge at the fyrſte
degree of the ſame Signe. but in theſe twoo tables the
arke of aſcenſion is accompted from the fyrſt degree of A-
ries, as from the common beginning, and eandeth at the
laſte degree of euery ſeuerall Signe. And now by this fyrſt
table if you examine y̆ former rule you ſhal find it to be tru.

Scholar

226   THE FOVRTHE TREATISE OF

A TABLE FOR THE DIVERSITIES

of Afcenfions for the firfte 6 Signes from the Equinoctiall to the Polare circle, accomptinge the
beginninge of euery arke, from the firfte
degree of Aries.

The elevation of the Pole.	Aries		Taurus		Gemini		Cancer		Leo		Virgo	
	Deg.	Min.	Deg.	Min.	Deg.	Min.	Deg.	Min.	Deg.	Min.	Deg.	Min.
0	27	54	57	48	90	0	122	12	152	6	180	0
1	27	42	57	26	89	34	121	50	151	54	180	0
2	27	30	57	4	89	8	121	28	151	42	180	0
3	27	17	56	52	88	41	121	6	151	29	180	0
4	27	5	56	20	88	15	120	44	151	17	180	0
5	26	53	55	57	87	49	120	21	151	5	180	0
8	26	16	54	50	86	30	119	14	150	28	180	0
10	25	51	54	5	85	36	118	29	150	3	180	0
11	25	38	53	42	85	9	118	6	149	30	180	0
15	24	46	52	9	83	19	116	33	148	58	180	0
20	23	39	50	6	80	54	114	30	147	51	180	0
25	22	27	47	54	78	18	112	18	145	39	180	0
30	21	9	45	32	75	28	109	56	145	21	180	0
35	19	41	42	52	72	16	107	16	143	55	180	0
40	18	4	39	49	68	36	104	13	142	16	180	0
45	16	10	36	25	64	14	100	37	140	22	180	0
50	13	52	31	47	58	47	96	11	138	4	180	0
55	11	1	26	6	52	37	90	30	135	23	180	0
60	7	16	18	12	42	8	82	36	131	28	180	0
65	2	4	5	48	21	8	70	10	126	15	180	0
66½	0	0	0	0	0	0	64	22	124	11	180	0

Scholar. I perceaue that the fyrfte line of numbres vnder
the fignes, againft the cypher o, doth reprefent the quantities of the Afcenfions in the righte fphere, and all the other
lynes doo declare the fpeciall quantities of feuerall afcenfions

A TABLE OF THE DIVERSITIES

of Afcenſions for the 6 foutherlye Signes, accomptinge the beginninge of thoſe
Afcenſions, from Aries firſte
degree.

Degrees of latitude.	Libra		Scorpius		Sagittari.		Capricor.		Aquarius		Piſces	
	Deg.	Min.	Deg.	Min.	Deg.	Min.	Deg.	Min.	Deg.	Min.	Deg.	Min.
0	207	54	237	48	270	0	302	12	332	6	360	0
1	208	6	238	10	270	26	302	34	332	18	360	0
2	208	18	238	32	270	52	302	56	332	30	360	0
3	208	31	238	54	271	18	303	18	332	43	360	0
4	208	43	239	16	271	45	303	40	332	55	360	0
5	208	55	239	39	272	21	304	33	333	7	360	0
8	209	32	240	46	273	30	305	40	333	44	360	0
10	209	57	241	32	274	24	305	55	334	9	360	0
11	210	10	241	54	274	51	306	18	334	22	360	0
15	211	2	243	27	276	41	307	51	335	14	360	0
20	212	9	245	30	279	6	309	54	336	21	360	0
25	213	21	247	42	281	42	312	6	337	33	360	0
30	214	39	250	4	284	32	314	28	338	51	360	0
35	216	5	251	44	287	44	317	8	340	17	360	0
40	217	44	255	47	291	24	320	11	341	56	360	0
45	219	38	259	23	295	46	323	47	343	50	360	0
50	222	56	263	49	301	13	328	13	346	8	360	0
55	224	47	269	30	308	23	333	54	348	59	360	0
60	228	32	277	24	318	52	341	48	352	44	360	0
65	233	45	289	50	338	52	354	12	357	56	360	0
66½	235	48	295	30	360	0	0	0	0	0	0	0

ons in eche of thoſe diſtinct latitudes, which be noted in the
firſt columne in both tables. Therfore now I maye perceaue
according to ẙ former rule, ẙ the greateſt nūbre of any doun
right column is ẙ higheſt nūbre in ẙ hed of ẙ ſame column,
ſo that

ſo that it may truely bee ſaide (as appeareth in this firſte table) that in eche Oblique ſphere the aſcenſions of the arkes
from Aries vnto the eand of Virgo, do abate ſtill and waxe
leſſe and leſſe, in reſpecte to their aſcenſions that they haue in
the Right ſphere.

Thre ſigni-
fications of
Aſcenſion.

Maſter. Thus you ſee, howe there may be accompted diuers formes of aſcenſions : firſte (as I ſayde at the beginning
of that definition) it maye ſignifie that degree certenlye of
the Equinoctiall, whiche dooth aſcende with anye ſigne or
parte thereof : as for example. in the latitude of 50 degrees,
the laſte degree of Aries hath for his aſcenſion the 13 degree
and 52 minute of the Equinoctiall, as by the firſte of theſe
twoo tables it dooth appeare : and in the ſame table it appea
reth, that the laſte degree of Taurus hathe for his aſcenſion
in the ſame latitude the 31 degree and 47 minut of the Equi
noctiall. And in the ſeconde ſignification, the aſcenſion of
Aries whole ſigne is that whole ark of 13 degrees and 52 minutes, and ſo the whole arke from the beginning of Aries,
to the eande of Taurus, hathe for his aſcenſion that whole
arke of 31 degrees, and 47 minutes of the Equinoctiall.
And in this ſignification dooth Iohn de ſacro Boſco vſe the
name of Aſcenſion, and in this ſenſe his rules be true : accordinge to whiche ſenſe I haue drawen to you certaine tables :
the firſte for the aſcenſions of the twelue Signes in the right
Sphere : the ſecond, for the aſcenſion of the Signes in 52 degrees of latitude : the thirde and fourthe are theſe twoo
tables laſt before, which for diuers latitudes doo dedare the
quantities of the Aſcenſions of al arkes of whole ſignes accompted from the beginning of Aries. The thyrde ſignification of aſcenſions is the quantitie of that arke of the Equi
noctiall whiche aſcendeth with anye certaine arke of the Zo
diake : as for example. that arke of the equinoctiall that aſcēdeth with any ſigne ſeuerally taken, is called the aſcenſion of
that ſigne. So haue you for euery ſigne certain ſeuerall arkes
of aſcenſion aſſigned, and ſet forthe here in diuers tables, ac
cording

cordinge to diuers eleuations of the Pole. And in this fignification muſt it be vnderſtande, when it is ſayde that any
ſigne hath a Right aſcenſion or an Oblique aſcenſion, for if
the arke of the Equinoctiall that riſeth with that ſigne, bee
greater then 30 degrees, then hathe that ſigne a Righte *A Ryghte*
aſcenſion : and if the arke of the Equinoctiall be leſſer then *aſcenſion.*
30 degres, then is that aſcenſion called an Oblique aſcenſion : *An Oblique*
but if the ſayd arke of the Equinoctiall be iuſte 30 degrees, *aſcenſion.*
then is it a Meane or Equall aſcenſion. *A meane*
    Scholar. Nowe do I better vnderſtande the vſe of theſe *aſcenſion.*
names then I dyd before : and alſo I perceaue howe the names of greater and leſſer portion are to be referred, not of
eche greater to eche leſſer, for ſo the aſcenſion of Taurus
myghte be accompted greater then the aſcention of Aries,
and leſſer then the aſcention of Gemini, in all climates with
out the Polare circle. And ſo one aſcenſion might be both
greater and leſſer, and therefore bothe ryghte and crooked
whiche is an abſurditie.
    Maſter. Thus hath ordre taught you, that wherof you wer
in doubt and manifeſtly approued that that ſeemed very ob
ſcure. Now therfore returne to your author again. And repete his other rules as he doth teache them.
    Scholar. His fifte rule is this : The arkes whiche followe *The fifte*
Libra, vnto the eande of Piſces, in an Oblique ſphere, doo *rule.*
increaſe their aſcenſions aboue the aſcenſions that they haue
in the Right ſphere in as muche as the portion of the Equi
noctiall is augmented. And the increaſe of thoſe aſcenſions
is agreeable in rate to the decreaſe of thoſe other aſcenſions
whiche ſuceede from Aries to Libra.
    Maſter. This rule muſte be vnderſtande of aſcenſions in
the ſeconde ſignification : and that may you trye by the later
of thoſe twoo tables which I gaue you laſte.
    Scholar. It appeareth ſo in deed. for Libra increaſeth from
207 degrees and 54 minutes, vnto 235 degres & 49 minutes.
And Scorpio frō 237 degrees & 48 minuts, vnto 295 degres
                        V.i.   and

and 36 minutes. likwaies Sagittarius from 270 degrees vnto
360 degrees. So dooth it appeare, that Libra dooth increaſe
betweene the Equinoctiall and the Polare circle, 27 degres,
and 54 minutes. And Scorpio increaſeth 57 degrees and 50
minutes. Alſo Sagittarius augmenteth by 90 degrees.
And now contrarye waies, Aries doth abate from 27 degres
and 54 minutes to nothinge. Taurus diminiſheth frome 57
degrees and 48 minuts vnto nothinge alſo. And Gemini
abateth from 90 to 0 : ſo dooth theſe three in decreaſe agree
with the other in increaſe exactly.

  Maſter. And ſo maye you iudge of the other three cou-
The ſixte ples. And therfore ſayth your author, that hereby it is ma-
rule. nifeſt, that two equall arkes lying one againſt the other, and
in an Oblique ſphere, haue their aſcenſions ioyntlye taken
togyther equall wythe the Aſenſions of the ſame arkes in
a Ryghte Sphere, ioyntlye taken alſo : for althoughe
thoſe arkes bee vnequall togyther, yet as muche as the
one abateth on the one ſyde, ſo muche the other increaſeth
on the other ſyde. and ſo bothe arkes in the Ryght ſphere
are equall to bothe thoſe arkes in any Oblique ſphere.

  Scholar. But I praye you, in what ſignification of aſcenſion is that rule to be vnderſtande?

  Maſter. In anye of thoſe twoo which be referred to arkes :
for the fyrſte can haue no place here, bicauſe it ſignifieth the
aſcenſion of one pointe only, and not of any arke as the o-
ther twoo do, and as this rule doth importe.

  Scholar. Then may I proue by examples in both ſortes of
tables. And firſte to beginne with thoſe tables that accompt
the whole arkes from the beginning of Aries, I fynd the aſcenſion of Aries in the head of the table, that is in the right
ſphere, to be 27 degrees & 54 minutes, & the aſcenſiō of Libra

$$\frac{\begin{array}{rr} 27 & 54 \\ 207 & 54 \end{array}}{\begin{array}{rr} 235 & 48 \end{array}}$$

(which is againſt it) 207 degres & 54 minuts. which both ioyned togither, make 235 degrees & 48 minuts. Now to proue y̆
like in an Oblique Sphere, I take the latitude of 40 degrees.
             and

18	4
217	44
235	48
7	16
228	32
235	48

and there I fynde for Aries his afcenfion 18 degrees and 4 minutes : and for Libra I fynde in the feconde table 217 degrees and 44 minutes : whiche both beyng added togither, do make 235 degrees and 48 minuts. that is precifely equall with the former afcenfions in the right fphere. Alfo in the eleuation of 60 degrees I trye the like, where Aries hath 7 degrees and 16 minutes, and Libra hath 228 degrees and 32 minutes, which by additiō amount to the fame fum as before.

Mafter. Attempt the lyke in the other tables.

Scholar. I take the arke of Aries afcenfion as before 27 degrees and 54 minutes : and the afcenfion of Libra (accomptyng only the arke of it from his owne beginninge) in lyke forte 27 degrees and 54 minutes. fo that both ioyned togither, make 55 degrees and 48 minutes. Then in the latitude of 55 degrees, I fynde for Aries 11 degrees and one mynute : and for Libra, 44 degrees and 47 minutes. and by additiō I find that they make the fame numbre as before.

27	54
27	54
55	48
11	1
44	47
55	48

Mafter. Make proofe in fome other arke.

Sholar. I take fyrfte the arke from the beginning of Leo, to the eande of the fame Signe, and fynd it to bee 29 degrees and 54 minutes in the ryght fphere : and fo for the Afcenfiō of the Signe of Aquarius, beyng equall to it, and agaynfte it in the Zodiake, I fynde the lyke noumbre, whiche make by addition 59 degrees and 48 minutes. Then in the latitude of 30 degrees I trye the lyke, and fynde for Leo 35 degrees and 25 minutes : and for Aquarius there dooth rife 24 degrees and 23 minutes : which make alfo togither the fame

29	54
29	54
59	48
35	25
24	23
59	48

fum of 59 degrees and 48 minutes. So in both thofe fignifications, whether I accompte feuerall arkes from feuerall beginnings, or generall arks from one generall beginning, the rule is founde true. Now refteth but one rule more of afcenfiōs in this author to be difcuffed, and that is this : that in an oblique fphere eche 2 arkes of the Zodiake being equal and equally diftaunt from any one of the Equinoctiall pointes, fhall haue equall afcenfions.

The 7 rule.

V.ij.   Ma-

Mafter. This rule is partly agreeable with the lafte rule, and partly feuerall, in as muche as euery contrarye arke is lyke diftaunte frome the one Equinoctiall pointe, as the fyrfte arke is frome the other Equinoctiall pointe. thys rule dooth agree (after a forte thoughe not proprely) wyth the other lafte before : but confideringe that Aries and Pifces as whole fignes haue lyke arkes, and are equallye dyftaunt from one Equinoctiall pointe, thoughe in backe ordre : for the eande of Aries is iufte equall in diftaunce from the precife Equinoctioll pointe, as the beginninge of Pifces is from the fame. And in this pointe thefe Signes haue thys feuenth rule as a fpeciall rule for theym and their Afcenfions. Lykewaies Taurus compared wyth Aquarius, Gemini with Capricorne, Cancer with Sagittarius, Leo with Scorpius, and Virgo with Libra, as this figure dooth

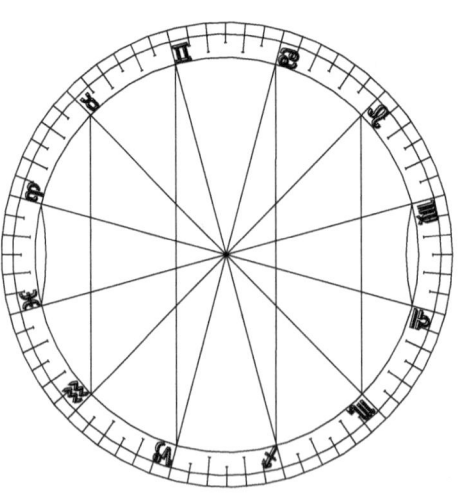

fhew exactly, althoughe in ẙ fame I haue marked alfo the contrary fignes that it might be a cō mon figure for bothe thofe rules, fo ẙ euery feueral fign hath 2 matches, with which it may be conferred, one of theym righte againfte him. and that comparifon is in the 6 rule : and the other leffe diftaũt, & ther conference belongeth to this 7 rule.
Schol-

Scholar. As this figure doth teache me what fignes may be confrerred togither, fo the tables before written doo declare the quantities of their afcenfions in thofe feueral latitudes : and the true meaning of bothe thofe rules, as well as of other, touchinge afcenfions.

Mafter. But this mufte you farther knowe, that thofe rules doo fpeake generallye of anye twoo arkes, whether they bee greater or leffer then a Signe, and doo not meane of Signes onlye.

Scholar. That mufte needes follow ordrely : for if Aries bee equall in afcenfion with Pifces, and Taurus equall in rifinge with Aquarius, then ioyntly Aries and Taurus muft needes be of one quantitie in afcenfion with Aquarius and Pifces, by compofition of proportions, as is taughte in Geometrye and Arithmetike alfo.

Mafter. Lykewaife (by refolution of propofitions) if al Aries be like in afcenfion with all Pifces, then the firft degre of Aries fhall afcende equallye with the lafte degree of Pifces : and the 20 degree of Aries, wyth the 10 degree of Pifces : & in lyke manner of eche other degree equally diftaunte from the Equinoctiall pointes : and fo lykewaies of euerye minute : for thefe rules of equalitie or inequalitie of Afcenfions of arkes, doo ferue as well for the arkes of degrees and mynutes, as for the arkes of whole Signes, or of greater quantities. Alfo this rule is general, that all arkes that afcende rightly, do defcende crookedly, be they great or fmall : and contrarye wayes, what arke fo euer afcendeth crookedlye, doth defcende righte : whereby it commeth to paffe, that alwaies the one figne counteruailyng with his contrary, there is euermore one halfe of the Zodiake aboue the Horizonte, as well as there is one halfe of the Equinoctiall aboue the fame. fo that when fo euer anye degree of the Zodiake doth fet in the wefte, the contrarye degree dooth rife in the eafte. Of this it foloweth, that in the longefte daye in the yeare there dooth rife but fyxe Sygnes, and in the

V .iij.   fhorteft

ſhorteſt daye there riſeth as manye ſignes.

Scholar. Thereof it maye ſeeme to come to paſſe, that in
aunciente tyme the day and the nyghte were euermore diui-
uided into 12 equall parts, (how longe or how ſhort ſo euer
they were) and thoſe partes were called Vnequall howers, of
*Howers
vnequall.* whiche yet manye men doo write, and doo call them howers
of the Planets : but as I iudg by the ordre of the aſcenſions,
euerye ſigne hathe not equall Aſcenſion, nor equall time in
riſyng, & therfore may thoſe howers be well called Vnequall,
which depend of the motion of the Zodiake, beeyng in it
ſelfe vnequall in his Aſcenſion.

Maſter. It is thought of ſome men to be a more apte reaſon to call thoſe howers vnequall, bicauſe not only the ſommer howers are vnequall to the winter howers, but alſo the
daye howers vnequall to the night howers.

*Naturall
howers.* Scholar. Iohn de ſacro Boſco doth call them naturall howers, and defineth them to be the meaſure of the tyme, in
whiche halfe a ſigne dooth aſcend.

Maſter. As the 6 ſignes that riſe in the daye or in the
nyghte keepe not one vniforme equalitye in their ryſynge,
ſo doth the Aſcenſions of the halfe ſignes differ more vnequallye : and by that meanes the howers of the daye can not
be equall togither, norther yet the howers of the nyght may
be called equall togither : wherefore other you muſt not allowe that definition, or els you muſt not parte the daye and
the nyght into equall partes.

Scholar. I knowe not what to ſaye to this, for norther
can I defende that definition, norther yet can I improue that
partition.

Maſter. Thoſe howers haue beene the occaſion of much
contention, and therfore were they wittilye reiected oute of
the daylye vſe, wherein they were ones common, and were
*Equall hou
res called
Equinoctial
howers.* lefte only to learned men, for learned vſes, and in their ſteed
other howers more certaine and equall were diuiſed, whiche
doo diuide the naturall day into 24 equall partes, and theſe
keepe

THE CASTLE OF KNOWLEDGE 235

keepe one iufte quantitie, how fo euer the Artificiall day do varye his quantitie.

Scholar. This I knowe well : but yet touchynge the fyrfte howers, called the Planet howers, I woulde gladlye vnder-ftande fome example for their exacte diuerfitie in fome one daye.

Mafter. You fhall haue anone one generall table for ma-ny dayes, namely for euerye fyxte daye in the yeare nighe hande, and that table fhall fuffice for the whole yeare : and yt fhall be calculate accordyng to that exact forme of diftin-ction of howers, by halfe Signes of the Zodiake : but in the meane ceafon, bicaufe you fhall not be ignorant of the vul-gare forme of vnequall howers, I haue heere fette forth an ordrelye partition of them, accordynge to the lengthe of euerye daye or nighte in the yeare, by increafe frome 12 mi-nutes to 12 minutes, for eche day or nyghte, from the fhor-teft daye, or nyghte of 1. minute of length, vnto the longeft daye or nyghte of 24 howers.

Scholar. But what if the longeft daye be not fo longe, as it is not with vs in Englande?

Mafter. The table doothe ferue for all places where the dayes be of fhorter lengthe : as by the ouermofte title and that fyrfte columne on the lefte hande you may perceaue.

Scholar. I was to negligente, that I did not confider that, for as it maye ferue for that daye in the yeare which is but 16 howers longe, (thoughe the longeft daye bee longer) fo maye it ferue for that place where the longeft daye is but 16 howers in quantitie.

Mafter. Yea and for the myddle of the earthe vnder the Equinoctial, where the longeft day is but 12 howers, fo that it ferueth from the Equinoctiall circle, vnto the Polare cir-cle, and for all Climates that be betweene them, as by the ho wers in the firfte columne you may perceaue. So that if you *The vfe of the table.* will knowe the quantitie of anye hower vnequall, or hower of the Planetes, after this forme : fyrft you mufte knowe the

V.iiij. iuft

236   THE FOVRTHE TREATISE OF

A TABLE FOR THE HOVRES OF

Planetes after the common forme.

Minutes.	0		12		24		36		48	
	Hour.	Minu.	Hour.	Minu.	Hour.	Minu.	Hour.	Minu.	Hour.	Minu.
0	0	0	0	1	0	2	0	3	0	4
1	0	5	0	6	0	7	0	8	0	9
2	0	10	0	11	0	12	0	13	0	14
3	0	15	0	16	0	17	0	18	0	19
4	0	20	0	21	0	22	0	23	0	24
5	0	25	0	26	0	27	0	28	0	29
6	0	30	0	31	0	32	0	33	0	34
7	0	35	0	36	0	37	0	38	0	39
8	0	40	0	41	0	42	0	43	0	44
9	0	45	0	46	0	47	0	48	0	49
10	0	50	0	51	0	52	0	53	0	54
11	0	55	0	56	0	57	0	58	0	59
12	1	0	1	1	1	2	1	3	1	4
13	1	5	1	6	1	7	1	8	1	9
14	1	10	1	11	1	12	1	13	1	14
15	1	15	1	16	1	17	1	18	1	19
16	1	20	1	21	1	22	1	23	1	24
17	1	25	1	26	1	27	1	28	1	29
18	1	30	1	31	1	32	1	33	1	34
19	1	35	1	36	1	37	1	38	1	39
20	1	40	1	41	1	42	1	43	1	44
21	1	45	1	46	1	47	1	48	1	49
22	1	50	1	51	1	52	1	53	1	54
23	1	55	1	56	1	57	1	58	1	59
24	2	0								

iuſt quantity of the day artificiall, from ſonne riſyng to ſon
ſettinge, and thereby alſo the quantitie of the nyghte : then
ſhall you ſeke the houres of their length in the firſt column,
vnder the title of howers : and if the daye or nyght haue any
minutes aboue thoſe euen howers, you ſhall ſeke them in the
higheſt

higheſt range of numbres, where they bee ſet from 12 to 12, and take that numbre of minutes that is nexte in quantitye to your minutes in the day propounded : and in the cōmon angle, againſte your howers and vnder your minutes, you ſhall fynde the iuſte quantitie of the minutes that make an hower vnequall, for that daye or nyght : but that muſt you vnderſtande ſeuerally.

Scholar. I were to groſſe headded if I wold make a doubt thereof. And bycauſe I will declare vnto you how I vnder- ſtande the vſe of it, I wyll by an example or twoo make it appeare. When the Artificiall daye is 14 howers longe, and 20 minutes, and the nyghte then is 9 howers longe and 40 minutes of neceſſitye : I woulde knowe the iuſte quantitye of the howers vnequall. Firſte therefore, in the fyrſte co- lomne I ſeeke oute the numbre of the howers, whiche is 14, then in the higheſt raunge of numbres I ſeeke the odde minutes, beinge 20, and bicauſe I fynde no ſuche numbre there, I take the nexte numbre whiche is 24, and by thoſe 2 numbres in their common angle againſte 14 towarde the righte hande, directly vnder the 24 minutes, I fynde 1, 12, whereby I vnderſtande, that eche vnequall hower is longer then the equall hower by 12 minutes that daye. and for the nyghte I fynde againſte 9 and vnder the numbre of 36 (whiche is nexte vnto 40) the iuſte quantitie of eche vne- qualle hower of the ſame nighte, to bee 0, 48, that is but 48 minutes : and ſo is the vnequall hower of the nyghte leſſer by twelue minutes, then is the equalle hower. And ſo bothe thoſe howers ioyned togither, doo make twoo howers, equall to twoo Equinoctiall or Equall howers. for ſo muche as the one is to lyttle, the other is to greate. Againe for an other triall, I take the artificiall daye to bee 8 howers and 36 minutes long, and therfor to know the quan titie of an vnequall hower, I ſeeke againſt 8, and vndernethe 36, wher I fynd 0, 43, which giueth me to vnderſtand that the vnequall hower that daye is only 43 minutes in quantity, &
the

the nyghte then beynge 15 howers long and 24 minutes, yeldeth his vnequall howers of 1 hower and 17 minutes longe :
whereby it is feene alfo, that fo muche is fupplied by the one
hower as was wantinge in the other. fo that euermore one
vnequall hower of the day ioined with an vnequal hower of
the nyghte, will make two howers equall to two equinoctiall howers.

Howers e-
quall, equi-
noctial, vul
gare and na
turall.

Scholar. You meane thofe common howers which we vfe
vulgarlye, whiche are called alfo of fome men Naturall howers, takinge that name of the Naturall daye, whiche they
diuide into 24 equall partes, (thoughe other men adfcribe
that name to Vnequal howers) and fo of their common vfe
ar they named Vulgare, lyke as they are called Equinoctiall
howers, bycaufe (as I haue learned) they depende of the reuolution of the Equinoctiall : and therefore keepe they one
conftante quantitie, eche beyng equall with other.

Mafter. You remembre it well. And as thefe are taken of
the motion of the Equinoctiall, and are nothng els but the
fpace or meafure of time wherein 15 degrees of the Equinoctiall do paffe the meridiane line, fo againe it feemeth to the
wifeft forte of men, that the Vnequall howers ought to bee
gathered by the motion of the Zodiake, whofe feuerall
forme of afcenfion for euery halfe figne, dooth make a feuerall and diftinct quantitie of Vnequall howers, and haue no
fewer fortes of differences, then there be diftincte and feuerall degrees or pointes, at whiche that arke of 15 degrees
maye beginne his afcenfion, as partly in this table folowing
it dooth appeare : where you may fee in the fyrfte columne
on the lefte hande, and in the lafte on the right hand, the degrees of the fignes fet : not euery one feuerally, but only frō
6 degrees to 6 degrees, whiche are fo mennye as may feeme
to fuffice for a conuenient diftinction of the feuerall diuerfities in fuch hours, namely in that latitude of 52 degres, for
whiche it is calculate. And nexte vnto thofe degrees in the
feconde columne, and in the lafte faue one, are fet the names
of

Vnequall
howers.

The decla-
ratiō of the
table.

of the 12 Signes in their conuenient ordre, that is to fay, in
the one parte the 6 Signes whiche be called north Signes, as
Aries, Taurus, Gemini, Cancer, Leo, and Virgo : and in
the other are fet the 6 fouth Signes, Libra, Scorpio, Sagittarius, Capricornus, Aquarius, & Pifces. And againft thofe
fignes and degrees ar fet the quantities of euery hower in the
daye for that time, when the Sonne is in any fuche degree of
thofe fignes. And for the better knowledge of the howers,
their names and numbres are fet forth in the head of the table : where alfo is fet a diftinction by diuerfitye of the daye
and nighte accordinglye as the Sonne is then in the fouthe
Signes or in the northe fignes.

  Scholar. I doo perceaue it to bee reafonable, that the firft
hower of the daye mufte be accompted that hower, in whofe
beginning the Sonne doth rife : fo that euery daye the fyrfte
hower is begonne with the afcenfion of that degree of anye
figne wherein the fonne is. And the firft hower of the night
is begonne with the afcenfion of that degree, which is oppofite or contrary to the place of the fonne : whiche place is
commonly called in latine Nadir Solis, althoughe in deede
the one woorde is an Arabike woorde, and not latine. And
after that firfte hower as the other howers of neceffitye doo
follow in ordre of numbre, fo their diftinction in quantitie
doth follow in this table : and the difference of them is agre
able to the diuerfitye of the afcenfion of eche halfe figne of
the Zodiake, as they doo followe in ordre. So that to come *Example.*
to an example, for declaration that I doo vnderftande that
table. yf I woulde knowe the quantitie of the vnequall howers, when the fonne is in Aries and in his fyrfte degree, I
muft entre the fyrfte parte of the table, where I fynde on the
lefte hande the Signes and their degrees : wherefore againfte
Aries and ỹ cyphar o, which betokeneth the very beginning
of the figne, I note all the howers as they followe in ordre :
whereby I perceaue that the fyrfte hower of the day is but 25
minutes of an equall hower in lengthe : the feconde hower

                  is

240   THE FOVRTHE TREATISE OF

A TABLE FOR THE DISTINCTI

calculate for the latitude of

Howers of the daye, for the northe Signes : and of the nyghte, for the fouth Signes.

Signes		1 / 7		2 / 8		3 / 9		4 / 10		5 / 11		6 / 12	
Hours		1	2	3	4	5	6	7	8	9	10	11	12
		H. M.	H. M.	H. M.	H. M.	H. M.	H. M.	H. M.	H. M.	H. M.	H. M.	H. M.	H. M.
0	♈	0 25	0 27	0 30	0 37	0 47	0 59	1 11	1 20	1 25	1 26	1 27	1 25
6		0 25	0 28	0 33	0 41	0 52	1 4	1 16	1 23	1 26	1 27	1 26	1 26
12		0 26	0 30	0 36	0 45	0 57	1 9	1 19	1 24	1 26	1 26	1 26	1 26
18		0 27	0 32	0 39	0 49	1 2	1 13	1 22	1 26	1 17	1 26	1 26	1 26
24		0 29	0 34	0 43	0 54	1 7	1 17	1 24	1 26	1 26	1 26	1 26	1 26
30	♉	0 30	0 37	0 47	0 59	1 12	1 20	1 25	1 26	1 27	1 25	1 26	1 26
6		0 33	0 41	0 52	1 4	1 16	1 23	1 26	1 27	1 26	1 26	1 26	1 26
12		0 36	0 45	0 57	1 9	1 19	1 24	1 26	1 26	1 26	1 26	1 26	1 27
18		0 39	0 49	1 2	1 13	1 22	1 26	1 27	1 26	1 26	1 26	1 26	1 26
24		0 43	0 54	1 7	1 17	1 24	1 26	1 26	1 26	1 26	1 26	1 27	1 26
0	♊	0 47	0 59	1 11	1 20	1 25	1 26	1 27	1 25	1 26	1 26	1 26	1 25
6		0 52	1 4	1 16	1 23	1 26	1 27	1 26	1 26	1 26	1 26	1 26	1 24
12		0 57	1 9	1 19	1 24	1 26	1 26	1 26	1 26	1 26	1 27	1 26	1 22
18		1 1	1 13	1 22	1 26	1 27	1 26	1 26	1 26	1 26	1 26	1 24	1 19
24		1 7	1 17	1 24	1 26	1 26	1 26	1 26	1 26	1 27	1 26	1 27	1 15
30	♋	1 11	1 20	1 25	1 26	1 27	1 25	1 26	1 26	1 26	1 25	1 21	1 11
6		1 16	1 23	1 26	1 27	1 26	1 26	1 26	1 26	1 26	1 24	1 17	1 7
12		1 19	1 24	1 26	1 26	1 26	1 26	1 26	1 27	1 16	1 22	1 13	1 2
18		1 22	1 26	1 27	1 26	1 26	1 26	1 26	1 26	1 24	1 19	1 9	0 57
24		1 24	1 26	1 26	1 26	1 26	1 26	1 27	1 26	1 27	1 15	1 4	0 52
0	♌	1 25	1 26	1 27	1 25	1 26	1 26	1 26	1 25	1 21	1 11	0 59	0 47
6		1 16	1 27	1 26	1 26	1 26	1 26	1 26	1 24	1 17	1 7	0 54	0 47
12		1 26	1 26	1 26	1 26	1 26	1 27	1 26	1 21	1 13	1 2	0 49	0 39
18		1 27	1 26	1 26	1 26	1 26	1 26	1 24	1 19	1 9	0 57	0 45	0 36
24		1 26	1 26	1 26	1 26	1 27	1 26	1 27	1 15	1 4	0 52	0 41	0 33
0	♍	1 27	1 25	1 26	1 26	1 26	1 25	1 22	1 11	0 59	0 47	0 37	0 30
6		1 26	1 26	1 26	1 26	1 26	1 24	1 17	1 7	0 54	0 47	0 34	0 33
12		1 26	1 26	1 26	1 27	1 26	1 22	1 13	1 2	0 49	0 39	0 32	0 27
18		1 26	1 26	1 26	1 26	1 24	1 19	1 9	0 57	0 43	0 36	0 34	0 26
24		1 26	1 26	1 27	1 26	1 27	1 15	1 4	0 52	0 41	0 33	0 28	0 25
30		1 26	1 26	1 26	1 25	1 21	1 11	0 59	0 47	0 37	0 30	0 27	0 25
		H. M.	H. M.	H. M.	H. M.	H. M.	H. M.	H. M.	H. M.	H. M.	H. M.	H. M.	H. M.

THE CASTLE OF KNOWLEDGE   241

ON OF THE VNEQVALL HOWERS
52 degrees.

Howers of the nyghte, for the northe Signes : and of the daye, for the fouth Signes.

| 7 | | 8 | | 9 | | 10 | | 11 | | 12 | | | |
| 1 | | 2 | | 3 | | 4 | | 5 | | 6 | | Signes | |
1	2	3	4	5	6	7	8	9	10	11	12	Hours	
H. M.	H. M.	H. M.	H. M.	H. M.	H. M.	H. M.	H. M.	H. M.	H. M.	H. M.	H. M.		
1 26	1 26	1 26	1 25	1 21	1 11	0 59	0 47	0 37	0 30	0 27	0 25	♎	0
1 26	1 26	1 26	1 24	1 17	1 7	0 54	0 47	0 34	0 33	0 26	0 24		6
1 16	1 27	1 26	1 22	1 13	1 2	0 49	0 39	0 32	0 27	0 25	0 25		12
1 26	1 26	1 24	1 19	1 9	0 57	0 45	0 36	0 34	0 26	0 25	0 25		18
1 27	1 26	1 27	1 15	1 4	0 52	0 41	0 33	0 28	0 25	0 24	0 26		24
1 26	1 25	1 21	1 11	0 59	0 47	0 37	0 03	0 27	0 25	0 25	0 27	♏	0
1 26	1 24	1 17	1 7	0 54	0 47	0 34	0 33	0 26	0 24	0 25	0 28		6
1 26	1 22	1 13	1 2	0 49	0 39	0 32	0 27	0 25	0 25	0 26	0 30		12
1 24	1 19	1 9	0 57	0 45	0 36	0 34	0 26	0 25	0 25	0 27	0 32		18
1 27	1 15	1 4	0 52	0 41	0 33	0 28	0 25	0 24	0 26	0 29	0 34		24
1 22	1 11	0 59	0 47	0 37	0 30	0 27	0 25	0 25	0 27	0 30	0 37	♐	0
1 17	1 7	0 54	0 47	0 34	0 33	0 26	0 24	0 25	0 28	0 33	0 41		6
1 13	1 2	0 49	0 39	0 32	0 27	0 25	0 25	0 26	0 30	0 36	0 45		12
1 9	0 57	0 45	0 36	0 34	0 26	0 25	0 25	0 27	0 32	0 39	0 49		18
1 4	0 52	0 41	0 33	0 28	0 25	0 24	0 26	0 29	0 34	0 43	0 54		24
0 59	0 47	0 37	0 30	0 27	0 25	0 25	0 27	0 30	0 37	0 47	0 59	♑	0
0 54	0 47	0 34	0 33	0 26	0 24	0 25	0 28	0 33	0 41	0 52	1 4		6
0 49	0 39	0 32	0 27	0 25	0 25	0 26	0 30	0 36	0 45	0 57	1 9		12
0 45	0 36	0 34	0 26	0 25	0 25	0 27	0 32	0 39	0 49	1 2	1 13		18
0 41	0 32	0 28	0 25	0 24	0 26	0 29	0 34	0 43	0 54	1 7	1 17		24
0 37	0 30	0 27	0 25	0 25	0 27	0 30	0 37	0 47	0 59	1 11	1 20	♒	0
0 34	0 33	0 26	0 24	0 25	0 28	0 33	0 41	0 52	1 4	1 16	1 23		6
0 32	0 27	0 25	0 25	0 26	0 30	0 36	0 45	0 57	1 9	1 19	1 24		12
0 34	0 26	0 25	0 25	0 27	0 32	0 39	0 49	1 2	1 13	1 21	1 26		18
0 28	0 25	0 24	0 26	0 29	0 34	0 43	0 54	1 7	1 17	1 24	1 26		24
0 27	0 26	0 25	0 27	0 30	0 37	0 47	0 59	1 11	1 20	1 25	1 26	♓	0
0 26	0 24	0 25	0 28	0 33	0 41	0 52	1 4	1 16	1 23	1 26	1 27		6
0 25	0 25	0 26	0 30	0 36	0 45	0 57	1 9	1 19	1 24	1 26	1 26		12
0 25	0 25	0 27	0 32	0 39	0 49	1 2	1 13	1 22	1 26	1 27	1 26		18
0 24	0 26	0 29	0 34	0 43	0 54	1 7	1 17	1 24	1 26	1 26	1 26		24
0 25	0 27	0 30	0 37	0 47	0 59	1 11	1 20	1 25	1 26	1 27	1 25		30
H. M.	H. M.	H. M.	H. M.	H. M.	H. M.	H. M.	H. M.	H. M.	H. M.	H. M.	H. M.		

Xi

is 27 minutes longe : the thirde hower 30 minutes, that is
halfe an equall hower iuſte : and in the ſame line goinge forwarde, the 12 and laſte hower of the daye is 1 hower and 25
minutes in lengthe. Then for the nighte the howers appeare
in the other parte of the table, where the firſte hower dooth
containe one equall or common hower, and 26 minutes : the
ſeconde hower and the third be of lyke quantitie, and ſo do
they afterwarde decreaſe vntyll the laſte hower of the nyght.

Exaumple. An other example : when the ſon is in the 10 degree of Cancer, bicauſe I can not fynde that degree in the table, I take
the degree nexte vnto it, whiche is the 12 degree, and proce-
dynge with it, I fynde the fyrſte vnequall hower to containe
1. equall hower, and 19 minutes : and the ſecond vnequall hower hath in it 1. equall hower and 24 minutes. Nowe for the
nyghte I look in the ſeconde parte of the table, and fynde
the fyrſte vnequall hower to bee but 49 minutes in lengthe,
and the ſeconde but 39 minutes. and ſo in ordre folowinge.
This muſte I doo when the Sonne is in anye of the northe
ſignes, but if the ſon be in any of the ſouth ſignes, thē muſt
we accompt the day howers in the ſecond part of the table, &
the howers of the night muſt be ſought in the firſte parte of
the table : in all other pointes I perceaue there is ſmall difference.

an ordre for Maſter. Yet by the way this maye you note, that if you
proportion. woulde deſire more preciſely to knowe the iuſte quantitie of
the howers, for anye ſuche degree of the Signes as is not
expreſſed in your table, you ſhall woorke by the rule of
proportion, to knowe the more exacte quantitie of the
vnequall howers. as for example : In the former worke where
you ſuppoſed the ſonne to be in the 10 degree of Cancer, bicauſe that degre is not found in the table, you muſt work by
proportion to know it, & that in this forme : firſte conſider
the howers againſt the next nūbre of degrees, as well beneth
your degre as alſo aboue the ſame, & marke the difference betweene them two, which difference ſhall alwaies be the ſecōd
                                                       numbre

numbre in the Golden rule : and the fyrſt noumbre of that
woorke ſhall alwaies be 6 degrees, bicauſe that is the ordina-
rye exceſſe in this table of eche two numbres next togither :
Now for the third numbre, you ſhall ſet the exceſſe of your
degrees proponed, aboue the leſſer degres in that table, next
beneth your ſaid numbre, which in this example is 4, for ſo
much is betwene 6 & 10. And the difference in howers in ẏ table is but 3 minutes : for againſt the 6 degree of Cancer, ther
is but one hower and 16 minutes : and againſt the 12 degre is
ſet one hower and 19 minutes. Therefore thus doo
I ſet thoſe numbres accordyng to the golden rule,
ſaying : If 6 degrees giue three minutes, then 4 degrees muſte yelde twoo minutes. thoſe two muſt
bee added to the leſſer numbre, and ſo dooth there ryſe
one hower and 18 minutes for the exacte quantitye of the
fyrſte vnequall hower, the Sonne beeynge in the tenthe
degree of Cancer.

Scholar. I praye you lette me prooue the ſame for the ſeconde hower of the nyght, where againſt the 6 degree I find
6 hower and 47 minutes : and againſte the 12 degree I ſee 6
hower and 39 minutes, heere the exceſſe is 8 minutes : then
ſette I the figures thus in the golden rule, and
ſay : If 6 yelde 8, then ſhall 4 giue $5\frac{1}{3}$: if I adde
theſe vnto the leſſer numbre of time, which is 39
minutes,

Maſter. You are to farre deceiued, and therefore I interrupt your woordes, for all thinges are to bee gouerned by
reaſon. So that if the howers do increaſe in quantitie, then
is it reaſonable to adde the parte proportionable to the leſſer numbre of tyme, as it was in the former example : but
in this example you ſee the time dooth not increaſe, but decreaſe, (ſeynge the tyme againſt 6 degrees is greater then the
tyme againſt 12 degrees) and therefore by good reaſon the
parte proportionable is to be abated from the greater, and
not to be added to the leſſer.

X.ij. Schol.

Schol. So is it reaſonable : therfore muſt I take that 5 ⅓ from 47, & then reſteth 41 ⅔, whiche is the preciſe quantitie of that vnequall hower. And nowe I thanke you, I am fully inſtructed touching that matter : ſo that for anye vnequall hower accordinge to the place of the ſonne in this latter table, and after the lengthe of the daye in the fyrſte table, I canne fynde oute the quantitie of eche vnequalle hower : but theſe twoo formes doo not make exactly one quantitye of howers vnequall.

Maſter. As in that you ſhall haue more exacter declaratiō hereafter. And for this preſent tyme I wyll ſay no more but that eche of both waies hath good vſes. And the fyrſt form whiche ſeemeth moſt plaine and leaſte artificiall, hathe comprobation of manye men, and namelye of Ptolemye in the ninth chapter of his ſecond boke of Almageſtes. but omittyng for a time that that remayneth touching howers, I will now ſpeake ſomwhat of the quantities of daies, in whiche matter you ſhall call to mynd, that the Naturall daye is not one with the Artificiall daye : for the firſte is commonly accompted from Sonne riſinge one daye, to Sonne riſing the nexte daye. but the ſeconde, that is the Artificiall daye, is reckened only from ſonne riſinge, to ſonne ſetting : ſo that there is no night accompted in the Artificiall daye, as there is in the Naturall daye.

Daies arti-
ficial and
Naturall.

Scholar. This I perceaue well inoughe: and farther alſo, that the Naturall daies are euer 24 howers longe, in all our knowen cuntries, but the Artificiall daies do increaſe and de creaſe diuerſely. And as I deſire to know the cauſes therof, ſo I do meruail how it cometh to paſſe, that in any cuntry or cli mat the naturall daies ſhuld differ.

Maſter. To the intente that we may proceede ordrely, we wyll begin with the one ſorte of daies, and ſo come to the talke of the other. And firſte as concerning Naturall dayes, I ſayde that they were cōmonly accompted from ſon ryſing to ſon ſetting : which deſcriptiō being true, what ſhal we ſay
of

of thofe northe and fouthe cuntries, where the Sonne continueth aboue the Horizont in fome places three weekes, in
other 6 weeks, and fo increafing tyll it extend to halfe a year.
in al which places if we call the naturall day ẙ fpace from fon
rifyng to Sonne rifyng again, then can not the naturall day
be of one quantitie to all nations, and fo fhuld thofe daies
naturall differ in nature, whiche were agaynfte nature vtter
lye : and therefore dyd I vfe that woorde commonlye in the
former defcription : but if I fhall define the naturall daye
exactlye, I mufte call it that iufte tyme in whiche the eight *The natu-*
Sphere or Firmamente dooth exactlye accomplyfhe his *rall daye.*
courfe, whiche tyme of naturall daye is the common meafure of all other tymes : and thys tyme is always equalle
in all places, howe be it accordynge to the former defcription, yf the retournynge of the Sonne bee accompted
frome anye one parte of the Meridiane lyne, to the fame
parte of the fayde lyne, then maye that defcription well extende to all partes of the worlde : for althoughe fome nations haue the Sonne in fyghte halfe a yeare togither, yet
dooth the fonne retourne to theyr meridiane lyne towarde
the fouthe, at the eand of 24 howers within a little, and in all
places lykewaies where the daye it not full 24 howers, the
fonne doothe retourne to their horizon, at the eand of 24
howers nygh hande.

    Scholar. I heare you fpeake in bothe thefe declarations,
with a doubtfull limitation of the 24 howers, as thoughe
that tyme were not the precife or iufte meafure of the naturall daye.

    Mafter. So fhall it appeare vnto you, yf you confider
that the fonne dooth euerye daye runne one degree almofte
towarde the eafte, accordynge to the fucceffion of the fignes, as before is mentioned : for if this daye the fonne be
in the fyrfte degree of Libra iuftely at noone, then tomo-
rowe at noone hee wyll bee in the feconde degree : and fo

                  X.iij.   the

the thirde daye hence in the thirde degree : and by the fame
reafon at the monethes eande, wyll the fonne haue paffed Libra cleerely, and bee in the beginninge of the nexte figne,
whiche is Scorpius : and therefore muft he be flacker in com
ming to the Meridian line, by fo muche time as ferueth for
the rifynge of all the figne of Libra in a Righte fphere.

Scholar. That tyme muft be an hower and 52 minutes.
for (as I remembre) the partes of the Equinoctiall whiche
doo ferue for the afcenfion of Libra, are 27 degrees and 54
minutes.

Mafter. As that is true, fo marke what is the difference
now for euerye day of that moneth, and then fhall you perceaue the difference of the Naturall dayes, as muche as dependeth of that caufe.

The firfte
caufe of di
uerfitye in
Naturall
dayes.

Scholar. For the fyrfte degree of Libra, the quantitye of
his afcention is 55 minutes of the equinoctiall, whiche maketh in time of an hower 3 minutes and $\frac{2}{3}$, and fo maye I fee
for diuers degrees at the beginninge of Libra, by the table
of the afcenfions in the Right fphere : but towarde the eande
of the fame figne, I fee 57 minutes agreeyng to the afcenfion
of one degree, whiche maketh fome difference in tyme alfo,
thoughe it bee fmall.

Mafter. Marke now about the middle of Scorpius, how
eche degree of the Zodiake hath one degree of the Equinoctiall agreeynge to his afcenfion, whiche maketh in tyme
4 minutes of an hower : and about the mydle of Sagittarius
one degree of the Zodiake hathe aunfwerable to him 64 or
65 minutes of the Equinoctiall. and fo in other diuers degrees of Signes fhall you fynd diuers quantities of their afcenfions, whereby it muft needes appeare, that if the Sonne
dyd moue forwarde in the Zodiake euery daye one degree
iuftlye, that the fonne fhoulde be 4 minutes after the 24 howers flacker then he was the daye before in touching the meridiane line, if there were not an other caufe of diuerfitye by
the fundrye quantities of the afcenfions.

Scholar.

THE CASTLE OF KNOWLEDGE   247

Scholar. This caufe is manifeft. And bicaufe I fee for fome degrees of the Zodiake but onlye 55 minutes of the Equinoctiall, whiche maketh in time 3 minutes and $\frac{2}{3}$: and for other degrees 65 minutes, whiche is 4 minutes and $\frac{1}{3}$: fo doth it appeare that the greateft difference is but $\frac{2}{3}$ partes of a minute : whiche is a fmall matter.

Mafter. Yet this fmall matter will caufe muche matter in Aftronomicall computations, though there were no more difference of diuerfitie in Naturall dayes but this only : but yet are there twoo other caufes in all Oblique fpheres, and but one in the Right fphere. The feconde common caufe in bothe fpheres, is the eccentricitye of the Sonne.

The fecond cause of vn equal daies naturall.

Scholar. What meane you thereby? for I doo not vnder-ftande that eccentricitye.

Mafter. It is a matter not agreeable for this treatife, but that by occafion I am moued to name it as a concurrente caufe touchinge inequalitye of naturall dayes : yet fomwhat to faye of it as may fuffice for this prefent, by example you fhall vnderftande both what eccentricitye is, and alfo howe it maye caufe diuerfitye in naturall dayes : for declaration whereof here in this fy-gure you fee two circles a greater and a leffer : the greater dooth betoken the eighte fphere or fir-mamente, and the leffer dooth reprefent the ec-centrike circle of the fphere of the Sonne.

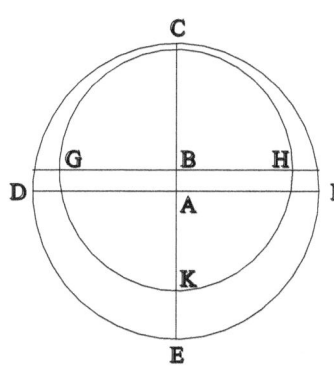

Thefe 2 circles as you fee, are eccentrike, for that they haue not one common centre, fith the centre of the greater circle is by A, and the centre of the lef-fer circle is by B. the diftaunce betweene A and B is the qua-

X.iiij.   titie

titie of their eccentricitye. Nowe maye you fee that eche circle is diuided into 4 quarters : and lykewife you may fe, that
the higher halfe of the leffer circle doth not fully anfwere to
halfe the greater circle : and againe the nether halfe of the
leffer circle doth occupy more then the halfe of the greater
circle. whereby it mufte needes bee euidente to all men, that
when the Sonne moueth in the higher part of his eccentrike
circle, hee doth moue flowlyer then he dooth in the nether
parte of the fame eccentrike : I meane in comparifon to the
Zodiake of the eyghte fphere : and thereby muft it appeare
that the Sonne doth not euerye daye moue lyke numbre of
minutes in the Zodiake : and you maye eafilye coniecture
hereby, that this is an other caufe of diuerfitye in the quan-

The thirde   titye of the naturall dayes. A thyrde diuerfitye is that whi-
cause   che is peculiare to euerye feuerall climate, and not common
of diuerfi-   to anye two on one fyde of the Equinoctiall, and that is the
tie of daies
Naturall.   obliquities of the Horizonte, yf the daye fhall bee accompted from fonne rifynge to fonne rifynge againe : but this
varietie is fo greate and fo diuers, that it is in manner infinite : and therfore doo Aftronomers reiecte the ordre of
accompt of daies, and recken the day from noone to none,
whiche accompte ferueth generally for all the partes of the
worlde, as if all Climates had one Horizont : for as in the
ryghte fphere bothe the Poles doo touche the Horizont, fo
the meridianes of euery climate and of all regions do paffe
by bothe the Poles of the worlde : and therefore all afcenfions accompted vnto that meridiane line, muft bee eftemed
as ryghte afcenfions, I meane afcenfions lyke vnto them that
be in the righte fphere.

Scholar. Nowe do I perceaue, that although there may
be affigned thre caufes of varietie in the naturall dayes, yet
one of them whiche is gathered by the obliquitie of the ho
rizonte in not regarded of Aftronomers, fith they doo accompt the beginning of the daye from ÿ noone fteede, and
the fonne beynge in the meridiane lyne. The fecond caufe by
the

the eccentricitie of the fonne I may coniuecture to appertain to a more higher fpeculation, then this treatife doth admit : but yet may be fomwhat vnderftande euen nowe by a fmall explication. The thirde caufe whiche dependeth of the diuerfitie of the afcenfions by obliquitye of the Horizonte, is peculiare to this treatife, and maye be gathered oute of the tables of afcenfions whiche ferue for the Ryghte fphere : of all whiche varieties at a time of more conuenient leafure, I will make for mine exercife a table at large. but in the meane ceafon I praye you, proceede as you haue begonne.

Mafter. Touching the diuerfities of Naturall dayes this maye fuffice : and for a common and meane quantitie you maye affigne 24 howers and 4 minutes, bicaufe that is the common nombre : for althoughe many be greater, yet manye other bee leffer. and this numbre is mofte nygheft the meane. Nowe touching Artificiall daies you fhall fynde no *The diuerfi tie of the artificiall daies.* fewer diuerfities : wherein although all the former three caufes be concurrent, yet the principall caufe is the obliquitie of the Horizont. And althoughe I haue twyfe before made mention of thofe daies, yet doth there reft more to be fayd of them. for in bothe places before I dyd briefly touche the caufes of diuerfitie of fuche Artificialle daies in diuers climates, and in the table of the diftinction of climates, I dyd fette forth the quantitie of the longeft daye in eche of them : and nowe will I fhew you fomwhat of the reafon of their inequalitie in anye one climate. Fyrft therfore to begin withal, you knowe that before the fonne in his naturall courfe can paffe the full of one degre, he is caried by the violence of the Starrye fkye rounde aboute the earthe. fo that in going betweene the firfte degree of Capricorne, and the fyrfte of Cancer, he dooth confume halfe a yeare, and therefore maketh aboue 182 reuolutions lyke fpirall circles, which are diuerflye parted by the Horizont, accordyng to the diuerfities of the eleuation of the Pole. As in the Ryght fphere they are all parted by the Horizont into two equall partes :

fo

ſo in euerye bowing Sphere, they are vnequally deuided by
the Horizont, ſo that where the north pole is eleuate aboue
the Horizont, there thoſe circles of the ſonnes reuolutions
which be from the equinoctiall northward, haue the greater
portion aboue the horizont, and the leſſer parte vnder the
ſame : and contrarye waies thoſe circles (or ſpires if you like
better ſo to call them) whiche be from the Equinoctiall to
the tropike of Caprocorne, and ſerue for explication of the
Sonnes motion, they haue their greater portion vnder the
Horizont, and the leſſer portion aboue the ſame. And com
paringe eche one of theſe to other, that circle whiche is far-
theſt towarde the ſouth, is moſte parte vnder the Horizont
of anye other. and euerye one of them the more it departeth from the ſouth and draweth toward the north, the grea
ter is his portion that is aboue the horizonte, and the leſſer
is that other portion whiche is vnder the ſame. wherfore the
middlemoſt bounde of thoſe two extremes, is iuſte halfe vn
der, and halfe aboue the Horizonte : and therfore the ſonne
beyng in it, doth make his abode iuſte lyke tyme aboue the
earthe, as he doth vnder it, and therby the daies and nights
are equall : but from thence towarde Cancer, the daye dooth
ſtill increaſe aboue the nighte : and from thence toward Capricorne, the daye dothe ſtill abate ſhorter then the nyghte :
which thinge will eaſilye appeare to the ſight, bothe by theſe

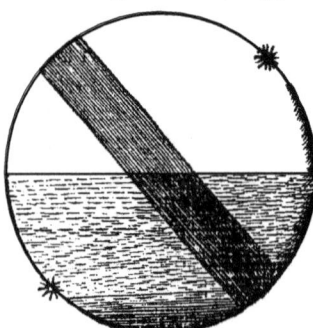

figures here drawen, and al
ſo by the diuers poſitions
of the materiall Sphere or
globe. And ſtyll the higher
that the Pole is eleuate a-
boue the Horizōt, the grea
ter parte of the northerlye
circles is aboue the Horizont, and the leſſer parte
of theym vnder the Hori-
zonte. And contrary waies
of

of the foutherlye circles, the greater portions of them are vnder the horizont, and the leffer portions aboue it. Nowe is it eafily perceaued, that feynge the fonne dooth kepe hys dailye courfe in one of thofe circles, then accorginglye as that circle in whiche the fonne doth moue, is parted by the horizonte, fo is the partition of the 24 howers into daye and nyghte agreeablye : fo that if the circle of the fonnes courfe be more vnder the horizont then aboue it, then fhall the nyghte be longer then the daye : and if the greater parte of the fonnes circle be aboue the horizont, then the day fhal exceede the nighte, in lyke proportion as the partes of the circles are in comparifon togither.

Scholar. Thefe diuers circles (I perceaue) are not in the fphere of the fonne, but are accompted in the eighte fphere betweene the two tropikes, fo that euery daye by the reuo-lution of the Firmament, the fonne is caried frome eafte to wefte rounde about the earthe, and by this violente motion doth defcribe a fpirall circle (as you call it) and not an exact circle : but yet maye it ferue in this cafe, as if it were a iufte circle : the difference is fo lytle of the fpace betweene the fpi-rall lynes in comparifon to their compaffe, whiche by the table of declination before expreffed, I geffe to bee in pro-portion fcarfe $\frac{1}{1000}$, which is no part notable in this cafe. And this farther I note : that two circles on contrary partes of the Equinoctiall equally diftaunt from it, are parted by the ho-rizont after one rate, and into lyke portions : but yet in fuch difference, that the parte of the one circle aboue ground, is equall to the parte of the other that is vnder ground : and fo contrary waies. wherby it foloweth, that the day of the one is equall to the nyghte of the other, and fo contrarye wayes alfo. Again feeyng that the fonne dothe defcend from Can-cer vnto Capricorne, by the fame circles of reuolution, by whiche he dydde afcende from Capricorne vnto Cancer, it muft needes follow that euery two dayes in the yeare equal-ly diftaunte from the longeft daye, or from the fhorteft, are

equall

equall in their artificiall daye, and in their nighte. Thefe generall thinges I maye eafilye gather : but howe I maye knowe
iuflye the quantitye of euerye Artificiall daye from other,
and the precife tyme of the fonne rifinge and fetting, I canne
not fo eafilye gather. wherefore if it pleafe you in thofe two
pointes I defyre your inftruction.

Mafter. Althoughe for this treatife the apteft forme be
by the vfe of the fphere and the due placinge of it, yet it is
harde to place the fphere fo well, and to vfe it fo aptlye, that
it myghte declare a iufte precifenes. and therfor after that I
haue taughte you the vfe of the Sphere for that point, I will
alfo by fupputation giue you a table fufficiente to declare
bothe vnto you for all partes vnder our parallele, and fomwhat more. Firfte for the vfe of the globe, you mufte fet it
accordinge to the latitude of the Region that you defire
to know the daies in, and then marke the degree of any figne
that the Sonne is in that daye, whofe quantitie you defire to
knowe : fette that degree iufte in the horizonte towarde the
eafte, and marke what degree of the equinoctiall is in the ho
rizonte at the fame tyme : then tourne the fphere weftwarde
tyll the degree of the fonne be iuft in the Horizonte againe
in the weft parte, and marke then what degree of the Equinoctiall doth lighte on the Horizont in the eafte parte, accomptynge trulye howe manye degrees bee betwixte thofe
twoo degrees which you haue marked, and that arke of the
Equinoctiall, is called the arke of that day : which you may
eafilye tourne into howers, accomptynge 15 degrees to an
hower, and for euery degree leffe then 15 accompting 4 minutes of an hower.

Scholar. This were eafye inough to doo, if I vfe the helpe
of the table that I fee in fome bookes, whiche teacheth eafily howe to tourne degrees of the Equinoctiall into partes
of tyme, as here in Orontius worke it is fette forthe. but I
dyd abbrydge it for my felfe as here appeareth : and bicaufe
the table was not extended aboue 60 degrees by Orontius,
                                                          I dyd

A TABLE FOR CONVERTINGE
degrees of the Equinoctiall into partes of tyme.

The ark of the Equino.	Partes of tyme.		The ark of the Equino.	Partes of tyme.		The ark of the Equino.	Partes of tyme.	
Degree	Houres	Minuts	Degree	Houres	Minuts	Degree	Houres	Minuts
1	0	4	75	5	0	205	13	40
2	0	8	80	5	20	210	14	0
3	0	12	85	5	40	215	14	20
4	0	16	90	6	0	220	14	40
5	0	20	95	6	20	225	15	0
6	0	24	100	6	40	230	15	20
7	0	28	105	7	0	235	15	40
8	0	32	110	7	20	240	16	0
9	0	36	115	7	40	245	16	20
10	0	40	120	8	0	250	16	40
11	0	44	125	8	20	255	17	0
12	0	48	130	8	40	260	17	20
13	0	52	135	9	0	265	17	40
14	0	56	140	9	20	270	18	0
15	1	0	145	9	40	275	18	20
20	1	20	150	10	0	280	18	40
25	1	40	155	10	20	285	19	0
30	2	0	160	10	40	290	19	20
35	2	20	165	11	0	295	19	40
40	2	40	170	11	20	300	20	0
45	3	0	175	11	40	305	20	20
50	3	20	180	12	0	315	21	40
55	3	40	185	12	20	330	22	0
60	4	0	190	12	40	340	22	40
65	4	20	195	13	0	350	23	20
70	4	40	200	13	20	360	24	0

I did for mine owne eafe make out the reft in this forme.

Maft. This is a table of to much eafe, and therfore doth ra
ther teache negligence, then anye thinge els. for him that li-
fteth to exercife his witte in readines of accompte, it is
an eafy matter to tourne degrees into howers without anye
tables, and therefore fuch tables myght well be fpared, & yet

Y.i.   manye

manye bokes are full of them : but if you lyſted, you might haue abbridged it more frome 15 vpwarde, takinge onlye euen 15 ſtyll. as thus. 15, 30, 45, 60, 75, &c. ſo ſeemeth all the reſte ſuperfluous, excepte your numbre of degrees in the daye arke, happen iuſte agreeable with ſome one of thoſe in the table : but nowe to procede, giue one example for declaration of your vnderſtandinge herein.

Exaumple.   Scholar. Then to begin I ſette the globe to the eleuation of 52 degrees, and confidre the place of the ſonne the 14 day of Auguſte, and fynde it to be by the Ephemerides, in the fyrſt beginning of Virgo, therefore do I ſet the beginning of Virgo in the verye horizont, and then do I ſee with it the 137 degree of the Equinoctiall in the ſame Horizont, whiche I doo marke : afterwarde I tourne the ſphere tyll the place of the ſonne be in the Horizont on the weſt part, and then in the eaſte parte I marke the degree of the Equinoctiall, whyche is 347 degrees. Nowe abatinge 137 oute of 347, there reſteth the whole daye arke, whiche

3 4 7
1 3 7
───
2 1 0

is 210 degrees, whiche make 14 howers, as by the former table is eaſily ſeene. wherfore I conclude that the 14 daye of Auguſt, the ſonne ſhineth 14 howers, and then muſte the nighte be but euen 10 howers, ſith bothe times make iuſt 24 howers : but yet I ſee not howe to knowe the howers of the ſonne ryſinge, and ſettinge.

Maſter. I am ſure you thinke that the Noone is the middle of the daye, and that the ſonne ſhyneth lyke ſpace beefore noone and after noone.

Scholar. That is moſte certaine.

Maſter. Then partinge the whole time of the ſonne ſhining, or of the artificial day into 2 equal parts, the one halfe doth limite the hower after none at which the ſon doth ſet.

Scholar. That is in the exaumple 7, and ſo muſte it needes be. And now I ſee by the ſame reaſon, the ſonne muſt ryſe 7 howers before noone, that is at 5 of the clocke in the mornynge.

Maſter.

Maſter. So is it. And for that eande that you maye haue a generall rule therein, euermore abate halfe the quantity of the daye from 12 howers, and then will the remainer declare the iuſte hower and minute of the ſonne riſynge.

Scholar. Then by your fauoure I will proue ones againe : *Exaumple.* wherfore I take the 16 daye of Iulye, the ſonne beyng in the 3 degree of Leo, which degree I ſette in the eaſte parte of the horizonte, and then doth there appeare in the ſame Hori- zonte the 98 and almoſt $\frac{1}{3}$ degree of the Equinoctiall : then turnynge the degree of the ſonne to the weſt part of the ho- rizonte, I fynde in the eaſte parte the 332 and $\frac{1}{3}$ almoſte of the equinoctiall : then ſubtrayinge the leſſer from the grea- $3\ 3\ 2\ \frac{1}{3}$ ter, there reſteth 234 : which I turne into partes of time, and $\underline{\ \ 9\ 8\ \frac{1}{3}}$ it dooth yelde 15 howers and 36 minutes. whiche is the iuſte $2\ 3\ 4$ length of that artificiall daye. and of it the one halfe is 7 ho- wers and 48 minutes : wherby I knowe that at 48 minutes, $1\ 2\quad 0$ after 7 of the clocke at nyghte, the ſonne ſetteth on that 16 $\underline{7\quad 4\ 8}$ daye of Iuly : and then abating ſo much from 12, there reſteth $4\quad {}^{1}\ 2$ 4 howers and 12 minutes : ſo that the ſonne riſynge appea- reth to be twelue minutes after 4. of the clocke in the mor- nynge. And nowe I thinke my ſelfe conninge inoughe in all this matter.

Maſter. Yet for more eaſe : after that you haue noted the degree of the Equinoctiall that dooth riſe with the place of the ſon, you may marke the degree that riſeth with the con- trarye point againſt the ſon : and abate then the fyrſt oute of the ſecond, and ſo accompliſh your woorke, as you did be- fore. for it is all one thinge, but that you need not to loke in cōtrary ſides of your ſphere for your worke. And this ſhall you note farther : that if the firſt aſcenſion of the place of the ſon be greater then the ſecond aſcenſion of the Nadir of the *A Cautele.* ſon, you ſhal put to the ſecond aſcenſion, 360 degrees, & then abate as you are taught before. As for example : the firſt day *Exaumple.* of February the ſon is by the Ephemerides in the 22 degree

Y.ij   of

of Aquarius, that degree I find in the Zodiak of my fpher, and I fette it iufte in the eafte parte of the Horizonte, and ther may I fe that the 343 $\frac{1}{3}$ degree of the Equinoctiall doth afcend at the fame inftant in the Horizont alfo : which I muft accompt for the true afcentiō of ỹ degre of Aquarius. Then tourne I to the 22 degree of Leo, beinge the Nadir of the fonne, and with it when it is fette in the Horizonte, I marke the 125 $\frac{3}{4}$ degree of the Equinoctiall to afcende. Nowe when I woulde fubtracte 343 $\frac{1}{3}$ out of 125 $\frac{3}{4}$, it will not be : and therfor I put vnto the leffer numbre 360, and fo it amounteth to 485 $\frac{3}{4}$, and then from it I abate 343 $\frac{1}{3}$, and there remaineth 142 $\frac{5}{12}$: whiche if you chaunge into partes of time, do make 9 howers and 30 minutes : and that is the quantitie of the fyrfte daye of Februarye.

Scholar. The halfe of that is 4 howers, and 45 minutes, whereby I knowe, that at the 45 minute that is $\frac{3}{4}$ of an hower after 4 of the docke the fonne fetteth : and rifeth in the mornynge 15 minutes, that is $\frac{1}{4}$ of an hower after 7 of the clocke. But why doo you adde thofe 360 degrees?

Mafter. Seeyng wee intende to abate the fyrfte afcenfion oute of the feconde, to thintente that their diftaunce maye bee knowen, feeynge the whole compaffe of the circle is but 360, from whiche if you abate the fyrfte afcenfion being the greateft numbre, then wyll there remaine the diftaunce betwene ỹ afcention & the end of the equinoctial : vnto which differēce you muft adde fo many degres as ỹ fecōd afcentiō requireth, as both reafon & practife wil declare vnto any mā.

Scholar. It is reafonable. Therfore now it may pleafe you to declare the fame woorke by exactnes of tables.

Mafter. Bicaufe you fhall not be driuen to feeke in the E-phemerides for the place of the Son, but that one table may ferue for it, as well as for the quantities of daies and other cō clufions alfo, I wil make the tables common for fundry vfes, whofe partes I will fyrfte declare, and after that will expreffe the vfes of them alfo.

The decalration of the tables.

In

THE CASTLE OF KNOWLEDGE　　257

THE TABLES OF QVANTITIES
of dayes Artificiall, and nightes, for all Englande.

Signes for the daye.				Eleuation of the Pole, or latitudes of Regions.					Signes for the nighte.			
daies of moneths.	degrees of signes.	daies of moneths.	degrees of signes.	51	52	53	54	55	daies of moneths.	degrees of signes.	daies of moneths.	degrees of signes.
10	♈ 0	13	♍ 30	12 0	12 0	12 0	12 0	12 0	0 ♎	13	30	10
11	1	12	29	12 4	12 4	12 4	12 4	12 4	1	14	29	9
12	2	11	28	12 8	12 8	12 8	12 9	12 9	2	15	28	8
13	3	10	27	12 12	12 12	12 12	12 14	12 14	3	16	27	7
14	4	9	26	12 16	12 16	12 16	12 18	12 18	4	17	26	6
15	5	8	25	12 20	12 20	12 21	12 22	12 23	5	18	25	5
16	6	7	24	12 24	12 24	12 26	12 26	12 28	6	19	24	4
17	7	6	23	12 28	12 28	12 30	12 30	12 32	7	20	23	3
18	8	5	22	12 32	12 32	12 34	12 35	12 36	8	21	22	2
19	9	4	21	12 36	12 36	12 38	12 40	12 40	9	22	21	1
20	10	3	20	12 40	12 40	12 42	12 44	12 44	10	23	20	29
21	11	2	19	12 44	12 44	12 46	12 48	12 49	11	24	19	28
22	12	1	18	12 48	12 48	12 50	12 52	12 54	12	25	18	27
23	13	31	17	12 52	12 54	12 54	12 56	12 58	13	26	17	26
24	14	30	16	12 54	12 58	12 59	13 2	13 3	14	27	16	35
25	15	29	15	12 58	13 2	13 4	13 6	13 8	15	28	15	24
26	16	28	14	13 2	13 6	13 8	13 10	13 12	16	29	14	23
27	17	26	13	13 6	13 10	13 12	13 14	13 17	17	30	13	22
28	18	25	12	13 10	13 14	13 16	13 18	13 22	18	1	12	21
29	19	24	11	13 14	13 18	13 20	13 22	13 26	19	2	11	20
30	20	23	10	13 18	13 22	13 4	13 27	13 32	20	3	10	19
31	21	22	9	13 22	13 26	13 28	13 32	13 36	21	4	9	18
1	22	21	8	13 26	13 30	13 32	13 36	13 40	22	5	8	17
2	23	20	7	13 30	13 34	13 36	13 40	13 44	23	6	7	16
3	24	19	6	13 34	13 38	13 40	13 44	13 48	24	7	6	15
4	25	18	5	13 38	13 42	13 44	13 48	13 52	25	8	5	14
5	26	17	4	13 42	13 46	13 49	13 53	13 57	26	9	4	13
6	27	16	3	13 46	13 50	13 54	13 58	14 2	27	10	3	12
7	28	15	2	13 50	13 52	13 58	14 2	14 6	28	11	2	11
9	29	14	1	13 52	13 56	14 2	14 6	14 11	29	12	1 ♓	10
10	30	13	0	13 56	14 0	14 6	14 10	14 16	30	13	0	9
				H.M.	H.M.	H.M.	H.M.	H.M.				

(Vertical month labels at left: MARCHE, SEPTEMBRE, SEPTEMBER, AVGVST, APRIEL. At right: SEPTEMBRE, MARCHE, OCTOBRE, FEBRVARYE.)

258   THE FOVRTHE TREATISE OF

The feconde parte of the table.

Signes for the daye.				Eleuation of the Pole, or latitudes of Regions.					Signes for the nighte.			
daies of moneths.	degrees of signes.	daies of moneths.	degrees of signes.	51	52	53	54	55	daies of moneths.	degrees of signes.	daies of moneths.	degrees of signes.
APRIEL. 10	♉ 0	13	30	13 56	14 0	14 6	14 10	14 16	♏ 0	13	30	9 *FEBRVRYE*
11	1	12	29	14 0	14 4	14 10	14 14	14 20	1	14	29	8
12	2	11	28	14 4	14 8	14 14	14 18	14 24	2	15	28	7
13	3	10	27	14 8	14 12	14 18	14 22	14 28	3	16	27	6
14	4	9	26	14 12	14 16	14 22	14 26	14 32	4	17	26	5
15	5	8	25	14 14	14 20	14 26	14 30	14 37	5	18	25	4
16	6	*AVGVST* 7	24	14 18	14 24	14 30	14 34	14 42	6	19 *OCTOBRE*	24	3
17	7	6	23	14 22	14 28	14 33	14 38	14 46	7	20	23	2
18	8	5	22	14 26	14 32	14 36	14 41	14 50	8	21	22	1
19	9	4	21	14 30	14 34	14 40	14 48	14 54	9	22	21	31
20	10	3	20	14 34	14 38	14 44	14 52	14 58	10	23	20	30
21	11	2	19	14 36	14 42	14 48	14 56	15 2	11	24	19	29
22	12	1	18	14 40	14 46	14 52	15 0	15 6	12	25	18	28
23	13	31	17	14 44	14 50	14 56	15 3	15 10	13	26	17	27
24	14	30	16	14 46	14 54	15 0	15 6	15 14	14	27	16	26
25	15	29	15	14 50	14 56	15 4	15 10	15 18	15	28	15	25
26	16	27	14	14 54	15 0	15 8	15 14	15 22	16	29	14	24
27	17	26	13	14 56	15 4	15 12	15 18	15 26	17	30	13	23
28	18	25	12	15 0	15 8	15 14	15 22	15 30	18	31	12	22
29	19	24	11	15 4	15 10	15 17	15 26	15 34	19	1	11	21
30	20	23	10	15 6	15 14	15 20	15 30	15 38	20	2	10	20
MAYE. 1	21	22	9	15 10	15 18	15 24	15 34	15 42	21	3 *NOVEMBRE*	9	19
2	22	21	8	15 12	15 20	15 28	15 37	15 45	22	4	8	18
3	23	20	7	15 16	15 24	15 32	15 40	15 48	23	5	7	17
4	24	19	6	15 18	15 28	15 36	15 44	15 52	24	6	6	16
6	25	18	5	15 22	15 30	15 39	15 47	15 56	25	7	5	15 *RYE*
7	26	17	4	15 24	15 34	15 41	15 50	16 0	26	8	4	14
IVLYE. 8	27	16	3	15 28	15 36	15 44	15 54	16 4	27	9	3	13 *IANVARYE*
9	28	15	2	15 30	15 40	15 47	15 57	16 7	28	10	2	12
10	29	♌ 14	1	15 34	15 42	15 50	16 0	16 10	29	11	♒ 1	11
11	30	13	0	15 36	15 44	15 54	16 4	16 14	30	12	0	10
				H.M.	H.M.	H.M.	H.M.	H.M.				

THE CASTLE OF KNOWLEDGE   259

The thyrde parte of the table.

Signes for the daye.				Eleuation of the Pole, or latitudes of Regions.					Signes for the nighte.			
daies of moneths.	degrees of signes.	daies of moneths.	degrees of signes.	51	52	53	54	55	daies of moneths.	degrees of signes.	daies of moneths.	degrees of signes.
11	Ⅱ 0	13	30	15 36	15 44	15 54	16 4	16 14	0 ♐	12	30	10
12	1	12	29	15 38	15 46	15 56	16 6	16 17	1	13	29	9
13	2	11	28	15 41	15 49	15 59	16 9	16 20	2	14	28	8
14	3	♌ 10	27	15 44	15 52	16 2	16 12	16 24	3	15	27	7
15	4	9	26	15 46	15 54	16 4	16 14	16 26	4	16	26	6
16	5	8	25	15 49	15 57	16 7	16 17	16 29	5	17	25	5
17	6	7	24	15 52	16 0	16 10	16 20	16 32	6	18	24	4
18	7	5	23	15 54	16 2	16 12	16 22	16 34	7	19	23	3
19	8	4	22	15 56	16 5	16 15	16 25	16 37	8	20	22	2
20	9	3	21	15 58	16 8	16 18	16 28	16 40	9	21	21	1
21	10	2	20	16 0	16 10	16 20	16 30	16 42	10	22	20	31
22	11	1	19	16 2	16 12	16 22	16 32	16 44	11	22	19	30
23	12	30	18	16 4	16 14	16 24	16 34	16 46	12	23	18	29
24	13	29	17	16 5	16 15	16 26	16 36	16 48	13	24	17	28
25	14	28	16	16 6	16 16	16 28	16 38	16 50	14	25	16	27
26	15	27	15	16 8	16 18	16 30	16 40	16 52	15	26	15	26
28	16	26	14	16 9	16 19	16 31	16 42	16 54	16	27	14	35
29	17	25	13	16 10	16 20	16 32	16 44	16 56	17	28	13	24
30	18	24	12	16 12	16 22	16 34	16 46	16 58	18	29	12	23
31	19	23	11	16 13	16 23	16 35	16 47	16 59	19	30	11	22
1	20	22	10	16 14	16 24	16 36	16 48	17 0	20	1	10	21
2	21	21	9	16 15	16 25	16 38	16 50	17 2	21	2	9	20
3	22	20	8	16 16	16 26	16 38	16 51	17 3	22	3	8	19
4	23	19	7	16 17	16 27	16 39	16 52	17 4	23	4	7	18
5	24	♋ 18	6	16 18	16 28	16 40	16 52	17 4	24	5	6	17
6	25	17	5	16 19	16 28	16 40	16 53	17 5	25	6	5	16
7	26	15	4	16 20	16 29	16 41	16 54	17 5	26	7	4	15
8	27	14	3	16 20	16 30	16 42	16 54	17 6	27	8	3	14
9	28	13	2	16 20	16 30	16 42	16 54	17 7	28	9	2	13
10	29	12 ♋	1	16 20	16 30	16 43	16 54	17 8	29	10	1 ♑	12
11	30	11	0	16 20	16 30	16 44	16 54	17 8	30	11	0	11
				H.M.	H.M.	H.M.	H.M.	H.M.				

Left margins: MAYE · IVLY · IVNE · IVNE
Right margins: IANVARYE · NOVEMBER · DECEMBRE · DECEMBRE

in the firſte columne are ſet the daies of the monthes, and in
the ſecond the degrees of the Signes in the Zodiake, in whi
che the ſonne is that daye : ſo likewaies the thirde and fourth
columne do ſerue for the like matter, ſeeing twiſe in the year
the daies are equall. And bicauſe at other 2 times in the year
the nights ar equall to thoſe daies, therfore on the right hād
of the table are ther 2 columnes of moneths, and other two
columnes of ſignes agreeable therto, in which thoſe nights
are equall with the daies of the monethes on the lefte hand,
and therfore ar the title ſet ouer the ſignes & moneths on the
lefte hand, ſignes for the day : and on the right hande ſignes
for the nighte : that is to ſaye, that if the moneth and ſigne
for which you ſeke, be on the left ſide of the table, then do the
numbres vnder the eleuation of the Pole declare the quantitie of the day : but if the monethes & ſignes be on the right
ſide, then is that quantitie the length of the night. and ouer
the 5 midle pillers, you ſe the title to be the Eleuation of the
Pole, or latitude of regions, whiche are there but only 5 ex-
preſſely ſet, namely 51, 52, 53, 54, & 55 : whiche may ſerue for all
Englād, from the ſouth ſea vnto Scotlād. And ſo may it do
for diuerſe of the northe partes of Europe and Aſia. Nowe
for the vſe of them, this is the ordre. When ſo euer you wold
know the quantitie of the daye Artificiall and of his night,
ſeeke out the day in the columnes on the right hande, or on
the lefte hand as it will chaunce, and by it in the next column
you may ſee the place of the Son in the Zodiake : then goyng right forth towarde the middle of your table tyll you
come directly vnder the column that ſerueth for your Region in latitude, there ſhall you finde 2 numbres : the firſt be
tokening howers, and the ſecond minutes of howers, which
declare the iuſte quantitie of the day for the moneths on the
lefte hande : or els if the moneth that you ſeeke for be on the
right hand, then do thoſe numbres of hower and minutes
betoken the quantitie of the nyghte.

　　Scholar. I perceaue it well, and ſe by reaſon it muſt nedes
be

be ſo : as for examples ſake. the 24 daye of Auguſte I deſire
to knowe the lengthe of the day and the place of the Sonne
in the Zodiake : wherfore fyndynge the ſaide 24 daye in the
fyrſte table of thoſe thre ryght againſt it, I may ſee the place
of the ſonne, whiche is then the 11 degree of Virgo : and
from thence proceedinge forth righte towarde the myddle
of the table, I fynde vnder the numbre of 52 degrees of latitude 13 howers and 18 minutes : whereby I perceaue that the
Artificiall daye from ſonne ryſynge to ſonne ſettinge, is ſo
longe with vs : and the nyght is the reſte of 24 howers, that
is 10 howers and 42 minutes. And the lyke quantities of
daye and nyght muſt needes be the 29 daye of marche, when
the ſonne is in the 19 degree of Aries. But on the 20 daye of
February, the ſonne beyng in the 11 degree of Piſces, that 13
howers and 18 minutes is the quantitie of the nyghte, and
the day then is but 10 howers and 42 minutes in length : and
ſo likewaies the ſeconde daye of Octobre, when the ſonne
is in the 19 degree of Libra.

Maſter. This is ſufficiente : for as you haue doone in this
ſo maye you doo in all other lyke. yet for the more certenty
I will proue you with one queſtion more : For London whi
che is ſuppoſed to be 51 degrees and 24 minutes in latitude,
I woulde knowe the quantitie of the daye Artificialle when
the ſonne is in the 28 degree of Scorpio.

Scholar. I fynde that ſigne of Scorpio in the ſecond table
on the right hand, and the 10 daye of Nouembre anſwering
vnto it. And bicauſe 24 minutes are leſſe then halfe a degre,
I do ſeeke the quantitie of the daye vnder 51 degrees rather
then vnder 52, and ſo fynde I 15 howers and 30 minutes : whiche in this caſe is the quantitie of the nyghte, as the title declareth that is ouer thoſe ſignes : therfore the lengthe of the
daye is 8 howers and 30 minutes.

Ma. You haue done well. But yet for an exacter preciſenes,
you may take the part proportionable for the odde minuts
of the eleuation, as thus. for the latitude of 51 degrees, the

*A cautele
for the part
proportio-
ble.*

day

daye is 8 howers and 30 minutes : and for 52 degrees, it were
8 howers and 20 minutes : fo are there 10 minutes difference betweene thofe two eleuations. Then faye by the Golden rule : If 60 minutes giue 10, what fhall 24 minutes giue?
and it will appeare to bee 4 minutes. Thofe 4 minutes mufte I abate frome the greater noumbre in
this example (and in all this worke wher the num
bres decreafe) and it will yelde 8 howers & 26 minutes : where as yf you did fynde the numbres to increafe,
then fhould you adde thofe partes porportionable vnto the
leffer numbre, as by proofe you may try, for that day when
the fonne is in the fecond degree of Leo.

Scholar. That is (by the fecond table) the 15 daye of Iuly,
and then is the daye in lengthe 15 howers and 30 minutes, in
the latitude of 51 degrees : but in the latitude of 52 degrees,
it is 15 howers and 40 minutes, fo it increafeth 10 minutes :
and therfore mufte I adde the parte proportionable (which
is 4 minutes as before) vnto 30. and fo haue I the true quantitie 34 minutes aboue 15 howers. And nowe I thinke I am
perfecte inoughe for all places betweene 51 degrees of latitude and 55 : but for other places I knowe no fuche waye.

Mafter. It were to longe a woorke to fette out all diuerfities of eleuations, and fcarfe agreeable for this treatife, wher
thefe thinges are but incidente, and not principall matters.
but at other times in more conuenient place it fhall be done
if I maye vnderftande this my labour to be profitably im-
ployed. And thē alfo will I make explicatiō of dyuers other
matters, whiche you did in your table at the beginning of
this treatife propounde, although at this time I thinke many of them lytle appertaining to this booke. But yet before
I eande this treatife, I muft fpeak fomwhat of twoo or three
matters more : And firfte of the chieffe Conftellations and
figures in the Starry fkye. For a ground you fhall note, that
the ftarres are not only in multitude infinite, but many of
them alfo fo fmal, that fcarfe any mans eye can difcern them.
                                                    wher-

Conftella-
tions.

wherefore to auoide confuſion, and to growe to a certenty,
the auncient Aſtronomers did note only 1022 ſtarres, wher
of the moſte parte they did aſſigne to certain limites, enclo
ſing them in figures of men, beaſtes, or other formes, and
accordinglye gaue them names, partly that they might the
more eaſily bee remembred, partlye for remembraunce of
ſome woorthy facte, and partly alſo for ſome notable ſigni
fication of the ſtarres comprehended in eche of them. All
whiche matters I will nowe ouerpaſſe, tyll a more conuenient place, and will repeate onlye their names and places
generally, diſtincting them accordynge to the accuſtomed
manner, into three ſortes : whereof the one ſorte are called
Northerlye conſtellations, the other ſorte Southerly conſtellations, and the third ſorte are the twelue ſignes, which
paſſe in the myddle betweene ſouthe and northe : for heere
in this place I meane not to referre ſouthe and north to the
Poles of the Equinoctiall, but as all learned men before me
haue doone, to the poles of the Zodiake. And ſo may the
Zodiake be accompted exactly in the myddle. But nowe to
beginne as Ptolemye doth, with the northerly conſtellations : The moſte northerly conſtellation is the leſſer Beare,
called Vrſa minor, and Cynoſura, and contayneth in it 7
ſtarres. This is the chiefe marke whereby mariners gouerne
their courſe in ſaylinge by nyghte, and namely by 2 ſtarres
in it, which many do call the Shafte, and other do name the
Guardas, after the Spaniſh tonge. Nigh vnto it is the greater Beare, called Vrſa maior, contayninge 27 ſtarres, wherof 7 are moſte notable, and are in latine named Plauſtrum,
and in engliſh Charles waine, which ſerueth alſo well in ſai-
lynge : and manye of the olde Greekes obſerued it onlye in
their nauigation, as the Sydonians and all the Phenicians
marked the leſſer Beare. Aboute theſe 2 Beares is there a
longe trace of 31 ſtarres, cōmonly called the Dragon. Then
foloweth Cepheus, whiche conſiſteth of 11 ſtarres.

  Bootes alſo is in the ſame coaſte, whome Proclus and o-
thers

The northe cõſtellatiõs.

1
Vrſa minor.

2
Vrſa maior.

3
Dragon.

4
Cepheus.

5
Bootes.

264   THE FOVRTHE TREATISE OF

thers doo name Arctophylax. and it hath 22 starres, beside
one very bryght starre called Arcturus, which standeth be-

6
*The northe
Croune.*

tweene Bootes legges. By Arctophylax ryghte hande, is
the northe Croune, called also Ariadnes Croune, and hath

7
Hercules.

in it 8 starres. Then foloweth Hercules, whom the greekes
doo call Engonasin, as it were the Kneeler, bicause of his
gesture : and it containeth 28 starres. By hys lefte hande,

8
Lyra.

is there an other constellation, whiche is called the Harpe,
in latine Lyra and Fidicula. and also Vultur cadens, that is
the fallynge Grype, it comprehendeth 10 starres. By it is

9
The Swan.

the Swanne, named Cygnus, and Auis generallye, as the
Greekes call it Ornis, whiche some men of to muche ouer-
syght do translate, Gallina a Hen : it consisteth of 17 starres.

10
Cassiopeia.

After it dooth Ptolomye recken Cassiopeia, whiche is by

11
Perseus.

Cepheus, and hath 13 starres. Nexte vnto hir is Perseus,
with Medusas headde, and it includeth 26 starres. Then

12
The Carter.

foloweth Erichthonius, with the Goate and the 2 Kyddes.
this constellation is also named Auriga the Cartar : and cō
taineth 14 starres with one in his right foote, which is common to Taurus also. An other constellation is there which
ioyneth heade to heade with Hercules, and is called of the

13
Serpētarius.

Greekes, Ophiuchus, and of the latines Serpentarius, that

14
The serpent.

is the manne with the Serpente, or Serpent bearer : and it
hathe 24 starres. Besyde the Serpent, which containeth 18
starres in him selfe, and is named of latines Anguis, and of
greekes Ophis. Then is there an other small constellation of 5 starres, a lytle southe of the swannes heade, and it is

15
The Dart.

named the Darte, Sagita or Telum in latine, and in greke
Oistos. By it towarde the southe, is the Egle, includynge

16
The Egle.

9 starres : hee is called not onlye Aquila in latine, but also
Vultur volans, and in greeke Aetos. Vnder it towarde the

17
Antinous.

south is a constellation harde adioyning named Antinous

18
The dolphin

in all tonges, and hath but 6 starres. A lyttle from it is the
Dolphine, whiche hath in it 10 starres.

19
*The Fore-
horse.*

Then foloweth the Forehorse, noted with 4 darke starres,
                                                        and

THE CASTLE OF KNOWLEDGE   265

and harde by him is the Flying horfe, named Pegafus : and *The Flying Horfe* doth confifte of 20 ftarres. Vnto him ioyneth Androme- *21* da, fo that hyr headde lyeth on the nauell of Pegafus, and *Andromeda* one ftarre is common to them bothe. This conftellation dothe containe 23 ftarres.

By hir lefte foot is ther a fmall conftellation of 4 ftarres, *22* which is commonly called the Triangle, and in latine Tri- *The triagle* angulus, but the greekes name it after one of their letters Delta and Deltoton. And thus haue I briefly reckened all the northely conftellations, excepte Berenices heare, of whiche I will fpeake lafte of all other. And therefore nowe nexte in due ordre mufte the 12 fignes followe : amongeft *1* whiche Aries occupieth the fyrfte place, and contayneth 13 *Aries.* ftarres. Then Taurus whiche is adorned with 33 ftarres, *2 Taurus.* wherof 5 be in his forhead and face, and are called of the *Water ftars* Greekes Hyades, and of the latines Succule : amongeft whi che, one is more notable then all the refte, and is called O- culus Tauri, the Bulles eye : but the Greekes call it Lampa- dias, and the latines Palilicium : the Arabitians Aldebaran. Other 6 ftarres (as Proclus numbreth them, though other accompt them 7) ar in the backe of this figne, and be called Vergiliae in Latin, and in Greeke Pleiades, and alfo Atlan- tides : they are named in englyfh the brood Henne, and the *The feuen* Seuen ftarres, yet they dufter fo nyghe togither, that it is *ftarres.* harde to numbre them truly. and therfore many do difagre in reckenynge them. *3*

After Taurus, Gemini do followe, whiche comprehend *Gemini.* 19 ftarres : of whiche twoo beare name as moft famous, and they are in their headdes : the formoft is named Appollos headde, and the nexte is called Hercules headde, bi- caufe thofe two Twinnes were fo named of fome men, yet other doo call them Caftor and Pollux. Before their for- *Propus.* mofte foote is there one fayr ftar (befide the 18,) which ther fore is named in greke Propus. After Gemini foloweth Ca *4 Cancer.* cer cotaining 8 ftars, befide a cloudy tract which is named ỹ *Crybbe,*

Z.i.   Manger

Aſſes.
5
Leo.

Manger or Crybbe. Other two ſtarres are called the Aſſes whiche ſeeme to ſtande at the Crybbe. Then the Lion is nexte, as a princely ſigne, in whome are 27 ſtarres, but two of them more notable then the reſte : the one is in the tayle, and therefore is called Cauda Leonis, the other in the breſt and is called the Baſilyſke or Kyngely ſtarre, and alſo the Lions harte, Cor Leonis in Latin, and Baſiliſcos in greke.

6
Virgo.

Nexte after Leo, cometh Virgo, garniſhed with 26 ſtarres, but one eſpecially glyſtereth aboue the reſte, and is called Spica Virginis, the Virgins ſpike.

A leſſer ſtarre there is alſo, whiche yet is notablye marked, and called Protrigetes, Præuindemiator.

7
Libre.

After Virgo cometh Libra, the ſigne of Iuſtice and equitie : but it is the leaſte ſigne in quantitie of all other in the Zodiake, for it occupieth ſcarſe halfe a ſigne in lengthe, and no meruaile, ſyth that cruell Scorpius dooth inuade ſo greate a portion, and preſſeth all that Sygne oute righte. yet hathe it 8 ſtarres, but not one out of the Scorpions clawes.

8
Scorpius.

Then Scorpius with his hooked tayle, and with his clawes doth reache ſo farre, that two full ſignes he taketh in length and 30 degrees almoſte in bredth, yet hath he but 21 ſtarres beſide thoſe whiche bee in his clawes, and are common to them & to Libra : amongeſt all which the principall is that, whiche is called the Scorpions harte, and is named of the Greekes Antares, and of Arabitians, Calbalatrab.

9
Sagittarius.

After him enſueth one of the Centaures lyke an archer on horſe backe, with manye fayre ſtarres, though they bee not of the greateſt : he hath in all 31. this ſigne is called Sagit

10
Capricorn.

tarius in latine, and in greeke Toxotes. Capricorn then ſo loweth with his monſtrous ſhape, nother fyſh nor fleſh, but myxed of both : a winterly ſigne and no waies pleaſant, but that he geueth hope of the cõfort of the Springe, bicauſe in it the ſonne beginneth to retourne to vs againe. hee hath in him 28 ſtarres of meane quantitye.

Aqua-

THE CASTLE OF KNOWLEDGE 267

Aquarius fo fafte dooth followe him at hande, that hee *Aquarius.* reacheth almofte as forwardlye as Capricorne, within leffe then 8 degrees : this figne hath in him 22 ftarres peculiare to him felfe,, althoughe Proclus name 4 of them in hys ryghte arme, to be the Water potte. But befyde thefe 22 *The water* ftarres, there are other 19, whiche in their dyuers and cro- *potte.* ked pofition doo make a forme of a Ryuer, and are called the Water whiche Aquarye fheddeth. With thefe 19 ftarres Ptolemye doth accompte one more, whiche is a bewtifull ftarre of the bryghteft forte, and is in the mouthe of the Southe fyfhe, fo that it is common to them bothe. this ftar is called of Arabitians Fomahant : fo that in all there are reckened in this figne, 42 ftarres.

Lafte of the 12 fignes commeth the Fyfhes, tyed by the *Pifces.* tayles with a common Lyne : the formofte Fyfhe hath but 8 ftarres, and his line hath 10. the latter Fifhe hath 11 ftarres, *The Lyne.* and his lyne hath but 5 : and where thofe two lines are knitte togyther, there is one ftarre more, whiche is called the Knotte, that is in Greeke named Syndefmos : fo that all the ftarres togither, of this figne, are 34.

Whether Proclus did miftake any thinge in this figne, I wifhe other to iudge, bicaufe I intended here not to intreat at large, and muche leffe to fcan other mennes writinges. And thus wyll I eande the 12 fignes of the Zodiake.

Nowe to diuerte vnto the fouthe fignes : fyrfte appeareth *The Whale.* the greate Whale, contayning 22 ftarres, whereof three bee mofte noted : the fyrfte in the nether chappe, whiche is in la tine called Mandibula ceti, and in Arabike Menkar. the fe- conde is called the Whales bellye, in Arabike Baten kaitos, and in Latine Venter Ceti. the thirde is the Whales tayle, named Cauda ceti in latine, and in Arabike Deneb kaitos. Nexte foloweth Orion, the Stormy figne, and hath diuers *Orion.* ftarres to the numbre of 38 : but the mofte notable are 6. the fyrfte is in his ryghte fhoulder, and is called by the

Z.ij.   Ara-

Arabitians Bed Algeuze. The fecond is in the lefte fhulder and is named Bellatrix. Other thre ftande as bullions fet in his gyrdle, and are called of manye englyfhe men the Golden yarde. Then is there in his lefte foote, a greate ftarre of the brighteft fort, which is named of Arabitians Algebar, and Rigel Algeuze. Befide thefe fixe there are other ftarres more notable for their forme then for their quantities. as the two ftarres which betoken his clubbe in his right hand, and 9 ftarres by his lefte hande, whiche reprefente a Lions fkynne : and other three doo limite his fworde, lying croffe his backe vnder his girdle.

3
The Riuer.

Betweene Orion and the Whale is there a greate tract of ftarres, whiche reprefent the forme of a Riuer : and therefore are they called the Ryuer. whiche fome more peculiarly name Eridanus, and other Nilus. Proclus calleth it Orions ryuer, bicaufe it beginneth at his lefte foote and hath one ftarre common with his foote, but befide that it hathe 34 ftarres : wherof the lafte is one of the greateft lyght.

4
The Hare.

By the beginninge of this Ryuer, vnder the feete of Orion is there a conftellation of 12 ftarres, named the Hare.

5
The great Dogge.

And after it toward the eafte is the greater Dogge, (of whō the Caniculare daies bear name) and is called of the grekes Sirius, and of the Latines Canis, hauing 18 ftarres, but one efpecially in bryghtnes more notable then anye of the reft, and that is in his mouthe, and is called peculiarlye Sirius and Canis, by the name of the whole Signe, and of the Arabians Alhabor. Northe almoft from this Dogge is ther a conftellation of 2 only ftarres named Canicula, the leffer Dogge : and in greeke Procyon, the fore dogge, whō Tully therfore calleth Antecanis, and other name him Precanis.

6
The leffer Dogge.

7
Argo the Shyppe.

At the tayle of the greater Dogge is the famous fhippe Argo, whiche comprehendeth 45 ftarres, wherof 8 bee bewtifull but one in efpeciall which is in the foote of the roother & is called Canopus, & of the Arabitians Suhel. This ftar is not feen in Englād, France, Germany nor Italy, & fcarfly in

the

THE CASTLE OF KNOWLEDGE   269

the moſte ſoutherly partes of Spaine. And here by the waye I will note a place in Proclus very much corrupted, whiche nowe I will only correct as I thinke good : and an other time will intreate more largely of it and of other mo. the wordes in Greeke are theſe.

ὀ δὲ ἐν ἄκρῳ τῷ πηδαλίῳ τῆς ἀργοῦσ κᾶμεν⊙ λαμπρόσ ἀεήρ.κάναβ⊙ ὀνο-
μάζεται,ὃ ρος μλίεν ῥοδίῳ μόλις θεωρητὸσ ὃσιν,ἡ παντελῶσ ἀφ᾽ ὑψιλῶν τόπων
ὁρατός. ἐν ἀλεξανδρ͂ εἰᾳ δὲ ἐςὶ παντελῶσ ∗ ἐυφανήσ. σχεδὲν γὰρ τίταρτον μέ-
ρ⊙ ζωδίσ ἐκ θῦ ὀρίζοντοσ μετεωρισμέν⊙ φαίνεται.

∗ ἀφανής
in all the
Greeke
bookes.

Stella vero illa ſplendida que in imo Argus gubernaculo ſita eſt, Canopus dicitur. ea in Rhodo vix conſpicitur, aut certe ab editis locis. In Alexandria vero prorſus ∗ conſpicua eſt, vtpote ſere quarta ſigni portione ſupra Horizontem euecta.

Non cer
nicur, naſ
tulit lati-
nus inter
pres, gre
ci codi-
cis erro-
rem imi-
tatus.

The bright ſtarre in the foote of the roother of Argus is called Canopus, whiche in the Rodes can ſcantely be ſeene, excepte it be from highe places : but in Alexandria it maye well be ſeene, for it doth riſe there nyghe a quarter of a ſigne aboue the Horizont.

Scholar. This is contrarye to the common tranſlation.

Maſter. And that common tranſlation is as contrary to common ſenſe, but therof an other time ſhall we talke, when I mynd to teache you the exacte ordre of aſcenſion for all theſe conſtellatiōs, and of their chiefe ſtarres alſo. And now to proceede as we began. Nexte after this ſhip ther foloweth the great Serpent whiche is called of the greekes and latines Hydra. it containeth 25 ſtarres, and ſtretcheth in greate lengthe by the ſpace of 3 whole ſignes. one ſtarre there is in it bryghter then the reſte, and that is named by the Arabians, Alphard.

8
The Ser-
pent of the
ſouthe.

On this Hydre there reſteth other 2 ſmall conſtellations, the one named the Cuppe, and the other the Rauen.

The Cuppe includeth ſeuen ſtarres all of one bygnes. This Cuppe ſtandeth on the Hydres backe, almoſte in the myddle of him.

9
The Cuppe

10
The Rauen.

The Rauen ſtandeth on the ſame Hydre, more nearer towarde the pointe of his tayle : and it is formed of 7 ſtarres
alſo, of whiche that which is in his lefte wing, is called in A-
rabike, Algorab.

11
The Cen-
taure.

The Cen-
taures ſpear.

Vnder the taile of this Hydre and thoſe twoo other ſmall
conſtellations, there ſtandeth the centaure Chiron, lyke a
lyghte horſeman with his chafinge ſtaffe : he hath in him 37
ſtarres, whereof 4 be in the garniſhe or penſile of his ſpear,
and them doth Proclus recken as a peculiare conſtellation,
and nameth it in greeke Thyrſolochus. And Ptolemy doth
recken thoſe ſtarres naming them to be in that ſpeare : wherfore I muſe howe Stofler ſeemed ſo ignoraunte herein, to
deny that Ptolemye doth make any mention of that ſpear,
and hym ſelfe deuiſeth oute of Ptolemye 6 wronge ſtarres
for that purpoſe : it appeareth hee was deceaued by the olde
tranſlation, where Clypeus is tranſlated for Haſta : that is,
ſhielde for ſpeare. whiche wrong tranſlation Schoner, Copernicus, and Eraſmus Rheinhold doo follow, and dyuers
other learned men, but againſt reaſon.

Scholar. I thinke it (as manye thinges els be) is receaued
by credite of authoritie, withoute diſquiſition of reaſon,
whiche blyndeth manye wittye man oftentymes.

Maſter. Yet is their faulte the more pardonable, if they
acknowledg their errour when thei be friendly admoniſhed :
but this is beſide our purpoſe at this time, therefore to returne : This Centaure with his righte hande dooth holde a
Wolfe, whiche is a ſeuerall conſtellation made of 19 ſtarres,
althoughe Hyginus and others doo recken fewer in him, as
they doo vntrulye in manye other. Vnder that beaſte towarde the ſouthe, harde vnder the Scorpions tayle, ſtandeth
the Altar, made of 7 Starres, of the meaneſt lyght : but it is
not ſeene in Englande aboue the Horizont. By this Altar
eaſtwarde betweene the two former feete of Sagittarye, there
is the Croune of the ſouthe, formed of 13 ſmall ſtarres : Pro
clus and Theon doo call it alſo Vraniſcus, as manye later
writers

12
The Wolfe.

13
The Alter.

14
The ſouthe
Croune.

writers in their tyme did name it : but Theon dooth farther
affirme that it hath 19 ftarres : whiche mufte feeme to bee an
errour, rather in the booke then in the author : wherein obferuation canne not healpe vs in Englande, fyth it rifeth not
aboue our horizont, but only toucheth it.

After it foloweth the Southe fyfhe, containynge 12 ftarres :
wherof one only is of the greateft lyght, and that is it which
ftandeth alfo for the eande of the water that runneth frome
Aquarius. This fyffhe lyeth betweene the conftellations of
Capricorne and Aquarye, fo that it is partely vnder them
both.

15
The fouthe Fifhe.

Thefe bee the Conftellations moft commonlye noted a-
mongeft auncient writers : howebeit one more there is named to lye betweene the Lions taile and Vrfa maior, whiche
is called Berenices heare, fome call it in latine Trica, and o-
ther Berenicis crines. Conon that famous aftronomer dyd
fyrfte name it, and Callimachus did dedare it, and therefore
doth Proclus adfcribe the fyrfte noting of them vnto Callimachus. The ftarres in it are 7, as Hyginus and Baffus do
accompt them : but they are verye darke, and therefore
Ptolemye doth numbre only thre of them, as the boundes
of that forme. Befyde thefe 50 conftellations, there bee a
greate numbre of ftarres, whiche be not affigned to any figure, but lye difperfedly about thofe other conftellations,
whereof 61 are in the northe parte of the fkye, and annexed
with the northerly fignes : and other 19 in the fouthe part of
the Zodiake, vnto whiche if you adde 337 whiche be in the
northe conftellations, and 316 in the fouthe conftellations,
with 292 in the Zodiake, fo haue you in all 1025 ftarres whiche be noted by Aftronomers, but in Ptolemyes accompte
there appeare but 1022, bicaufe he doth not accompte anye
ftarre of Berenices heare, but called it the Traces of heare.
Thefe ftarres be not of one quantity, but fom much brigh-
ter then other, and therefore are they diftincte into diuers
meafures of lyght, and namely 8, whiche are called the firft

16
Berenices heare.

6   1
1   9
3   3   7
3   1   6
2   9   2
————
1   0   2   5

Z.iiij.   greatnes

greatnes, the feconde, the thirde, the fourthe, the fyfte an the fyxte, vnder whiche they are that be called Cloudy ftarres : and a leffer forte yet named Darke ftarres : of all which, and the meafure of their quantitie, I will at an other tyme fpeak more fullye, for this place and time agreeth euell with the matter, and that muche worfe, then at the beginning it fee-med to doo.

Scholar. There remaine yet manye tytles vntouched of them whiche I gathered.

Mafter. And manye of theym fmally agreeable for this treatife, but doo more aptly appertaine to Cofmography, and therefore ought to be referued for that worke : faue that fome of them are peculiare for the Theorike of Planetes, and yet will I lightly touch them in fewe words, for fo much as may feeme to healpe to this treatife.

Howe the numbre of fpheres is knowen.

Scholar. I remember at the beginninge you promifed to fhewe a caufe why you name but 8 fpheres, where as other men do accompte more : and alfo how it may appeare, that there are fo manye, for the eyes can fee but one only, whiche is the firmament.

The Moone.

Mafter. Your felfe fayde, you had marked (as many ma-riners, yea and all men do almofte) that the Moone dothe euerye daye runne eaftwarde notably, fo that in a weeke fhee paffeth a quarter of the fkye in that courfe, and in 15 daies fhe runneth halfe the compaffe of the fkye, and fo in a mo-neth fhe retourneth to the fonne againe, hauinge paffed all

The Sonne.

the circuit of heauen. fo of the Sonne you haue vnderftand that in a yeare he trauerfeth ouer all the lengthe of the Zo-diake, contrary to the courfe of the Firmament, whereby it mufte needes appeare vnto you, that feeynge the fonne and the moone haue courfes diftinct from the Fixed ftarres, thei mufte needes haue diftincte fpheres alfo, wherein they doo moue, and accomplifhe their courfes.

Scholar. I remembre I haue hearde it often repeated as a principle in nature, that one fymple body can haue but one
fymple

fymple motion. and therfore where diuers motions bee, it mufte needes followe that there are diuers bodyes as theyr workers, whiche you in this talke do call fpheres.

Mafter. As you may thinke that their fpheres are diftinct from the Firmament by reafon of their feuerall motions, fo are they diftincte a fonder by the fame reafon.

Scholar. It is mofte certaine.

Mafter. Then if by good obferuation it haue bene pro-ued, that there be 5 other ftarres whiche haue their motions all diftincte from the Starry fkye, and eche of them frome their fellowes, it will appeare reafonable that euerye one of them hath a feuerall fphere perculiare for him felfe, and for his priuate motion.

Scholar. It will followe of neceffitye.

Mafter. Then I will beginne with your felfe for one of them, whiche I am fure you can not but marke, as all men, yea the verye Plowmen doo. And that is Venus, whiche I *Venus.* dare faye, you haue marked in the euenynge to fet after the fon, & then is fhe named the euenyng ftar, & yet doth fhe not at al times fhine like fpace after fon fetting, but fome times more & fomtime leffe. And if you marke hir well, then fhall you perceaue, that the fyrfte nyghte that fhe appeareth, fhee fhyneth leffe time then fhe dothe the feconde nyght, and fo increafeth the tyme of hir fhyninge for a fpace, and then dothe fhee abate againe by lyttle and lyttle, tyll fhe ioyne with the fonne, and then appeareth no more at euenynge, but fhortly after will fhe fhowe in the mornynge before the fonne ryfynge, and increafe the time of hir fhining by litle and lytle, tyll fhe comme to the fartheft of hir diftaunce frō the fonne, and then will fhe abate againe in lyke manner, till fhe come within the beames of the fonne, and leefe hir ap-pearynge for a tyme.

Scholar. This is mofte certaine and knowen of all men vulgarly, althoughe fewe men doo confidre the caufe ther-of : but nowe I doo remembre, what you taught me of the
                                                        afcen-

aſcenſions poeticall (as they be named) and namely of that whiche you thought meter to bee called apparition, whoſe contrary you called Occultation : ſo that when Venus doth ſhyne at euenynge after ſonne ſettinge, ſhe dothe riſe as ſom tearme it, with a ſonnely ryſinge : and when ſhee is hydden againe, ſhe is ſet with a ſonnely ſettinge. but that you iudge Apparition and Occultation more apter tearmes.

Maſter. You doo not geſſe muche amyſſe. And to the intent that you may conſidre this matter the better, I think it good that you do marke hyr motion the more diligently hereafter : as in this preſente moneth of Septembre, at the beginning of the moneth ſhe was about 36 degrees behynde the ſonne, and ſo ſhoulde ſhe ſhine almoſte 2 howers and a halfe after the ſonne, as it myghte appeare by the degrees of diſtaunce. but conſideringe the obliquitie of the Zodiake, and the latitude of Venus at that time, ſhe didde ſcarſe ſhine three quarters of an hower after the ſonne.

Scholar. This talke is to obſcure for me yet.

Maſter. I knowe it ryghte well. but yet I thoughte good to admoniſh you in that matter, leaſt at any time you ſhuld fynde the doubte, when you ſhall haue no opportunity to aſke councell therein : but now to proceede. before the eand of the ſame moneth of Septembre, the ſayde Planete wyll be cleane hydde with the ſonne beames : for within 2 dayes after (I meane the ſecond daye of Octobre) ſhe doth ioyne with the ſonne by coniunction. And frome that daye forwarde the ſonne doth outgo hir ſo faſte, that by the 13 daye of Octobre, ſhe wyll be out of his beames againe, and ryſe almoſte an hower and a quarter before the ſonne. And at the eande of Nouembre, ſhe will be 46 degrees behind the ſonne, in ordre of the ſignes, and yet ſhall ſhe riſe 4 howers and more before the ſonne, where as the numbre of degrees are equall to lyttle more then three howers. but the obliquities of the Horizont, doth make all the diuerſitie in this. excepte a meane trifle by the latitude of Venus. And thus

may

may you marke Venus in all that moneth, and in Decembre alſo vnto the eande of the yeare : but then dooth ſhe a-
bate her diſtaunce againe, wherby it is eaſye to vnderſtande
that ſhe hathe a ſeuerall motion from the ſonne, and a ſeuerall ſphere alſo.

Scholar. In Venus it doth appeare nowe eaſye inoughe
to conſidre, as well as in the Sonne and Moone : but is it as
eaſye in the other four Planetes?

Maſter. Yea in deede, for three of them which bee moſte
higheſt, if you lyſte to learne to knowe them, and to marke
their courſes : but Mercury is not ſo well marked, bicauſe *Mercury.*
he doth alwaies keepe his courſe high about the ſonne, and
therfore his obſeruation requireth greate diligence, and his
courſes appeare moſt ſtraunge, yet bothe he and Venus do
accompliſhe their courſe in a yeare with the ſonne : but Sa- *Saturne.*
turne is ſo ſlacke a mouer, that you ſhall not well perceaue
his motion vnder 4 moneths. in which time he doth moue
about 4 degrees : ſo that if you marke his place at any time,
and within 4 monethes after that time yf you do marke him
againe, you ſhall perceaue that hee is gone 4 degrees eaſtwarde, whiche you maye marke by the fixed ſtarres aboute
that place : but if you doo after a whole yeare marke hys
place, then ſhall you perceaue well and manifeſtly, that hee
is gone eaſtwarde 12 degrees, and ſomwhat more : as for example. The fyrſte daye of Septembre, the laſte yeare 1555,
Saturne was in the 12 degree of Aries, and this year of 1556
we ſee him to be in the 26 degree of the ſame ſigne, wherby
it dothe appeare, that he hathe moued 14 degrees eaſtwarde
in that yeare ſpace. And if you will haue farther proofe : In
the yeare of our Lorde 1549, the laſte daye of Nouembre,
Saturne was ſeene in the 26 degree of Capricorne, and this
yeare of 1556 the fyrſte of Septembre, the ſame ſtarre was
in the 26 degree of Aries : wherby it maye bee knowen that
hee hath moued three whole ſignes (whiche is a quarter of
the Zodiake) in 7 year ſpace. And ſo in leſſe then 30 years,
he

hee dothe go about the whole Zodiake.

Iupiter.   Iupiter hath a fwyfter courfe, for he paffeth the circuite of heauen in leffe then 12 yeares. fo doth he euery yeare run ouer one figne, and euery two moneths he paffeth 5 degres.

Mars.   Mars is yet fwyfter in courfe then hee, and compaffeth all the Zodiake in 2 yeare, and euery moneth paffeth halfe a figne. wherby for this point, he is more eafy to be marked, then anye of the other. but yet are his motions difficulte to marke in other pointes : but this may fuffice for tryall that he moueth eaftwarde, as all the other Planetes do : and ther fore muft he be iudged, as all the other alfo oughte to haue feuerall fpheres in whiche they moue. And although theyr fpheres can not bee feene, yet in as muche as their ftarres maye be fo well perceaued, it mufte needes follow, that they haue fpheres alfo : except we fhuld come to that abfurditie to faye, that they moue in the Ayer as byrdes do, or as fy-fhes in the water : whiche were to muche repugnante to any one ordrely motion, and much more difagreyng to fo ma-ny diuers motions as are in the Planetes, but namely in Mars and Mercury. And to the intend that you may know them the better, it fhall bee good that you learne their true places by the Ephemerides, and accuftome your felfe to loke for them, and to marke their bignes and colours how they differ from other ftarres. whiche is fpoken by waye of exhortation only, and not propouned as anye peece of this booke, but an other time I will inftructe you better therein.

Scholar. But in the meane time, howe fhall I know whe-ther there by anye more fpheres or no?

Mafter. There is thoughte to be in the 8 fphere or Fir-*Of the nith* mament, two other motions, whiche be difagreeable from *and tenthe* all other mouinges before mentioned, and therfore many *fphere.* thinke that they mufte of neceffitye confeffe 2 other fpheres from whiche thofe motions muft proceede peculiarly.

Scholar. What motions are thofe, and howe are they knowen?

Ma-

Mafter. Fyrfte there is one notable obferuation by conference of learned men in diuerfe ages, concernyng the Equinoctiall pointes, and lyke waies concerning thofe Tropicall pointes, that the Sonne toucheth twife euery yeare : for about the incarnation of Chrift, the equinoctiall point or inftaunte happened aboute the 25 daye of Marche, and nowe it is aboute the tenthe of the fame moneth, whyche difagreemente dooth ryfe partly by the miffe ordre in the Leape yeares, but mofte principallye thoroughe the anticipation of the Equinoctiall tearmes. For althoughe the Sonne doo at the yeares eande retourne to the fame poynte in the Starrye fkye where hee was at the beginninge of the fame yeare, yet is he not exactlye fo nighe vnto the Equinoctiall pointe as he was before, but doth ouer runne it euery yeare, and thereby in continuaunce of tyme it cometh to paffe, that men may fenfibly perceaue that the ftars are runne eaftward from that equinoctiall point.

Scholar. This feemeth fomething obfcure, excepte you can declare it more plainely.

Mafter. Do you not confidre betwene the fonne and the moone, that when fhe doth ioyne with him by coniunction and then ouerpaffeth him by her fwyfte motion, that when fhe retourneth againe to the fame place where fhe dyd leaue the fonne, fhe doth not fynde him there, but fhe muft ouer go that place, beefore fhee canne ouertake the Sonne againe, by reafon that the fonne dydde moue forwarde after the moone in the fame courfe, though muche more flowly : So likewaies when the Sonne departeth frome anye ftarre in the fkye, in the verye inftaunt of the equinoctiall equalitye, and in the very point of the interfection of the Equi noctiall and the Ecliptike line, where of neceffity that equalitie muft happen : if the fonne retourning after a year vnto that Equinoctiall pointe, do not fynde the ftarre there precifely, whiche he lefte there, but that he mufte ouer run that point, before he cā come again to ẙ faid ftar, may not we yea

&.i.   and

and muſt not we ſaye, that the ſtarre is moued forwarde in
his courſe eaſtwarde, as all the Planetes doo moue? Howe
bee it the quantitie is ſo lyttle, that it is not perceaued by
ſyghte alone, nother yet by inſtrumentes, in leſſe then an
hundreth yeare, ſo that no one man is hable to marke anye
greate diuerſitie in hys owne age, but muſt be fayne to con
ferre with other men that hathe made obſeruations longe
beefore and written them : ſo dydde Ptolemye conferre
his obſeruations, with Hipparchus obſeruatiōs, and found
that from Hipparchus tyme vnto his owne age, the Fixed
ſtarres were moued forward from the Eqinoctiall pointe,
two degrees, and 40 minutes : whereby he dyd coniecture,
that they moued euery hundreth yeare one degre, ſyth the
tyme betwene their 2 obſeruations was 265 yeare : and after
the like rate was the ſame motiō found by conference of the
obſeruations of Timochares & Hipparchus. what other mē
ſay for more preciſenes herin ſyth their tyme, I wil in ỹ The
orikes dedare vnto you : but all agree herein, that the ſtarres
do moue vniformly with all their ſphere eaſtward as the Pla
netes doo. wherefore many aſſigne that motion as peculiar
to the eight ſphere, and the daily motion from eaſte to weſt
they appoint to the nynth ſphere. Other men perceauinge
that the ſtarres doo alſo aſcende northwarde, and deſcende
againe ſouthwarde, doo aſſigne a certaine motion, whiche
is named by them Motus trepidationis, and they note it
to bee peculiare for the eighte ſphere, and the other motion laſte named before, they accompte to be propre to the
nynthe ſphere, and then of neceſſitye it foloweth, that a
tenthe ſphere (as they ſaye) muſte be aſſigned for the dayly motion.

Scholar. If it be true that there be ſuche varieties of motions, then it ſeemeth reaſonable to aſſigne ſo many ſpheres
as there be motions ſeuerall.

Maſter. Although you thinke ſo now, you may be perſua
ded

ded peraduenture to thinke the contrary hereafter, as moſt
wiſe men in that arte do.

Scholar. But in the meane ceaſon what ſhall I thinke?

Maſter. Thinke well on that that you haue learned, and
labour to be expert in all that, by often conference of your
learnynge, with the praĉtiſe of the globe, and ſo ſhall you be
apte to bee inſtruĉted in all the reſte the more eaſilye. for it
will requyre a witte ſomewhat readye, and praĉtiſed in theſe
former matters.

Scholar. I wyll then prepare me a Sphere (without which
I ſee I can doo lytle good herein) and ſo will I praĉtiſe theſe
former leſſons, that I truſte to be as readye in them, as any
auditor in framynge of accompte.

Maſter. By that means ſhall all other thinges in thys
arte appeare eaſye vnto you, whiche nowe myght ſeeme vn-
timely put forth, if I ſhoulde offer to teache them, as the
motions of the Sonne, Moone, and other Planetes, with
their eccentrikes, equantes, differentes and Epicycles.

Scholar. In deede I thinke this to harde yet, but of the
progreſſion, retrogradation, and ſtation of the Planetes,
and alſo of the eclipſes of the Sonne and Moone, I knowe
that Iohn de ſacro Boſco dyd write ſomwhat, and ſo myght
you brieflye nowe do.

Maſter. His woordes are ſhorte and therefore obſcure,
and ſo ſhould my wordes be. beſide that, it is a diſordrely
forme to put the carte before the horſe : I meane to write of
the paſſions of the Planets, before I haue ſufficiently taught
the full ordre of their motion. Therefore I will ſaye in
fewe wordes, that the reaſons of the paſſions canne not bee
taughte aptely, before the Theorikes of theyr motions.
but for the contentation of your mynde, I maye define after a
forte the eclipſes of bothe the Sonne and Moone : wher- *The E-*
of the fyrſte is but an appearaunte and a countrefete E- *clipſe of*
clipſe : and is no wante nor loſſe of the lyghte in the *the Sonne.*
Sonne it ſelfe, but is an impedimente, that hys lyghte
&.ij.   dooth

dooth not or can not extende vnto vs, by reafon that the
moone doth runne beetwene him and our fighte. And this
Eclipfe as it hydeth the fonne from vs for a time, fo in fom
partes of the earthe at the felfe fame inftaunte he is not anye
whitte eclipfed, but fhyneth cleerely and wholly. And therfore is that eclipfe called no Generall eclipfe, whiche fhould
extende to all the worlde, namely for that hemifpherye, but
is particulare for fome one climate, and yet not vniuerfall

The E-
clipfe of
the moone.

to all that climate, but contrarye waies the eclipfe of the
moone is a true eclipfe in deede : for there is no thinge that
runneth betweene our fyghte and her, and fo hydeth frome
vs her light, but fhe leefeth her light certainly. As if a glaffe
that ftandeth in the Sonne, doo receaue the lyghte of the
Sonne, and doo cafte beames (as wee maye fee) frome
hym, tyll fome cloude or fome other darke bodye paffe
betweene the Sonne and it, and then it leefeth hys lyght
cleerely, and hathe no lyghte but hys owne bryghtneffe,
whiche canne caft no beames, nother deferue anye name
of lyghte, in comparifon to the lyghte that it hadde of
the Sonne : So the Moone kepynge hyr courfe tyll fhee
bee at the full, that is to faye, in the contrarye poynte
of the Zodiake to the Sonne, and that then fhe bee with
out all latitude, and runne ryghte vnder the Ecliptike
lyne in the Zodiake, then dooth fhee lyghte directly in
the fhaddowe of the earthe, and therefore canne not receaue the lyghte of the fonne, but leefeth it for the time,
howe bee it not always a lyke. for fometime fhee commeth whollye withoute the fhaddowe of the earthe, and
then is fhee whollye eclipfed : at other times fhee commeth
but partely into the fhaddowe, and that fome tymes in
the ouer parte, and fometime in the nether parte, wherby fhee is eclipfed partly, and not vniuerfallye : for if the
mone paffe by the northe or ouer part of the fhaddow, and
touche it with anye parte of hir felfe, then is that parte
                                                           eclipfed

THE CASTLE OF KNOWLEDGE   281

eclipfed of neceffity, which is the fouthe part of the moone
or the nether part of her. And again if the mone do touch
the nether parte of the fhaddowe whiche is nexte to the Horizonte, then is the hygher or northerlye parte of the
Moone eclipfed. To tell you nowe of the Eclipticall
pointes, whiche be commonly called the Headde and the
Tayle of the Dragon, it were verye vntymely, and harde
for you brieflye to conceaue, and therefore I do willingly
omitte them.

Scholar. Yet this I perceaue by you, that the fonne is not
darkened in him felfe, but is hydde by the moone from vs,
whiche happeneth diuerflye : for fometyme all the Sonne
is hyd, and fometyme the hygher part only, and at other
times, the nether parte onlye. of all whiche formes, I may
fee examples on euerye common Almanach after a groffe

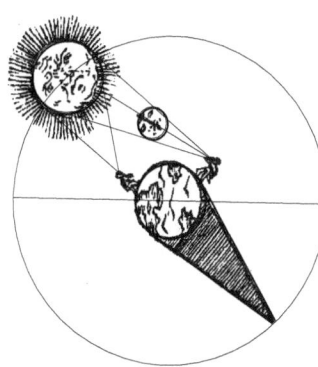

fort : but this Figure doth
more aptlye expreffe the
caufe thereof : where the
Moone dooth appeare to
be betweene any one Re-
gion and the Sonne, and
therefore hydeth the Son
frome the inhabitauntes
of that place : but in o-
ther Regions there ap-
peareth no fuche lette of
the Moone, but that they
maye fully fee the Sonne.
And other Nations bee-
tweene them, fee parte, and leefe other parte.

And thys I perceaue maye bee confidered dyuerfelye,
in as muche as anye bee nygher to theym that fee the
whole Sonne, or nygher to thofe that fee hys E-
clipfe.

&.iij.   Ma-

Mafter. There is in that nighnes double confideration :
one is of diftaunce betwene eafte and wefte, and ẙ other is of
diftaunce betweene fouthe and north. for when any nation
doth perceaue the higher cantle of the fonne endipfed, then
they that dwell more northerly, (vnder the fame meridian)
do leefe more of the fonne, and iudge that edipfe the greater : and contrary waies they that dwell directly towarde the
fouthe, the farther fouth they dwell, the leffer doth the part
edipfed appeare to them to be, tyll at lengthe vnto them
that dwell more fouthe there appeareth no edipfe at all.
The feconde confideration betwixte eafte and wefte, dooth
caufe only diuerfity in time of the Edipfe, but not in form :
& that is cōmon alfo for the edipfe of the Moone, but fo is
not the firft confideration, but ferueth for the fonnes edipfe
onlye.

Scholar. As for the edipfe of the mone, I thinke the former figures which you did fhewe me, do comprehende all
varieties of formes fufficiently, whiche be thefe two, for the

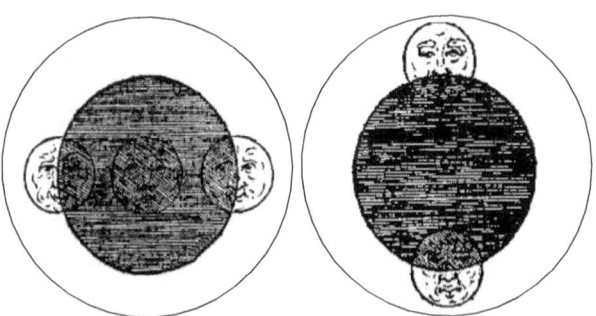

other two do reprefent thofe falfe formes, that do follow of
certaine falfe figures of the earthe and therfor do not ferue
here in place of true doctrine.

Mafter. This may you now alfo confidre, that although
the edipfe of the fonne is not general to all nations, bicaufe
                                                it is

it is not a true eclipſe or wante of lyghte, but onlye an appearaunte eclipſe, yet the eclipſe of the moone is a very E-
clipſe in deede, that is to ſaye, a wante of lyghte in hir ſelfe,
& therfore who ſo euer doth ſee her, dooth ſee alſo hir eclipſe
exactlye as it is : and it appeareth vniformlye to them all,
thoughe at that time the moone be not, nor canne not bee
aboue the horizonte to all people : and therefore vnto them
that haue the moone vnder their horizont, it is accompted
none eclipſe. And that is the cauſe why many eclipſes of the
ſonne and moone alſo are not noted in the common Ephemerides and Almanachs, bicauſe they appeare in ſuch time
as the Planet eclipſed, is vnder the Horizonte of that region
for whiche the Almanach or Ephemerides is written. farther more this is to be conſidered as a very truth and moſt
vnfallible, that the eclipſe of the ſon can neuer happen but
at the verye chaunge of the moone, for at other times ſhee
is ſo far in ordre of hir courſe from the ſonne, that ſhee can
not hyde any parte of him from anye nation in earth. And
for the eclipſe of the moone, the time of oppoſition or full
moone doth ſerue only. for the ſhaddowe of the earth whiche alwaye runeth towarde the Nadir of the ſonne directly,
can not touche the moone, excepte ſhe be verye nighe vnto
the ſame place. And that is the cauſe why the eclipſe of the
ſonne whiche happened at the deathe of Chriſt, may not be
accompted a naturall eclipſe, for ſo muche as it happened
in the time of the full moone, when it is not poſſible by natures ordre, that anye ſuche eclipſe ſhoulde happen. And
therfore dyd Dionyſe ẏ Areopagite beyng in Alexandria,
and Apollophanes his companiō, not only wonder at this
ſtraung and vnnaturall eclipſe, but concluded that it could
not happen without ſome meruailous cauſe, and a wonderfull immutation of natures workes.

  Scholar. So dooth our author of the ſphere note it, affirming that Dionyſe dyd ſay then : Other doth the God
of nature ſuffre now, or els the whole frame of the world ſhall

                                        &.iiij.   now

284   THE FOVRTHE TREATISE OF

nowe be diffolued.

Mafter. With this good claufe did he eande his booke,
and fo wyll we with the fame eande clofe vp our talke. Learnynge this good vfe in this naturall arte, that it leadeth mē
wonderfully to the knowledge of God, and his highe my-
fteries. as not only by example of thefe twoo philofophers
here it doth appear, but by the teftimonies of the fcriptures
in fundry places.

Scholar. This was that Dionyfe, whome Sainéte Paule
dyd conuerte afterwarde at Athenes, and rather muche bicaufe he hadde in remembraunce that miraculous Eclipfe.

Mafter. So maye wee gather manye argumentes by lyke
maters againft infideles and falfe Chriftians alfo : but that
frute will I referue for an other place : and for this prefente
will only faye, that there was neuer any good Aftronomer,
that denyed the Maieftie and prouidence of God, though
many other denyed bothe : but nowe farewell for a time : I
am dryuen to omytte teachinge of Aftrononye, and mufte
of force go learne fome lawe.

Scholar. The god that is author of true Aftronomye,
and made all the heauens for men to beholde, keepe you in
healthe and cleare from all trouble, that you maye, as you
mynde, accomplyffhe your workes, and finifh well and fpe-
dily, the frutes of your ftudye.

Mafter. Amen, and Amen.

THE CASTLE OF KNOWLEDGE 285

The titles of the fourthe Treatise.

What occasions moued men fyrste to iudge the forme of the worlde to be rounde, and namely three principall reasons thereof.   *1*

That the heauens are rounde in forme contrarye to the errour of Lactantius Firmianus, whiche thoughte it to bee flatte, and his opinion confuted by diuers reasons, namely by the vewe of the starres, by aptenes of mouynge, by reason of capacytie, and auoyding of emptines.   *2*

That the Firmament doth moue, thoughe Lactantius thought the contrarye : and howe it maye be proued, especially by the Milkye waye. And that the starres doo not mooue as byrdes in the ayer, or as fyshes in the water.   *3*

That the heauens are not cornered, nother of manye angles.   *4*

That all thinges shewe greater then they be, thorough vapours, and therfore the starres with the Sonne and Moone doo appeare greatest nigh vnto the Horizont.   *5*

Dyuerse opinions of the forme of the earthe : some thinkinge it to be of Cubike forme, other iudginge it Rygge formed, other affirmynge it to be plaine, other deeminge it hollowe as a dyshe, and other esteemynge it longe and rounde, like a piller or roller : all whiche beyng sufficiently confuted, it is full proued, that the earthe is iustly rounde in shape.   *6*

Then followe diuerse reasons, approuynge the water to be round, and a declaration with proofe why the water dooth not, nother can not ouerronne the whole face of the earth.   *7*

That the earthe and water togither doo make but one rounde Globe, and haue therefore one common centre.   *8*

That the earth is but as a pricke in comparison to the Skye, which is approued by foure dyuers argumentes.   *9*

The distaunce of euerye sphere frome the centre of the earthe, with an ordre to trye the quantities of the Sonne and Moone &c. in comparison to the earthe.   *10*

That the earthe is in the myddle of the worlde, and the contrary opinions repeated and confuted by sondry proofes.   *11*

That the earthe dooth not moue from the centre of the worlde.   *12*

A briefe rehersall of the parallele cirdes, with an instruction howe to fynde the distaunce of the Tropikes, and the greatest declination of the sonne, and of euerye degree of the Zodiake from the Equinoctiall cirde.   *13*

That the Arctike and Antarctike cirdes are not permanente, but mutable, accordynge to the chaunge of the regions, and so their quantities varieth, and their distaunce altereth, in respet to thother paraleles : and their ordre chaungeth diuersly.   *14*

The Zones beynge immutable, ought not to be distinct by the Arctike and Antarctike cirdes whiche are mutable, but rather by the Polare cirdes whiche perseuere styll, and keepe their quantities, their distaunce and their ordre vniformly.   *15*

That

16   That there ar no Zones vninhabitable other for heat or could, but may be and are alſo inhabited, as it is well knowen.

17   The Zodiake is named of the twelue Signes, whiche ſignes are taken in diuers ſignifications. and howe any ſtarre or Planete is named to bee in any ſigne. alſo what is the longitude, latitude and declinatiō of any ſtarres or Planetes.

18   The Colures, what they be, and howe many in numbre, and whereof they take their name.

19   The Horizonte celeſtiall and terreſtriall, howe they be diſtinĉte : where Proclus ſentence is reprehended, and thre ſeuerall tables ſet forth for diſtin ĉtion of howers, according to diſtaunce of myles from eaſte to weſte, and that for diuerſe climates.

20   The ordre and numbre of the Climates, with the eleuation of the Pole and the quantities of the longeſt daie in eche of them.

21   Of aſcention Aſtronomicall and Poeticall, and how euery one of them is diſtinĉte. with certaine rules of aſcention Aſtronomicall, and tables for the ſame, bothe in the Ryghte ſphere, and alſo in diuers Oblique ſpheres. with an examination of the rules of Iohn de ſacro Boſco.

22   The diſtinĉtion of howers into howres equall, and howers vnequall : and that howers vnequall be conſidered in twoo diuers ſortes, with tables ſette forthe for eche ſorte, concerninge their quantities.

23   Of daies Artificiall and Naturall. and what are the cauſes of diuerſi tie in eche of them, with tables for the quantities of the ſame : and a decal ration of the Sonne ryſinge and ſettinge.

24   The names of the conſtellations, with the numbre of their ſtarres.

25   A briefe declaration of the motions of the Planetes, and conſequent ly a reaſonable proofe for the numbre of their ſpheres. And farther what occaſion there was, that men ſhould imagine the ninthe and tenth ſphere to be, Where as there can none be ſeene aboue the eight ſphere.

26   A ſhorte explication of the eclipſes of the Sonne and the Moone.

Though faultes ofte times doo muche abounde,
When men doo leaſte ſuche chaunce ſuſpecte:
Yet good redreſſe maye ſoone be founde,
If faultes bee ſpied and full detecte.
But who that will in woorke proceede,
And ſeeke not firſte the faultes tamend,
I promiſe him ſmalle gaine in deede,
Thoughe truthe to ſeeke hee doo pretend:
Therefore amend if thou wilt ſpeede
Theſe faultes, ere thou on me doo reade.

The fyrſte numbre ſignifieth the page, the ſecond the lyne of the page.

9.28, ſphere which is. 10.12, eight ſphere. 10.29, proofe of my wordes and in the meane ceaſon to procede as I began: you muſt. 17. 17, doth. 18.1, the ſemicircle. 18.15, ϲρϊϕετου. 21.7, κύκλοσ. 23.10, ἰσημερίασ. 24, in the fiσ gure H, muſt be ſet by the mydle lyne againſt G. 25.26, χειμερινὴν. 27.8, κύκλων. 29.17, moueth or runneth. 30.7, οὖ꜔σ.32.33, there 2 circles. 33.22, drawen. 34.21, declareth. 36.18, and thorough. 41.17, they do. 56.12, to the colures. 57.35, their formes. 63.34, by their qualities. 68.17, call the latitude. 80.22, round aboute. 89.35, accordyngly. 97.20, at home. 103, in the margent, lib.3,c.24. 106.11, although. 106.33, heauen. 111.6, moſt apte of all other. 114.31, the rygge. 114.32. the one.116. in the margent, the reσ proofe. 117.21, inſtaunte. 121.19, the fifte parte. 121.20. the fifte parte. 124, in the margent is the lyne wronge ſette. 136.18, that is by D.136.24, that is by B. 145, and 146, the foure figures are not well placed in ordre, for the firſt ſhould be the thyrd, the ſeconde ſhoulde be fyrſte, and the third ought to bee ſecond. 147, ſet D vpon the greateſt ſhaddow, and E vpon the myddlemoſt. 153.11,33 minutes. 171.4, ſlowly. 172.8, ζύτων. 177.9, Arcturus is in libra ꝛc. aboue 31 degrees. 180.35, And H ꝸ L the 2. extreme points on the earth, vnto whiche ꝛc. 186.23 ſtand. 189.5, at an other time. 192, in the figure of the climates, B and D ſhould ſtand lower againſt the double lyne, which is the Equinoctiall. 194.23. conſidre. 207, the line in the example is wronge placed. 212.1, differeth not in this table the fyrſt. 212.16, 180 degres. 233.16, of proportions. 245.22, the daye is not. 248.20, reiect that ordre. 248.33, is not regarded. 260.10, the titles ſette. 266.12, protrygetes. 270.3. ryghte wynge. 272.1. fifte and the.

Imprinted at London by Reginalde
Wolfe, Anno Domini, 1556.

www.ingramcontent.com/pod-product-compliance
Lightning Source LLC
Chambersburg PA
CBHW051441170526
45166CB00001B/74